Concrete Materials and Technology

The design and implementation of high-quality concrete demands an underlying knowledge of concrete fundamentals as well as its constituent materials, and in various formulations. Starting with the basics, *Concrete Materials and Technology: A Practical Guide* examines the production and chemistry of cement, as well as the different types and their applications. Quality control processes and numerous methods for testing are presented and explained in detail. This book presents the fundamentals of concrete technology and serves as a useful guide for civil engineering students, project managers, concrete quality control managers and technicians.

Features:

- Explains the basics of different components and applications for different types of concrete
- Presents numerous methods for testing of concrete

Concrete Materials and Technology
A Practical Guide

Kambiz Janamian and José B. Aguiar

CRC Press
Taylor & Francis Group
Boca Raton London New York

CRC Press is an imprint of the
Taylor & Francis Group, an **informa** business

First edition published 2024
by CRC Press
6000 Broken Sound Parkway NW, Suite 300, Boca Raton, FL 33487-2742

and by CRC Press
4 Park Square, Milton Park, Abingdon, Oxon, OX14 4RN

CRC Press is an imprint of Taylor & Francis Group, LLC

Library of Congress Cataloging-in-Publication Data
Names: Janamian, Kambiz, author. | Aguiar, J. (José), author.
Title: Concrete materials and technology : a practical guide /
Kambiz Janamian and José Aguiar.
Description: First edition. | Boca Raton : CRC Press, [2024] |
Includes references and index. |
Identifiers: LCCN 2022062013 (print) | LCCN 2022062014 (ebook) |
ISBN 9781032470184 (hardback) | ISBN 9781032470191 (paperback) |
ISBN 9781003384243 (ebook)
Subjects: LCSH: Concrete.
Classification: LCC TA439.J34 2024 (print) | LCC TA439 (ebook) |
DDC 620.1/36—dc23/eng/20230123
LC record available at https://lccn.loc.gov/2022062013
LC ebook record available at https://lccn.loc.gov/2022062014

ISBN: 978-1-032-47018-4 (hbk)
ISBN: 978-1-032-47019-1 (pbk)
ISBN: 978-1-003-38424-3 (ebk)

DOI: 10.1201/9781003384243

Typeset in Times
by codeMantra

I dedicate this to:

Dear Parinaz, my incentive through this life

Contents

Preface..xv
Authors..xvii

Chapter 1 Introduction to Concrete Technology...1

    1.1   Concrete Constituent Materials...1
    1.2   Water Reduction in Concrete Production.............................4
    1.3   Concrete Workability 4
    1.4   Slump Test 6
    1.5   Hydration Reaction...7
    1.6   Concrete Curing 9
    1.7   Compressive Strength... 11
    1.8   Water-to-Binder and Water-to-Cement Ratio......................... 16
    1.9   Tensile and Flexural Strength of Concrete 17
    1.10 Elastic Module of Concrete 17
    1.11 Concrete Permeability and Water Tight Structures................ 19
    1.12 Durability and Effect of the Environment on Concrete 21
           1.12.1 Alkali Aggregate Reaction.................................22
           1.12.2 Carbonation Reaction......................................24
           1.12.3 Chloride Attack...25
           1.12.4 Sulfate Attack...27
           1.12.5 Effect of Freeze Thaw Cycle 29
    References 31

Chapter 2 Portland Cement...35

    2.1   What Is the Portland Cement?..35
    2.2   Cement History..37
    2.3   Cement Raw Materials 38
    2.4   Cement Production 39
    2.5   Compositions of Portland Cement.......................................41
    2.6   Types of Portland Cement 44
    2.7   Other Types of Cement..47
    2.8   Hydration Reaction..49
    2.9   Mineral Additives (Supplementary Cementitious Materials).... 50
    2.10 Blended Cement...51
    2.11 Quality Control of Cement 52
           2.11.1 Fineness of Cement 54
           2.11.2 Particle Size Distribution 55
           2.11.3 Cement Health...57
           2.11.4 Setting Time of Cement 58
           2.11.5 Cement Compressive Strength 60

          2.11.6 Specific Gravity of Cement 62
          2.11.7 Cement Heat of Hydration 63
    2.12 Analysis of Cement 64
          2.12.1 Cement Analysis Formulas 67
    2.13 Cement Transportation 69
    2.14 Cement Storage 70
    References 71

Chapter 3 Mineral Additives (Supplementary Cementitious Materials) 75
    3.1   What Are the Supplementary Cementitious Materials? 75
    3.2   Reaction with Cement and Water 77
    3.3   Silica Fume 78
          3.3.1   Silica Fume Specification 79
          3.3.2   Effect of Silica Fume on the Properties of
                  Fresh Concrete 80
          3.3.3   Effect of Silica Fume on the Properties of
                  Hardened Concrete 80
          3.3.4   Use of Silica Gel or Silica Slurry 81
    3.4   Fly Ash 82
          3.4.1   Fly Ash Specification 83
          3.4.2   Effect of Fly Ash on the Properties of
                  Fresh Concrete 84
          3.4.3   Effect of Fly Ash on the Properties of
                  Hardened Concrete 85
    3.5   Ground Granulated Blast Furnace Slag (GGBS) 85
          3.5.1   GGBS Specification 86
          3.5.2   Effect of GGBS on the Properties of Fresh Concrete 87
          3.5.3   Effect of GGBS on the Properties of
                  Hardened Concrete 88
    3.6   Natural Pozzolans 88
          3.6.1   Natural Pozzolans Specification 89
          3.6.2   Effect of Natural Pozzolans on the Properties of
                  Fresh Concrete 90
          3.6.3   Effect of Natural Pozzolans on the Properties of
                  Hardened Concrete 91
    3.7   Comparison Between Different Supplementary
          Cementitious Materials 92
    3.8   Use of Mineral Additives in Concrete Production 93
    References 94

Chapter 4 Aggregates 97
    4.1   Types of Natural Stones 98
    4.2   Aggregates and the Density of Concrete 101

4.3   Aggregate Size... 108
       4.3.1   Coarse Aggregates 109
       4.3.2   Maximum Size of Coarse Aggregate...................... 109
       4.3.3   Fine Aggregates... 111
4.4   Production of Aggregates 113
       4.4.1   Natural Aggregates.. 113
       4.4.2   Crushed Aggregates 114
4.5   Aggregates Test and Quality Control 116
       4.5.1   Abrasion Resistance 116
       4.5.2   Sieve Analysis Test.. 116
       4.5.3   Density and Water Absorption of Aggregates.......... 119
       4.5.4   Fillers and Passing by Sieve No. 200 125
       4.5.5   Harmful Materials in Aggregates 129
References 130

Chapter 5   Chemical Admixtures 133

5.1   The Reason for Using Concrete Admixtures 133
5.2   Accelerator Admixtures 134
       5.2.1   Set Accelerators.. 135
       5.2.2   Hardening Accelerators.. 137
5.3   Retarder Admixtures 138
       5.3.1   Dosage of Retarder Admixtures.............................. 139
5.4   Plasticizers and Super-Plasticizers 141
       5.4.1   Admixtures in ASTM Standard.............................. 143
       5.4.2   Chemical Bases of Plasticizers and
               Super-Plasticizers 143
       5.4.3   Mini Slump Test.. 148
       5.4.4   Marsh Funnel Test... 152
       5.4.5   Evaluation of Water Reduction Rate 154
5.5   Air-Entraining Admixtures 156
5.6   Water Proofing Admixtures 159
5.7   Curing Compounds 161
5.8   Viscosity Modifier Admixtures... 163
5.9   Pumping Aid Admixtures 164
5.10   Foaming Agent Admixtures... 165
References 167

Chapter 6   Water for Concrete 171

6.1   Water for Concrete Production... 172
6.2   Water for Concrete Curing 174
6.3   Test of Water.. 175
6.4   Decrease the Amount of Water for Concrete
       Production and Curing 175
References 177

Chapter 7 Testing of Concrete 179

　　　　　　7.1　Tests for Fresh Concrete 179
　　　　　　　　　　7.1.1　Slump Test 180
　　　　　　　　　　7.1.2　Flow Table Test 181
　　　　　　　　　　7.1.3　Slump Flow Test 182
　　　　　　　　　　7.1.4　Rheometer for Concrete 183
　　　　　　　　　　7.1.5　Temperature of Fresh Concrete 184
　　　　　　　　　　7.1.6　Density of Fresh Concrete 185
　　　　　　　　　　7.1.7　Air Content of Fresh Concrete 187
　　　　　　7.2　Tests for Hardened Concrete 188
　　　　　　　　　　7.2.1　Compressive Strength Test 189
　　　　　　　　　　7.2.2　Concrete Elastic Modules and Poisson's
　　　　　　　　　　　　　　Ratio Test 192
　　　　　　　　　　7.2.3　Density of Hardened Concrete 195
　　　　　　　　　　7.2.4　Permeability of Concrete 196
　　　　　　7.3　Test of Concrete in the Structures 197
　　　　　　　　　　7.3.1　Ultrasonic Test 198
　　　　　　　　　　7.3.2　Schmidt Hammer Test 200
　　　　　　　　　　7.3.3　Concrete Core Test 202
　　　　　　References 204

Chapter 8 Durability of Concrete Structures 209

　　　　　　8.1　Parameters Affecting Concrete Durability 209
　　　　　　8.2　Durability Against Carbonation 211
　　　　　　8.3　Durability Against Chloride Ion 213
　　　　　　8.4　Durability Against Sulfate Ion 214
　　　　　　8.5　Durability Against Freeze Thaw 216
　　　　　　8.6　Example for the Durability of a Concrete Structure 217
　　　　　　8.7　First Project: Bridge Deck in North Europe 217
　　　　　　8.8　Second Project: Commercial Building in a Big City
　　　　　　　　　Near the South China Sea 219
　　　　　　References 221

Chapter 9 Concrete Mix Design 225

　　　　　　9.1　The Goals of Concrete Mix Design 226
　　　　　　9.2　The Step-by-Step Method for Concrete Mix Design 226
　　　　　　　　　　9.2.1　Step (1): Specify Standard Deviation 226
　　　　　　　　　　9.2.2　Step (2): Specify Mix Design
　　　　　　　　　　　　　　Compressive Strength 228
　　　　　　　　　　9.2.3　Step (3): Specify Percentage of Each
　　　　　　　　　　　　　　Aggregate in Concrete 228
　　　　　　　　　　9.2.4　Step (4): Specify Fineness Module of
　　　　　　　　　　　　　　Total Aggregates 229

9.2.5   Step (5): Specify Water-to-Binder Ratio 230
9.2.6   Step (6): Specify Free Water for Concrete 233
9.2.7   Step (7): Specify the Amount of Portland
         Cement and Other Binders 236
9.2.8   Step (8): Specify the Total Volume of Aggregates 237
9.2.9   Step (9): Specify the Weight of Aggregates in
         Saturated Surface Dry Conditions 238
9.2.10   Step (10): Specify the Real Weight of
         Aggregates and Water in Concrete...........................238
9.3   Example 1 for Concrete Mix Design.....................................239
9.4   Example 2 for Concrete Mix Design.....................................246
9.5   Example 3 for Concrete Mix Design.....................................254
9.6   Implementation of Mix Design in the Projects 260
References 261

Chapter 10   Production, Transportation, and Implementation of Concrete.........263
10.1   Production of Concrete in the Laboratory.............................264
10.2   Production of Concrete in the Batching Plant......................267
10.3   Concrete Transportation With Truck Mixers 276
10.4   Other Instruments for Concrete Transportation 277
10.5   Concrete Pumping 278
10.6   Compaction of Concrete..280
10.7   Smoothing the Surface of Concrete Elements......................282
10.8   Cold Joint in Concrete 283
10.9   Curing of Concrete 284
10.10   Concrete Recycling System 286
References 287

Chapter 11   Usage of Fibers in Concrete 289
11.1   Steel Fibers 289
11.2   Glass Fibers 292
11.3   Artificial Fibers 294
         11.3.1   Polypropylene Fibers..295
11.4   Natural Fibers 297
References 298

Chapter 12   Hot and Cold Weather Concreting 301
12.1   Calculating Concrete Temperature According to the
         Constituent Materials Temperature 301
12.2   Definitions of Hot Weather Conditions for Concrete 302
12.3   Cement Hydration Reaction at Hot Weather Conditions.......304
12.4   The Effects of Hot Weather on the Properties of Concrete........304
12.5   Considerations for Hot Weather Concreting 305

12.6   Concrete Cracking at Hot Weather Conditions 309
12.7   Chemical Admixtures for Hot Weather Concreting............. 312
12.8   Calculations for Concrete Temperature at
       Hot Weather Conditions 313
12.9   Definitions of Cold Weather Conditions for Concrete 315
12.10 Cement Hydration Reaction at Cold Weather Conditions 316
12.11 The Effects of Cold Weather on the Properties of Concrete.......317
12.12 Considerations for Cold Weather Concreting........................ 318
12.13 Chemical Admixtures for Cold Weather Concreting 321
12.14 Calculations for Concrete Temperature at Cold
       Weather Conditions 322
References 324

Chapter 13 Concrete for Special Purposes 327

13.1   Self-Compacting Concrete 327
       13.1.1 Definitions of SCC 327
       13.1.2 Considerations for SCC Production and
                 Implementation.. 328
       13.1.3 Tests for Checking the Properties of SCC 331
       13.1.4 Example for SCC Mix Design and
                 Implementation..335
13.2   Watertight Concrete..340
       13.2.1 Definitions of Waterproof Concrete and
                 Watertight Structure 340
       13.2.2 Considerations for Watertight Concrete
                 Production and Implementation 341
       13.2.3 Tests for Checking the Properties of
                 Watertight Concrete 342
       13.2.4 Example for Watertight Concrete Mix Design
                 and Implementation...344
13.3   High-Strength Concrete...347
       13.3.1 Definitions of High-Strength Concrete 348
       13.3.2 Considerations for High-Strength Concrete
                 Production and Implementation 350
       13.3.3 Tests for Checking the Properties of
                 High-Strength Concrete 351
       13.3.4 Example for High-Strength Concrete Mix
                 Design and Implementation 352
13.4   Ultra-High-Strength Concrete 354
       13.4.1 Definitions of Ultra-High-Strength Concrete........... 355
       13.4.2 Considerations for Ultra-High-Strength Concrete
                 Production and Implementation 356
       13.4.3 Tests for Checking the Properties of
                 Ultra-High-Strength Concrete.................................357
       13.4.4 Example for Ultra-High-Strength Concrete
                 Mix Design and Implementation............................. 360

13.5 Mass Concrete 361
    13.5.1 Definitions for Mass Concrete 361
    13.5.2 Considerations for Mass Concrete Production
        and Implementation 362
    13.5.3 Tests for Checking the Properties of
        Mass Concrete 363
    13.5.4 Example for Mass Concrete Mix Design and
        Implementation 363
13.6 Precast Concrete 366
    13.6.1 Definitions of Precast Concrete 366
    13.6.2 Considerations for Precast Concrete
        Production and Implementation 366
    13.6.3 Tests for Checking the Properties of
        Precast Concrete 367
    13.6.4 Example for Precast Concrete Mix Design and
        Implementation 368
References 370

Index 373

Preface

Concrete technology is one of the most important subjects for civil engineers. The usage of concrete in all kinds of projects is growing every day. So, the design of high-quality and high-performance concrete is very important. After the production of concrete, quality control and implementation are also notable.

To design, control, and implement high-quality concrete, we should start with the knowledge about concrete constituent materials. Their quality control and all combinations of them. After that, we should continue with the concrete mix design and the process to ensure the quality of the concrete. Finally, the implementation process for concrete should be considered.

There are many books on the concrete technology subject in the market. The difference between this book is the practical view to the concrete technology. So, the book can be useful for civil engineering students, project managers, concrete quality control managers and technicians, and all of the engineers who should work with concrete in construction projects.

The first chapter of the book is an introduction to concrete technology. The reader will know about some expressions in concrete technology that we are going to use them in the future chapters of the book. So, this part is very important for the starters.

Chapter 2 is about the Portland cement, the most important material of concrete. The reader will study about the history and production of cement, chemistry of cement, different types of cement, and the quality control process.

Chapter 3 is about the mineral admixtures. Like Portland cement, they are another kind of binders in concrete. The use of these materials in concrete is very important for the durability purpose. So, it is necessary for a concrete technologist to know about them.

Chapter 4 is about the aggregates, the texture of concrete. These materials are very important for the rheology and behavior of fresh concrete, and they are also important for the quality of hardened concrete. Quality control of aggregates is another subject explained in this chapter.

Chapter 5 is about the chemical admixtures in concrete. These materials are very important for the production of high-quality and modern concrete. So, it is necessary to know their specification and usage.

Chapter 6 is about the water for concrete. The specification of suitable water for concrete will be explained in this chapter.

In Chapter 7, the reader will study about the most important tests for fresh and hardened concrete. The testing of concrete in the structures is also described in the last part of the chapter.

Chapter 8 is about the durability of concrete structures. This subject is very important today, because of the protection of resources in the environment. So, the production and implementation of durable concrete will be discussed.

Chapter 9 is about the concrete mix design. As mentioned before, the production of high-quality concrete depends on the concrete mix design. In fact, we should mix all of the raw materials with the specific amount to prepare a suitable concrete.

In Chapter 10, the production, transportation and implementation of concrete are discussed. So, the reader will study about the most advanced techniques of concrete usage in the structural projects.

Chapter 11 is about the different types of fibers suitable for concrete.

The hot and cold weather conditions could be very harsh for concrete. So, the implementation of concrete in these conditions is very important. This subject is explained in Chapter 12.

Chapter 13 is about the special kinds of concrete for special purposes. There are too many kinds of special concrete in the structures. Some of the most useful ones are explained in this chapter.

According to the abovementioned, this book can be used as a concrete technology handbook in most types of construction projects.

Kambiz Janamian
José B. Aguiar
2023

Authors

Kambiz Janamian is an experienced civil engineer and concrete technologist. He worked for more than 10 years in ready mixed concrete plants as the QC and development supervisor and consultant. He also worked as a concrete admixture formulator and researcher for many years. He was the supervisor for many joint projects between the concrete industry and universities. His research is related to concrete mix design, concrete admixtures, PCE super-plasticizers, ultra-high-performance concrete, and many other subjects related to the concrete technology. He published five books on concrete technology with the subjects of high-performance concrete, shrinkage and cracks in concrete, concrete admixtures, concrete mix design, and using of plasticizers and super-plasticizers.

José B. Aguiar is an Associate Professor with habilitation at the Department of Civil Engineering of University of Minho, Portugal. He gained his BSc in Civil Engineering in 1982 and received his PhD in civil engineering in 1990. He has over 250 publications. His main areas of interest include durability of concrete, concrete-polymer composites, incorporation of wastes in concrete, and energy efficiency of buildings. He published a book *A Comprehensive Method for Concrete Mix Design, 2020, Materials Research Forum.*

.

1 Introduction to Concrete Technology

Concrete is the most important construction material with a high amount of usage in the structures, because:

- It is very simple to find its constituent materials everywhere.
- Fresh concrete is a flexible material. So, you can shape it with all types of forms.
- The compressive strength of hardened concrete is very good. We can make a concrete with more than 300 MPa compressive strength nowadays.
- When we use concrete combined with steel bars or steel fibers, the tensile and flexural strength of the mix will be very good.

So, it is very important for a civil engineer to know about this magic material. The technique of making and controlling the quality of this material defined as the concrete technology.

Concrete technology is the technique for the preparation of high-quality constituent materials and mixing them. It seems that it is a simple work. But really it is not. The importance of choosing suitable constituent materials and good proportions is one of the most advanced techniques in civil engineering.

This chapter is the start for a concrete technologist. You should start with some of the most important expressions and definitions of concrete technology. This fundamental knowledge is necessary to continue the other subjects of the book.

Every page of this chapter contains many important definitions and concepts that you should learn about the concrete technology. We will use these concepts in the following chapters too many times. So, for beginners, this chapter is the base of other chapters.

Let's start our journey with the concrete constituent materials.

1.1 CONCRETE CONSTITUENT MATERIALS

Concrete is a mixture of below materials:

- Portland cement (Figure 1.1): It is a kind of powder, which contains calcium silicates and calcium aluminate chemicals. This is the main binder in concrete, which reacts with water to harden. This is also the main material for the smoothness of concrete. We can use about 300–600 kg of Portland cement in 1 m³ of concrete which is about 10%–25% by weight of concrete and about 9%–18% by volume of concrete.

DOI: 10.1201/9781003384243-1

FIGURE 1.1   Portland cement. (Photograph by the author.)

- Water: The material which is necessary for the hydration reaction of cement chemicals. We can use about 110–250 kg of water in 1 m³ of concrete which is about 5%–10% by weight and volume of concrete.
- Aggregates (Figure 1.2): They are parts of stone with different sizes from the biggest size, which is about 25 mm, to the finest size, which is less than 0.075 mm. Aggregates are the texture and structure of concrete. We can use about 1700–1900 kg of aggregates in 1 m³ of concrete, which is about 70%–80% by weight of concrete and about 60%–70% by volume of concrete.
- Mineral admixtures (Figure 1.3): These materials act as the helping binder of concrete. Most of them are the by-products of other industries, which we can use to increase the performance of concrete, especially in the subject of durability. Most of the times, we will replace a defined portion of Portland cement with these materials. For example, between 6% and 70% of cement can be replaced with these materials. So, the amount of use is about 20–250 kg of 1 m³ of concrete which is about 1%–10% by weight of concrete and about 1%–10% by volume of concrete.

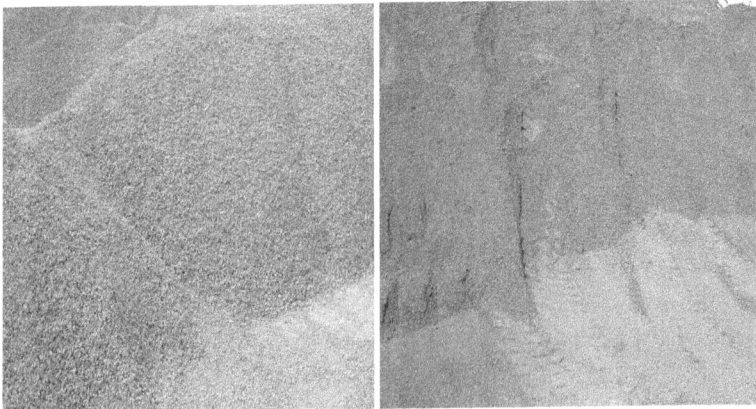

FIGURE 1.2   Coarse aggregates (left) and fine aggregates (right). (Photograph by the author.)

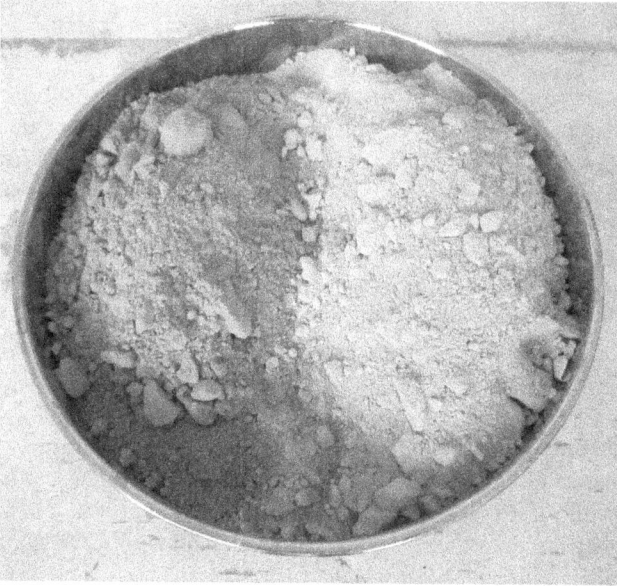

FIGURE 1.3   The most active mineral admixture (Silica fume). (Photograph by the author.)

- Chemical admixtures (Figure 1.4): These are special chemicals that we can use for special purposes like better workability, air entraining, accelerating or retarding the hardness of concrete, and many other purposes. We can use a little amount of these chemicals in concrete. Most of the times, the amount of use is less than 2% by weight of total binder (Portland cement + mineral admixtures) which is less than 0.5% by weight of concrete and less than 1% by volume of concrete. Although the amount of use for these chemicals is a little, the effect of them is too much on the quality and performance of concrete. So, in modern concrete technology, using chemical admixtures is crucial.

FIGURE 1.4   Different chemical admixtures. (Photograph by the author.)

1.2 WATER REDUCTION IN CONCRETE PRODUCTION

As mentioned before, the reason for concrete hardening is the hydration reaction between Portland cement chemical ingredients and water. This is in fact the reason for all of the properties of hardened concrete. To complete this reaction, we need a special amount of water which depends on the type of cement and usage of other binders (mineral admixtures). We can say roughly that this amount of water is about 25%–30% by weight of cement.

On the other hand, if we make a concrete with this amount of water, the concrete will be very dry. Most of the times, if we would like to make a workable concrete without using any chemical admixture, we need water about 60% by weight of cement. This is a huge difference! So, what will happen to our concrete with this high amount of water?

The excess water would evaporate from the concrete. This moving of water through the texture of concrete during its hardening will cause some micro-cracks inside the concrete. So, the concrete will be porous, and it will be less hard for loading. So, we should try to use less water in the production process of concrete. How can we use less water for concrete production with workability consideration?

The answer is: using plasticizers and/or super-plasticizers. By using these materials, we can make concrete with the amount of water less than 30% by weight of cement and good workability. We will talk about plasticizers and super-plasticizers in the later chapters of this book.

The most important advantages of water reduction in concrete production are:

- Increase the compressive strength of concrete: As mentioned before, using less water in concrete will cause less micro-cracks in the texture of concrete. So, we will have more compressive strength.
- Decrease the permeability of concrete: Less micro-cracks in the texture of concrete means less permeability. The water and aggressive ions like chloride or sulfate cannot infiltrate into the concrete and they cannot destroy concrete or the steel bars.
- Better bonding between concrete and steel bars: Because of the cohesion of concrete with less water, the bonding between concrete and steel bars will be better.
- Decrease the shrinkage and probability for cracks (Figure 1.5): Shrinkage is the reduction of volume of concrete during the time. There are too many reasons for concrete shrinkage. The problem is cracking which will happen because of the shrinkage on the surface and even sometimes inside the concrete. Reduction of water in concrete will help the stability in volume and decrease cracking.

1.3 CONCRETE WORKABILITY

Workability is defined as the ability of concrete for flowing, pumping, finishing, and all of the work that we need to do for the implementation of concrete. Workability is one of the specifications of fresh concrete. A good and workable concrete will cause better compaction and implementation. So, the final structure will have better quality.

FIGURE 1.5 Cracks on the surface of the concrete. (Photograph by the author.)

Several factors have effects on the workability of concrete:

- Concrete flowability: One of the most important factors affecting workability is the flowability of concrete. The more flowable concrete means the more workable. But as mentioned before, to produce a high-quality concrete we should use plasticizers and super-plasticizers for the flowability of concrete.
- Type and amount of cement in concrete: There are several types of cement to use in concrete with different specifications. Some of them are accelerated cements and the others could be retarded. These types of cement can change the need of concrete for water or super-plasticizer to flow. So, they could act on the workability of concrete. On the other hand, if we use more cement in 1 m³ of concrete, the need for water will increase. So, the cement type and the amount of cement in concrete are the factors that can change the workability performance of concrete.
- Types of aggregates and their gradation: We will talk about different types of aggregates in the later chapters. If we use crushed aggregates versus natural aggregates, they could decrease the concrete workability. Because, by using crushed aggregates, concrete will be harsh and it could be difficult to work with it. Another important factor is the amount of fillers in the fine aggregates. Fillers are the finest part of the aggregates that can be very important for the workability properties of concrete. Especially, the pumpability of concrete depends on the amount of fillers in the fine aggregates.
- Amount of air in concrete: When we produce any kind of concrete, about 1%–2% of its volume is air bubbles. As these bubbles have circular shape, they can act positively for the workability of concrete. So, more air bubbles in the texture of

concrete will cause better workability. But do not forget that on the other hand, more air bubbles in the texture of concrete will cause drastically decrease in the compressive strength and other mechanical properties of concrete.

- Amount of water in concrete: As you know, increasing the amount of water in concrete will cause decreasing the compressive strength. But on the other hand, it can cause better workability properties. So, the optimum dosage of water in concrete is very important. We can optimize the amount of water by using a special super-plasticizer for any purpose.
- Concrete and ambient temperature: The water demand and dosage of super-plasticizer will increase at hot weather conditions. On the other hand, when we transport concrete from the production plant to the final project, the ambient and concrete temperature will be very important. Because in this case, hot weather will cause more evaporation of water and also rapid hydration reaction of cement, and these two reasons will cause decrease in workability during the transportation time.
- Using plasticizers and super-plasticizers: As mentioned before, one of the most important reasons for using plasticizers and super-plasticizers is their effect on the workability of concrete. They can improve the flowability and sometimes the cohesion of concrete for better workability and pumping.
- Transportation of concrete: The transportation time, especially in hot weather conditions, is very important. If the transportation time will increase, the workability will decrease drastically. The other important factor is the type of mixer and the mixing speed for the transportation of concrete. More mixing speed will cause a more rapid hydration reaction and it will cause decrease in flowability and workability.

1.4   SLUMP TEST

Slump test is one of the most important tests to measure the flowability of concrete. This test procedure is according to ASTM C143. Although we can say that this is not an accurate test for the flowability of concrete, it can be useful in many cases.

To do the test, we need a cone with a height of 300 mm and lower role diameter of 200 mm and upper role diameter of 100 mm (Figures 1.6 and 1.7). We should pour

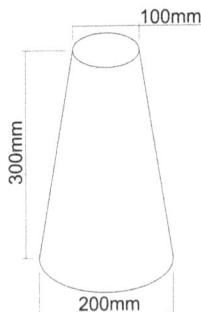

FIGURE 1.6   The dimensions of the slump cone. (Photograph created by the author.)

FIGURE 1.7   Picture of slump test tools. (Photograph by the author.)

the cone with concrete in three layers. After pouring each layer, we should compact the layer with a 16 mm rebar with 15 strokes. Then the upper limit of the cone should be cleaned of concrete. Finally, we should pull up the cone slowly and let the concrete fall down. The difference between the height of the cone (300 mm) and fallen concrete is the slump of concrete, which is an index for the flowability. The more flowable concrete means a higher number of slump.

For the slump less than 120 mm, we can say that the concrete is a stiff one and the test is accurate enough. For the slump between 120 and 200 mm, the concrete is a good flowable concrete. But the test accuracy is not good enough. For the slump more than 200 mm, the test accuracy is not good and we should use other tests for good measurement of the flowability. In fact, a concrete with the slump more than 200 mm is not a normal concrete. We can call these kinds of concrete an easy compacting concrete or self-compacting concrete.

1.5   HYDRATION REACTION

The hydration reaction is a chemical reaction between Portland cement compositions and water which causes the hardening of concrete. This reaction is an exothermic chemical reaction. The heat released by this reaction is called hydration heat.

All of the compositions of cement can react with water. But the most important ones for all of the mechanical properties of concrete are C_3S (tri calcium silicate) and C_2S (di calcium silicate). We will talk about the compositions of cement in Chapter 2. But you have to know that the most important compositions of any kind of Portland cement are C_3S and C_2S. You can see the hydration reaction for these compositions as follows:

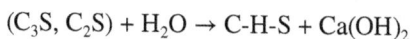

$$(C_3S, C_2S) + H_2O \rightarrow C\text{-}H\text{-}S + Ca(OH)_2$$

As you see, the reaction of C_3S and C_2S with water produces C-H-S (Calcium Hydrate Silicate) and $Ca(OH)_2$.

The exact chemical structure of C-H-S is not known. But we know that this chemical is responsible for the compressive strength and other mechanical properties of concrete. So, if we can increase the amount of C-H-S in the hydration reaction, we can increase the quality and performance of concrete. Now the question is: How can we do it?

To increase the amount of C-H-S in concrete, we should optimize the hydration reaction. To do it, we should:

- Use high-quality cement with a good production process: Like any other material, high-quality cement will cause a suitable hydration reaction.
- Optimize the temperature of concrete during hydration: Concrete temperature can increase or decrease the hydration reaction speed, especially in the earlier stage.
- Use less water in concrete production: This will cause optimizing the amount of water for the hydration reaction.
- Use high-quality plasticizers and super-plasticizers: Some of these chemicals can improve the hydration reaction in addition to reducing water.
- Cure the concrete during the hydration process: It means that we should control the moisture and temperature during time.

About the hydration heat, you should know that any kind of cement can release different amount of heat during the hydration reaction. We will talk about different types of cement later. Some kinds of cement are accelerated ones and the others are retarded ones. So, accelerated types of cement can release more heat during the hydration reaction and retarded cements can release less hydration heat.

It is very important to control the hydration heat in some projects because this heat can damage the quality of concrete. For example, in mass concrete projects like mass foundations (Figure 1.8) or structure of concrete dams, the problem of hydration heat could be very important. To control the heat of hydration we should follow the below considerations:

- Use retarded types of cement: This will cause a decrease in the speed of the hydration reaction.
- Use less cement in $1 m^3$ of concrete: It means the minimum cement for concrete according to the mechanical properties.
- Use retarder chemical admixtures: These chemicals can control the heat of hydration.
- Use other types of binders: Binders like slag or fly ash beside the Portland cement can decrease the heat of hydration.

On the other hand, in some cases, we need to increase the heat of hydration. For example, in winter and in cold weather conditions (Figure 1.9), to reduce the risk of icing, we need more heat in the concrete hydration reaction. For this reason, you should consider below points:

FIGURE 1.8 A mass foundation. (Photograph by the author.)

FIGURE 1.9 Cold weather construction. ("Constructing natural gas line in winter, Finland" by Jukka Isokoski.)

- Use more cement in $1\,\text{m}^3$ of concrete: You can increase the amount of heat by the hydration reaction.
- Use accelerated types of cement: These types of cement can release more heat, especially in the earlier stage.
- Use accelerator chemical admixtures: These chemicals can accelerate the hydration reaction and increase the amount of heat in the earlier stage.

1.6 CONCRETE CURING

Curing is the protection of finished concrete against drying and high changing in temperature. To achieve maximum performance of any concrete, curing is critical. For perfect curing, you should consider three points:

- Moisture: the first important point is the protection of concrete against drying. Hydration reaction needs enough amount of water during the time. On the other hand, ambient conditions like temperature and wind will dry the surface of the concrete element. So, we should protect the moisture inside the concrete.
- Temperature: We will discuss the temperature effect on concrete later. For now, you should know that we should control the temperature of concrete to prevent freezing and on the other hand, we should control the amount of heat inside the concrete to prevent cracking. The best temperature for concrete is between 15°C and 25°C.
- Time: We should do all of the considerations above, as long as it is possible. In some texts, the best time for concrete curing is one week. But, really it is very hard to do the considerations of curing for one week. So, at least we should do it for 48 hours and continue as much as possible.

Now, we should talk about the exact tasks for concrete curing. We will talk about this subject later. But, here we will list some of the most important tasks for concrete curing:

- Use of curing compound admixtures (Figure 1.10): We should spray these chemicals on the surface of the concrete element. So, the moisture evaporation will control. In fact, these chemicals make a film layer on the surface of concrete and this film will prevent evaporation.

FIGURE 1.10   The implementation of curing compound chemicals. ("Spraying the curing compound" by Robyn McKinley.)

- Use of covers: If we cover the surface of the concrete with plastic layers, we can prevent the evaporation of water from the concrete.
- Use of water jet (Figure 1.11) or fogging: With this task, we can secure the amount of moisture for the hydration reaction.
- Control the temperature of concrete in hot weather conditions: By using cold water, we can control the temperature of concrete in summer.
- Control the temperature of concrete in cold weather conditions: We can cover the surface of the concrete with blankets (Figure 1.12) and using of heaters to control the temperature of concrete in winter. This process is very important for the curing of concrete in cold weather conditions.

1.7 COMPRESSIVE STRENGTH

Compressive strength is the most important mechanical property of concrete. As concrete tensile and flexural strength is low, we need the compressive strength of concrete for the design of concrete structures.

The important question is: What is strength? And what is compressive strength?

FIGURE 1.11 Water curing of a concrete column. (An Iraqi construction worker" by Jim Gordon.)

FIGURE 1.12   Suitable blankets to control the temperature of concrete in winter. (Photograph by the author.)

Strength is the maximum amount of force on one unit of area that a material can tolerate without failure.

Compressive strength is the maximum amount of compressive force on one unit of area that concrete can tolerate without failure.

For concrete, as the age of concrete will increase, the compressive strength will raise. So, this is very important to know the age of concrete for the amount of compressive strength.

The most important age is 28 days because for most kinds of concrete, the increase of compressive strength after 28 days is very little. We can check the growth of compressive strength at other ages like 3, 7, or 11 days. On the other hand, we can check the growth of compressive strength at the ages like 42 or 90 days. You can see the common growing pattern of compressive strength for concrete during the time in Figure 1.13.

You should know that the growing pattern of concrete depends on several factors like the type of cement, use of mineral additives and the amount of them, type of mineral additive, and use of chemical admixtures specially accelerators or retarders. In Figure 1.13 we can see a common pattern for a concrete with only ASTM type-I cement and without any mineral or chemical admixture.

Now, we would like to see, how can we measure the compressive strength of concrete? To measure the compressive strength of concrete we should follow these steps:

- Prepare enough amount of concrete as sample. For example, we can give a sample from a truck mixer that transports concrete to the project. The amount of concrete should be enough for pouring the molds that we are going to use.
- Choose the appropriate kind of mold that we would like to use. The standard mold according to ASTM is the 15×30 cm cylinder mold (Figure 1.14). But you can use 10×20 cm cylinders or $15 \times 15 \times 15$ cm cubes instead (Figure 1.15). If you use the latter mold, you should convert the compressive strength to the equivalent for the 15×30 cm cylinder. In different countries all over the world, the more common kind of mold is different.

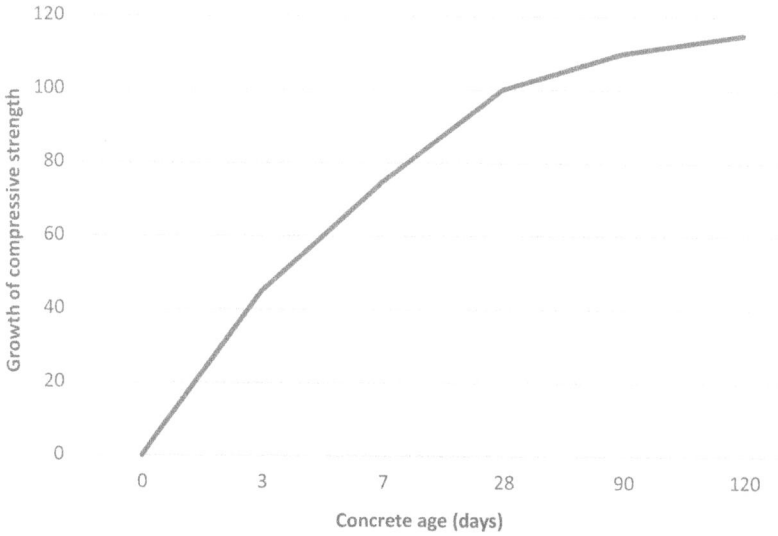

FIGURE 1.13   Common pattern for concrete compressive strength growing. (Graph created by the author.)

FIGURE 1.14   Standard 15 × 30 cm cylinder mold and specimen. (Photograph by the author.)

- Pour the molds according to the ASTM C39 in different layers with good compaction.
- Release the molds after 24 hours and put the specimens of hardened concrete into the water according to ASTM C39.
- In the appropriate age, for example, 28 days, bring the specimen out of the water, and let the surface dry. Now the specimen is ready for the compressive strength test.

FIGURE 1.15   Standard $15 \times 15 \times 15$ cm cube mold and specimen. (Photograph by the author.)

- Put the specimen under the hydraulic test machine (Figure 1.16) to load it until failure. You can calculate the compressive strength by dividing the final force by the surface area of the specimen.

As mentioned before, if we use other kinds of molds instead of the 15×30 cm cylinder, we should convert the compressive strength to the equivalent standard cylinder. There are too many suggestions for this reason in different texts and standards. For example, to convert the $15 \times 15 \times 15$ cm cube compressive strength to the standard 15×30 cm cylinder you can use Table 1.1.

Example 1.1: Calculate the concrete compressive strength for an ASTM standard cylinder specimen that failed with the maximum force of 82.2 tons.

$$\text{Area} = 7.5 \times 7.5 \times 3.14 = 176.6 \text{ cm}^2$$

$$\text{Compressive strength} = (82.2 \times 1000)/176.6 = 465 \text{ kg/cm}^2 = 46.5 \text{ MPa}$$

Example 1.2: Calculate the concrete standard cylinder compressive strength for a $15 \times 15 \times 15$ cube specimen which failed with the maximum force of 101.6 tons.

$$\text{Area} = 15 \times 15 = 225 \text{ cm}^2$$

$$\text{Cube compressive strength} = (101.6 \times 1000)/225 = 452 \text{ kg/cm}^2$$

FIGURE 1.16   Hydraulic concrete test machine. (Photograph by the author.)

According to Table 1.1 we have:

Standard cylinder compressive strength = 452 − 50 = 402 kg/cm² = 40.2 MPa

From a compressive strength point of view, we can use four types of concrete:

- Low-strength concrete: Compressive strength less than 30 MPa
- Normal strength concrete: Compressive strength between 30 and 50 MPa
- High-strength concrete: Compressive strength of more than 50–90 MPa
- Ultra-high strength concrete: Compressive strength of more than 90 MPa

TABLE 1.1
Rough Conversion of 15 × 15 × 15 cm Cube Compressive Strength to the Standard 15 × 30 cm Cylinder

15 × 15 × 15 cm cube compressive strength (MPa)	30	35	40	45	50	55
15 × 30 cm cylinder compressive strength (MPa)	25	30	35	40	45	50

The common compressive strength for concrete in different countries of the world could be different. For example, in developed countries, the common compressive strength of concrete is about 50 MPa. But in some other countries, the common compressive strength could be about 25 MPa.

The use of low-strength concrete is a waste of materials and resources because by using the same materials and using the beneficiary of high concrete technology, we can make high-strength concrete. We will talk about this subject later. But for now, it is very important to use a concrete with higher strength and performance.

1.8   WATER-TO-BINDER AND WATER-TO-CEMENT RATIO

As mentioned before, using less water in concrete production means higher strength and performance. But the real subject is not only less water. It depends on the water-to-binder and/or water-to-cement ratio.

In this concept, the cement is only pure Portland cement and the binder is the sum of Portland cement and mineral additives like silica fume, fly ash, ground granulated blast furnace slag (GGBS), and natural pozzolans.

The complete concept is: If we have lower water-to-cement (w/c) or water-to-binder (w/b) ratio, then we will have higher strength concrete as can be seen in Figure 1.17.

To decrease w/c ratio, we can increase the amount of cement or total binder and we can decrease the amount of water. We have restrictions to increase the amount of cement. Because of the hydration heat, cracking, economical reasons, and low activity of high amount of cement in concrete, we can use a maximum of 500 kg of pure Portland cement and a maximum of 700 kg of total binder in 1 m³ of concrete. So, the best thing that we can do is the reduction of water in concrete production.

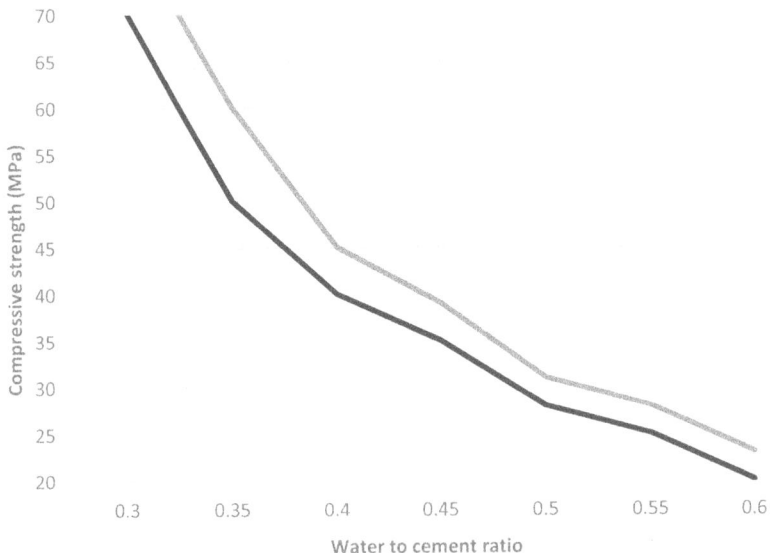

FIGURE 1.17   Effect of w/c on the compressive strength for different types of cement. (Graph created by the author.)

On the other hand, if we make concrete with 300–400 kg of pure Portland cement and water-to-cement ratio of about 0.55–0.6, then we will have concrete with 80–100 mm slump which is not a good slump for the implementation of concrete. So, what should we do to produce a concrete with w/c of 0.4 and lower?

The answer is using high-performance super-plasticizers. These chemicals will let the concrete flow without increasing the amount of water. The average dosage of plasticizers and super-plasticizers is between 0.3% and 2% by weight of binder, which can be between 1 and 10 kg of admixture in 1 m³ of concrete.

In fact, the amount of use for these chemical admixtures is very low, but the effect of them is very high in concrete production. We can say that the production of high-quality and performance concrete is possible only because of the high technology of these chemical admixtures.

1.9 TENSILE AND FLEXURAL STRENGTH OF CONCRETE

The most important mechanical property of concrete is compressive strength because the compressive strength of concrete can be high enough to use it in concrete structures. We can make concrete with a compressive strength of more than 300 MPa. So, for compressive strength, there are no restrictions.

On the other hand, for tensile strength, concrete is a weak material. We can say that the tensile strength of a normal concrete is about 8%–12% of the compressive strength. So, for a 50 MPa concrete that is a good concrete, we will have only about 5 MPa of tensile strength which is very low. So, for the design of concrete structures, we will not trust on the tensile strength of concrete. In fact, we assume that the tensile strength of concrete is zero. We only trust on the compressive strength.

As the tensile strength of concrete is very low, we can say that the flexural strength of concrete is also very low. Because, when we have an element under the positive flexural load, we will have compression in the upper layer and tension in the lower layer (Figure 1.18). Because of the lack of tension, we will have the failure of the element at the point of tension failure.

1.10 ELASTIC MODULE OF CONCRETE

Elastic module or young module is the stress-to-strain ratio of a solid concrete when the Hooks law is indefeasible. In fact, it shows the stiffness of concrete. So, when the compressive strength of concrete will increase, the elastic module also should increase.

To calculate the elastic module of concrete, we need the stress-strain curve. Then we should calculate the slope of the line tangent to the linear part of the stress-strain curve. You can see a sample of stress-strain curve for concrete in Figure 1.19.

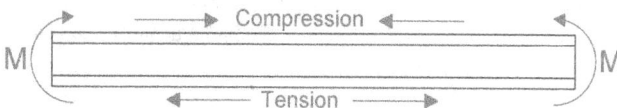

FIGURE 1.18 Correlation between flexural strength and tensile strength. (Graph created by the author.)

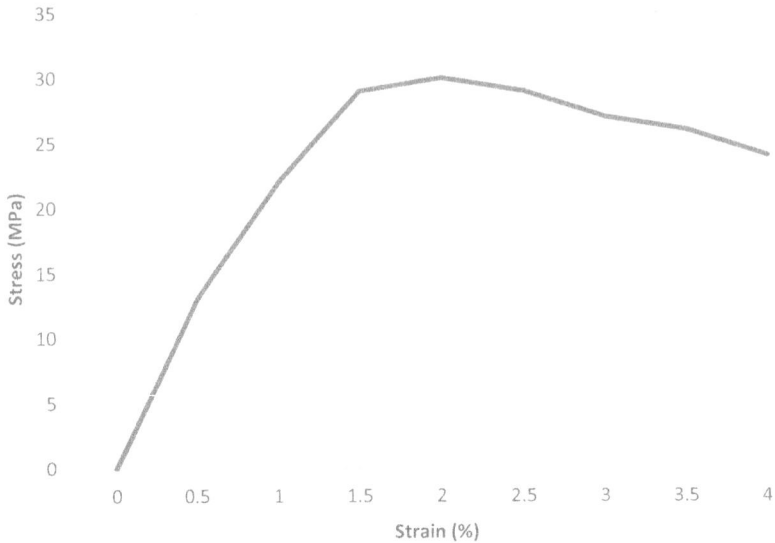

FIGURE 1.19   Stress-strain curve for a C30 concrete. (Graph created by the author.)

To calculate the elastic module for concretes with a compressive strength less than 50 MPa we can use below equation:

$$E = 5000\sqrt{f'c}$$

In this equation, E is the elastic module of concrete in MPa, and f'c is the 28 days standard cylinder compressive strength of concrete in MPa.

Another concept that is important to be known is the ductility of concrete. You should know that, if the compressive strength of concrete increases, the ductility will decrease. So, we can say, high strength concrete could be a brittle material.

For the design of concrete structures, using brittle materials is not recommended, especially in the seismic zones of the world because of the failure type of these materials. But you have to know three comments for this problem:

- The ductility of concrete with a compressive strength of less than 70 MPa is good enough to use as a ductile construction material.
- For the concretes with a compressive strength of more than 70 MPa, we can control the ductility by using suitable steel bars and steel fibers. So, these concrete also can be used as a good construction material.
- We cannot use concrete as a construction material without using steel bars or steel fibers, because of concrete weakness against tension. So, by using steel bars and/or steel fibers with concrete, we will have a very good construction material which we call reinforced concrete.

You can see a typical stress-strain curve for different compressive strength types of concrete in Figure 1.20.

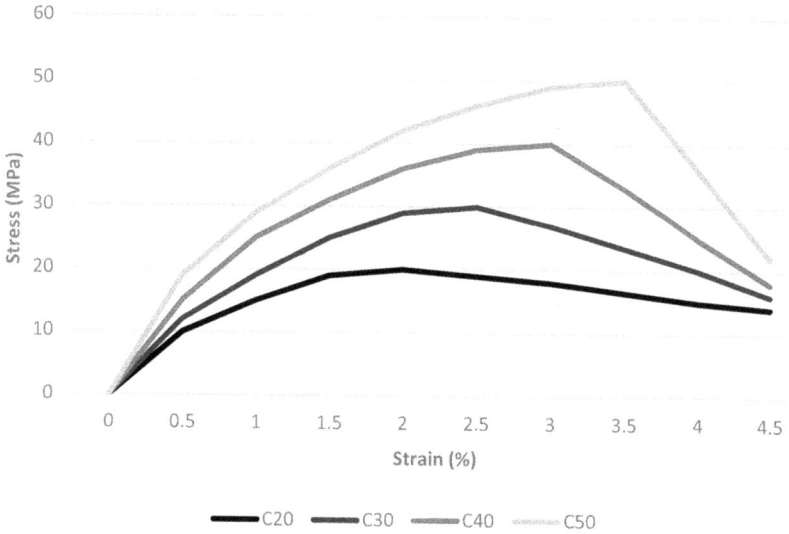

FIGURE 1.20   Typical stress-strain curve for different types of concrete. (Graph created by the author.)

As you can see in Figure 1.20, for higher compressive strength, after failure the stress will decrease with higher amount for the same strain. So, the material is more brittle. But as described before, we can control it by using steel bars or steel fibers.

1.11   CONCRETE PERMEABILITY AND WATER TIGHT STRUCTURES

The permeability of concrete is one of the key elements for the quality and performance of concrete. As mentioned before, to make a high-quality concrete, we should reduce w/c or w/b ratio. If we make this kind of concrete, then we are going to make an impermeable concrete.

For an impermeable concrete, water and other chemicals that can be very harmful to concrete cannot infiltrate the concrete. We would like to make an impermeable concrete for two reasons:

- For water-tight structures: Some of the concrete structures need to be water-tight. For example, swimming pools or water storages, which should be water tight. For these kinds of structures, we need to decrease the permeability of concrete.
- To increase the durability of structures: We will talk about the durability in the next part. You should know that, if aggressive chemical ions can go inside the concrete, they can reduce the mechanical behavior of the concrete structure. They can destroy the steel bars or the concrete itself. So, to control the infiltration of chemical ions into the concrete, we need to decrease the permeability.

There are too many points that we should consider to make an impermeable concrete. You can see some of them below:

- Reduce the water-to-binder or water-to-cement ratio: It can minimize the micropores inside the concrete. This is the best way to control the permeability of concrete.
- Use of mineral additives: They can reduce the micropores inside the concrete. These materials can reduce w/b and they can also produce more C-H-S and make a compacted concrete microstructure. The best powder additive for this reason is the silica fume.
- Use of plasticizers and super-plasticizers: These chemicals can reduce the w/b and also, and they can help to make a flowable concrete. One of the most important points to make an impermeable concrete is a good compacted concrete. If we have a good plastic flowable concrete made with a good super-plasticizer, we can guarantee good compaction of concrete.
- Use of PVC water stop tapes (Figure 1.21) in the joints: If you have any joint during concreting, you should use the water stop tapes to control the infiltration of water from the joints.
- Use suitable vibrator machines: As mentioned before, the compaction of concrete is very important and good vibration is one of the key elements for the best compaction.
- Prevention of cracking: If you make a good quality concrete with good implementation, we expect to have an impermeable concrete. But if you have any kind of crack on the surface of concrete structure, water and other chemicals can infiltrate from these cracks into the concrete. So, the concrete cannot be impermeable. To control cracking, you should consider all of the mentioned in the curing part of this session. You should cure the concrete as good as possible to control cracking.
- Use of waterproofing admixtures: You can find these kinds of chemicals in liquid and solid form. They are zooming on the pores inside concrete. They try to fill these pores to control the permeability. If you would like to

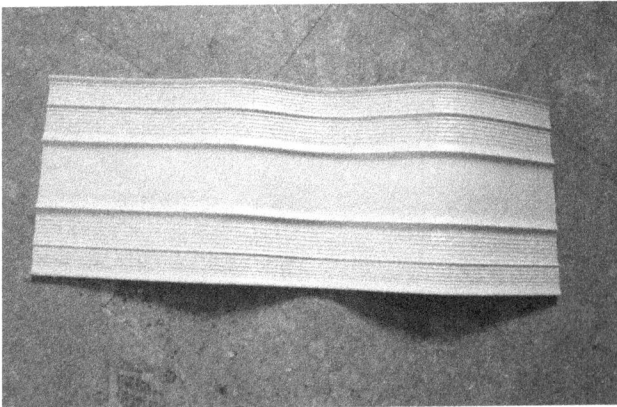

FIGURE 1.21   PVC water stop tape for the joints. (Photograph by the author.)

use these chemical admixtures, you should consider other points that we mentioned before to make a watertight concrete. The most important mistake is thinking of only using a waterproofing admixture.

1.12 DURABILITY AND EFFECT OF THE ENVIRONMENT ON CONCRETE

One of the most important concepts in concrete technology is the durability of concrete structures. What does it mean?

Durability is the perpetuity of the structure against environmental destroyable effects over time. As we are using the natural resources as the construction materials to build the structure, we should consider the durability due to sustainable development.

Sustainable development is a kind of development without destroying the natural resources and the environment. We should consider the sustainable development in all kinds of industries. In concrete industry, it means that we should produce the best quality concrete with the highest durability that is possible and by using minimum natural resources, without destroying the environment. To make a concrete with the minimum usage of natural resources and without destroying the environment, we should consider below points:

- Use minimum amount of water to control the consumption of drinkable water: To do that, we should use plasticizers and super-plasticizers to reduce the amount of water in concrete mix.
- Use minimum amount of Portland cement to control the air pollution: As the production of one ton of Portland cement will produce about one ton of CO_2, we should control the usage of pure Portland cement. To do that, we should replace a percentage of pure Portland cement with the mineral additives which are waste materials, and also by using plasticizers and super-plasticizers we can reduce the amount of water and w/c. So, we can use less Portland cement for the same mechanical properties.
- Use of mineral additives to increase the durability of concrete: As these materials are by-products of some industries, by using them, we can help to control pollution from the waste materials.
- Use of recycled materials if it is possible: There are too many trials for the usage of recycled aggregates in concrete. Sometimes the durability of concrete made with the recycled aggregates is low. So, the usage of recycled aggregates is restricted for some projects. But the researches are going on and we should try to use these materials again for the production of concrete. Because they can increase the pollution of our Earth.

Instead of making an environmental-friendly concrete, we should try to make a durable concrete for the structures. We will talk about durability in later parts of this book. But for now, we will discuss five environmental conditions that could affect the concrete and destroy the steel bars or concrete itself. Also, we will discuss the prevention methods to control the durability of concrete against these aggressive environmental conditions.

1.12.1 ALKALI AGGREGATE REACTION

This is a reaction between some active minerals in some kinds of aggregates and the alkali materials (sodium and potassium oxides) in cement. This reaction needs also a little amount of moisture to occur. It will cause expansion inside concrete and severe cracking will happen. So, the control of this reaction is very important due to concrete durability.

As you can see in Figures 1.22 and 1.23, the cracks of alkali aggregate reaction (AAR) are very dangerous for the durability and even stability of the concrete structure. These cracks will happen within about 2 or 3 years after the construction. So, this is a long-term reaction that proceeds slowly inside the concrete element. Because of this type of progress, we call this reaction also "concrete cancer." We will see destroying of the structural element from inside without any exterior motivator. This is exactly like the progress of cancer in the body.

We have two types of alkali aggregate reactions:

- Alkali silica reaction: This reaction will happen between the active silica minerals inside some kinds of aggregates and the alkalis of cement beside moisture.
- Alkali carbonate reaction: This reaction will happen between the active carbonate minerals inside some kinds of aggregates and the alkalis of cement beside moisture.

    The alkali silica reaction (ASR) is more dangerous than the alkali carbonate reaction because there are more aggregates with active silica materials. Also, the reaction for active silica is more powerful and more rapid than for the active carbonate.

    With attention to the above mentioned, we realize that the control of AAR is very important at the time of concrete production. To control this reaction, we can consider below points:

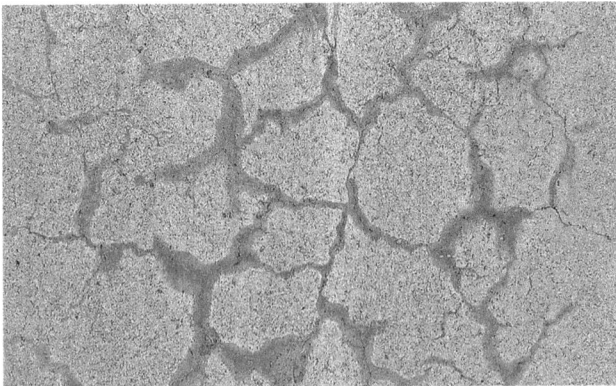

FIGURE 1.22 Cracks caused by alkali aggregate reaction. ("Surface of a concrete pillar of the building of the National Gallery of Canada at Ottawa".)

FIGURE 1.23 Alkali aggregate reaction. ("Bc rueckreise 025 swartz bay 2ndary efflorescence alkali silica reactions" by Achim Hering.)

- Checking the aggregates: The first and most important point is testing the aggregates for the potential of active materials needed for the AAR. There are several test methods for this reason. For example, we can test the aggregates by ASTM C227 test method.

  If the aggregates are potentially active, we should not use them in concrete production. But sometimes, we can only use one type of aggregate or all of the local aggregates are potentially active. In this case we should consider other points.
- Use of low alkali cement: Low alkali cement is a Portland cement with total alkali less than 0.6%. If we make a concrete with low alkali cement, we do not have the danger of AAR in the future even with the potentially activated aggregates. Because the amount of alkali is not enough to start and progress of reaction. Sometimes, there is no low alkali cement in a country or region, because the production process for low alkali cement depends on the raw materials of the cement factory. So, access to the suitable raw materials for low alkali cement in some regions is not possible.
- Use of blended cement: Blended cement means the use of Portland cement mixed with mineral additives with a defined percentage of mixing. When we use blended cement, we will decrease the amount of total alkali in the concrete mix. So, we can control the AAR.

- Reduce the permeability of concrete: As mentioned before, moisture is one of the critical elements for the AAR. So, if the moisture cannot infiltrate into the concrete, the reaction will control. We can reduce the permeability of any kind of concrete with the considerations mentioned in the previous session.

1.12.2 Carbonation Reaction

Carbonation reaction is a reaction between carbon dioxide in the air and water which produce carbonic acid and concrete. This acid will penetrate inside concrete and will reach the steel bars. As concrete itself is an alkali material, it can protect the steel bars from corrosion. But when the carbonic acid penetrates the concrete, it will decrease the pH around the steel bars and the corrosion will start. This corrosion of steel bars will cause the expansion of the bar and this expansion will cause the cracking of concrete. So, we can see the signs of corroded steel bars from outside of the concrete element.

The most important element for the carbonation reaction is enough amount of carbon dioxide in the air. So, the places for this reaction are the big crowded cities with high air pollution.

The second element for the carbonation reaction is enough amount of moisture. So, if the humidity of air in a crowded city will be high, the danger of carbonation reaction will be very high. But it does not mean that only in humid climate conditions we will have the carbonation attack. With only three to five times of raining in a year, we can see a carbonation reaction in concrete structures (Figure 1.24).

To reduce the carbonation attack effect in concrete structures, we can consider below points:

- Reduce the permeability of concrete: As mentioned before, the carbonation reaction needs water and carbon dioxide to penetrate into the concrete element. If we reduce the permeability of concrete, we can control the

FIGURE 1.24 Steel bar corrosion in concrete structures. ("This is one of a series of 46 pictures, which compares two bridges. The second bridge (picture 46) crosses a canal inside the City of Amsterdam" by Achim Hering.)

penetration of water and carbon dioxide into the concrete. So, the carbonation reaction will control. To reduce the permeability of concrete, you can see previous section of this chapter about permeability.

- Increase the concrete cover on the steel bars: We know that the carbonation itself does not have any effect on the concrete. It will attack the steel bars inside the concrete. If we increase the concrete cover on the steel bars, water and carbon dioxide will arrive to the steel bars later. So, the carbonation reaction can control.

The carbonation attack in urban structures like bridges has a significant effect on the durability of these structures. So, the control of this reaction in these kinds of structures is very important.

1.12.3 CHLORIDE ATTACK

One of the most dangerous chemical ions for concrete structures is the chloride ion. This will attack the steel bars and corrosion will start like an electrochemical cell (Figure 1.25). Some parts of the steel bars will be anode and the other parts act as the cathode. This kind of corrosion will happen rapidly in concrete structures. So, the prevention of this attack is more important than the other ones mentioned before.

Chloride attack will produce FeOH and also ferrite oxide which causes the expansion of steel bars and the concrete covering the steel bars.

How can a concrete structure expose to the chloride ion?

Concrete structures can expose to the chloride ion in two main ways:

- Deicing salt: As you know, in winter and at the time of snowing, to melt the ice, we are using salt or mixing of salt and sand. All kinds of these salts contain chloride ions. This ion from the surface of roads can move with the

FIGURE 1.25 Corrosion of the steel bars. ("This is one of a series of 46 pictures, which compares two bridges. The first bridge, (pictures 1–45), is part of the Canadian freeway QEW (Queen Elizabeth Way)" by Achim Hering.)

wheels of vehicles and can splash to the concrete elements like the columns of parking lots or bridges and even to the viaducts of the bridges or many other types of concrete structures. So, when we are using the deicing salts at cold climate locations, we should consider the chloride attack on the structural elements which may be exposed to the chloride ion.

- Chloride of sea water: Sea water contains high concentration of chloride. So, when a structure is exposed to sea water, the danger of chloride attack will be very high. On the other hand, the chloride ion can evaporate into the air from the sea water and go so far from the sea. This chloride can attack the structures that are not contacted to the sea water directly. This kind of chloride ion is called airborne chloride. The severity of the chloride attack from airborne ions is not comparable with the attack from direct sea water. But we should consider the danger of chloride attack in the structure even tens of kilometers far from the seaside.

As mentioned before, the control of chloride attack to the concrete structures is very important, especially in the seaside structures. So, to avoid corrosion due to the chloride attack we should consider below points:

- Reduce the permeability of concrete: To prevent the penetration of the chloride ion inside the concrete element, we should reduce the permeability of concrete. To do that, we should consider the points mentioned in the earlier section of this chapter about permeability. But, you should pay more attention to this part for the chloride attack. Because it is more dangerous than the carbonation reaction. For example, the use of a concrete with the minimum w/b ratio is more important.
- Use of mineral additives, especially the silica fume: Many studies showed that the use of mineral additives, especially the silica fume, besides reducing the w/b ratio in the concrete mix, can control the chloride attack in the structures. So, the use of silica fume in the seaside structures in many countries is mandatory. We will talk about the use of these mineral additives in concrete in the future. But, to control the chloride attack we should use these additives in nearly the maximum dosage.
- Curing: Concrete curing is one of the most important points for the reduction of permeability. But we point it here again because this is very important for the control of the chloride attack in concrete structures. Chloride ions can attack the steel bars even from a very thin crack on the surface of the concrete.
- Increase the concrete cover on the steel bars: To retard arriving of the chloride ion to the steel bars, we should increase the concrete cover on the steel bars. If the quality of concrete will be high enough, it can prevent the attack of chloride ions on the steel bars.
- Use of liquid epoxy on the surface of the steel bars: As the control of chloride attack is very important, some new methods were developed for this reason. One of these methods is using epoxy on the surface of the steel bars. This epoxy layer will prevent the formation of the electrochemical cell. So, it can prevent corrosion.

FIGURE 1.26 Corrosion of the steel bars. ("This is one of a series of 46 pictures, which compares two bridges. The second bridge (picture 46) crosses a canal inside the City of Amsterdam" by Achim Hering.)

If we implement all of the above mentioned about the quality of concrete, there is no need to use the epoxy layer on the surface of steel bars.

- Use of special cover on the surface of concrete: We can use special types of liquid admixtures which can implement on the surface of the concrete. These materials can prevent the penetration of chloride ions into the concrete. So, they can control the corrosion of the steel bars. If we implement all of the above mentioned about the quality of concrete, there is no need to use these covers on the surface of the concrete. Sometimes, for some very important structures, we can use these materials for more assurance about the safety of the structure against the chloride attack.
- Cathodic protection: This is a technique to control steel bar corrosion. In this technique, we should make the steel bars as the cathode in an electrochemical cell and use a sacrificial metal as anode which will corrode. This technique can protect steel bars from corrosion. But, it is not simple to use that in all kinds of structures. So, as mentioned before, the best we can do is the production and implementation of a high-quality concrete to protect the structure from corrosion (Figure 1.26).

1.12.4 SULFATE ATTACK

Sulfate ion is another dangerous chemical for concrete. Unlike the other chemicals that attack the steel bars, sulfate only attacks the concrete itself. So, the corrosion will start from the surface of the concrete.

The concrete elements could expose to the sulfate ion in two ways:

- Sulfate ion from the soil: Soil in some locations could contain a high amount of sulfate ions to attack the concrete. It will attack the structures that are in contact with the soil. Like several foundations or some columns. So, we

should test the soil to check the amount of sulfate, when we would like to start a structure that will be in contact with the soil.
* Sulfate ion from water: Sea water and also under-ground water, if it is in contact with the sulfate soil, could contain high amount of sulfate ions. It can attack the structures that are in contact with this kind of water. For example, offshore structures are in danger of chloride and sulfate attack.

The mechanism of sulfate attack in concrete is the reaction of three calcium aluminate (C_3A), which is one of the main compounds of Portland cement, with the sulfate ion. The real chemical reaction is very complicated and depends on the type of sulfate salt (calcium sulfate, sodium sulfate, magnesium sulfate). But, the main compounds produced due to the sulfate attack are the ettringite and thaumasite which are the products of a complicated chemical reaction. In this case, moisture is needed for the completion of the chemical reaction.

The main important factor for the sensitivity of concrete against sulfate ions is the amount of C_3A in the Portland cement or total binder that we are going to use for concrete production. C_3A is one of the main compounds of cement. We will talk about the compounds of Portland cement in Chapter 2. There are different amounts of C_3A in different types of Portland cement (between 3% and 12%). So, the most resistant cement against sulfate ion is the cement with the lowest amount of C_3A.

In the case of a sulfate attack, the main body of concrete will start to destroy (Figure 1.27). So, the loading capacity of the concrete element will decrease dramatically. But there is no corrosion in the steel bars directly because of the sulfate attack.

In the case of offshore structures, chloride and sulfate will attack the concrete elements together. So, high attention is needed for the quality of concrete to protect it from corrosion.

As mentioned before, the protection against sulfate attack is also very important because this kind of corrosion can decrease the loading capacity of the structure. So,

FIGURE 1.27   Cracks on concrete that can be caused by sulfate attack. ("Concrete texture".)

for the protection of concrete structures against sulfate attack, you should consider below points:

- Reduce the permeability of concrete: As mentioned before, sulfate reaction in concrete needs moisture to accomplish. So, if we can reduce the permeability of concrete, we can remove the water from the inner parts of the concrete element. So, we can protect it from corrosion. Although the corrosion will start from the surface of the concrete element, we can retard the start of corrosion in the inner parts of the concrete element.
- Reduce the w/b ratio: Reduction of w/b ratio, besides reducing the permeability of concrete, can prevent the travel of water during the body of concrete element. So, it is very important to produce a concrete with w/b ratio as low as possible.
- Use of ASTM Type II or V cement: The choice of cement in the case of a sulfate attack is very important. We will talk about the ASTM's different types of Portland cement in Chapter 2. For now, you should know that all types of Portland cement have more than 8% of C_3A instead of type II which has between 5% and 8% of C_3A and type V which has less than 5% of C_3A. We call type V Portland cement the sulfate-resistant cement also. For moderate exposure to sulfate we can use type II Portland cement and for severe exposure to the sulfate ion we should use type V Portland cement. The definition of moderate and severe exposure conditions in different texts and standards are different. But we will talk about this issue in the chapter on durability.

  Another important issue is the use of suitable cement in the condition of chloride and sulfate attack together. In this case, the best cement is type II Portland cement. Because recent researches showed that a low amount of C_3A will cause more mobility for the chloride ion inside the concrete body. So, the use of type V cement because of very low amount of C_3A is not good. On the other hand, the use of type I cement with a very high amount of C_3A is not good because of the sulfate ion. The best cement for this case is type II cement with a moderate amount of C_3A, which can control the mobility of the chloride ion inside concrete and the resistance of concrete against the sulfate ion together.
- Use of mineral additives: Mineral additives as a percent of the total binder in concrete mix can reduce the amount of total C_3A in the binder. So, it can prevent corrosion due to the sulfate ion. All kinds of mineral additives can improve the resistance of concrete against sulfate ions. But, recent researches showed that the best one is the GGBS. It has a very good effect on the resistance of concrete against sulfate ions.

1.12.5 EFFECT OF FREEZE THAW CYCLE

One of the environmental conditions which can destroy the concrete is the freeze thaw cycle. If we have the minus temperature at night and plus temperature at day in a region, and if we have enough moisture inside the concrete elements, then we will have freeze-thaw cycles in the concrete elements.

Water inside the concrete element will freeze at night. This freezing will cause a pressure inside concrete because of the expansion of water during the freezing process. Then in the morning and by growing the temperature more than zero, the ice starts to melt. This process will repeat during the next night and day. It can happen several times during the cold season in a region for a concrete element. The pressure inside the concrete and the repeating process will cause severe destruction of the concrete element.

A good example of this process is the concrete tables beside the gardens or beside the streets. There is enough moisture inside these concrete elements because of the watering of flowers and plants in the garden, or the splash of water caused by moving of the vehicles which cause always exposure of these elements to water. If the variation of temperature will happen in the region of these tables, the freeze-thaw cycles can destroy them as you can see in Figure 1.28.

To reduce the effect of freeze-thaw cycles on the concrete element, you should consider below points:

- Reduce the permeability of concrete: If we reduce the permeability of concrete, water cannot infiltrate into the concrete element and if there is no moisture inside concrete, the freeze thaw cannot progress inside the concrete element. So, the element will protect against the destroying effect of the freeze-thaw cycle.

  To reduce the permeability of concrete you should see the permeability session in this chapter. If the concrete elements are precast, like tables, control of the concrete production process and curing and other implementation considerations will be more easily. So, we can produce high-quality precast concrete elements.

- Use of air-entraining admixtures: We will talk about air-entraining admixtures later in this book. For now, these chemical admixtures can entrain air bubbles with the similar shape and similar size during all parts of the concrete elements body. These air bubbles will act as a safety valve to control the pressure of water freezing in the freeze-thaw cycle. So, they can reduce the pressure inside the concrete and the destroying effect of freeze-thaw cycles will be controlled.

FIGURE 1.28   Freeze-thaw cycles effect on the tables beside a street. (Photograph by the author.)

REFERENCES

Aitcin P.C. *High Performance Concrete*, E&FN SPON, 2004.

American Society for Testing and Materials, Standard Practice for Sampling Freshly Mixed Concrete, ASTM C172-99.

American Society for Testing and Materials, Standard Practice for Making and Curing Concrete Test Specimens in the Laboratory, ASTM C192-00.

American Society for Testing and Materials, Standard Practice for Capping Cylindrical Concrete Specimens, ASTM C617-98.

American Society for Testing and Materials, Standard Specification for Ready-Mixed Concrete, ASTM C94-00.

American Society for Testing and Materials, Standard Specification for Flow Table for use in Test of Hydraulic Cement, ASTM C230-98.

American Society for Testing and Materials, Standard Specification for Air-Entraining Admixture for Concrete, ASTM C260-00.

American Society for Testing and Materials, Standard Specification for Light-weight Aggregates for Structural Concrete, ASTM C330-00.

American Society for Testing and Materials, Standard Test Method for Compressive strength of Cylindrical Concrete Specimens, ASTM C39-01.

American Society for Testing and Materials, Standard Test Method for Flexural strength of Concrete, ASTM C78-00.

American Society for Testing and Materials, Standard Test Method for Slump of Hydraulic Cement Concrete, ASTM C143-00.

American Society for Testing and Materials, Standard Test Method for Air Content of Freshly Mixed Concrete by the Volumetric Method, ASTM C173-01.

American Society for Testing and Materials, Standard Test Method for Potential Alkali Reactivity of Cement-Aggregate Combination, ASTM C227-97.

American Society for Testing and Materials, Standard Test Method for Air Content of Freshly Mixed Concrete by the Pressure Method, ASTM C231-97.

American Society for Testing and Materials, Standard Test Method for Potential Alkali Silica Reactivity of Aggregates (Chemical Method), ASTM C289-94.

American Society for Testing and Materials, Standard Test Method for Static Modules of Elasticity and Poisson's Ratio of Concrete in Compression, ASTM C469-94.

American Society for Testing and Materials, Standard Test Method for Length Change of Concrete Due to Alkali-Carbonate Rock Reaction, ASTM C1105-95.

American Society for Testing and Materials, Standard Test Method for Potential Alkali Reactivity of Aggregates (Mortar Bar Method), ASTM C1260-94.

American Society for Testing and Materials, Standard Test Method for Flow of Hydraulic Cement Mortar, ASTM C1437-99.

American Society for Testing and Materials, Standard Test Method for Compressive Strength of Hydraulic Cement Mortars, ASTM C109-99.

Bertolini L, Elsener B, Pedeferri P, Polder R, *Corrosion of Steel in Concrete, Prevention, Diagnosis, Repair*, Wiley-VCH, 2004.

Connor Jerome J, Faraji Susan, *Fundamentals of Structural Engineering*, Springer, 2016.

Devanath, "Concrete texture." Retrieved from: https://pixnio.com/textures-and-patterns/concrete-texture/dust-stone-dry-concrete-grey-monochrome-texture-pattern-old.

European Standard Organization, Admixtures for Concrete, Mortar and Grout Test Methods, EN480 Series.

European Standard Organization, Concrete-Part 1: Specification, Performance, Production and Conformity, EN206-1, 2000.

European Standard Organization, Testing Fresh Concrete, EN12450 Series.

European Standard Organization, Testing Hardened Concrete, EN12390 Series.

Gjorv E. Odd, *Durability Design of Concrete Structures in Severe Environments*, Taylor & Francis, 2009.

Gjorv Odd E, *Durability Design of Concrete Structures*, Taylor & Francis, 2009.

Gordon, Jim, "An Iraqi construction worker (foreground) wets down the burlap covering freshly poured concrete pillars, as part of the concrete curing process, as he helps work on the Sumail Primary Health Clinic being built in Dahuk, Dahuk Province, Iraq (IRQ), by Iraqi contractors and sub-contractors as an US Army (USA) Corps of Engineers (USACE) designed and managed C type clinic (the largest and most capable of all health clinics) during Operation IRAQI FREEDOM." Retrieved from: https://nara.getarchive. net/media/an-iraqi-construction-worker-foreground-wets-down-the-burlap-covering-freshly–887a6f.

Hauschild Michael, Rosenbaum Ralph K, Olsen Sting Irving, *Life Cycle Assessment, Theory and Practice*, Springer, 2018.

Heinrichs Harald, Martens Pim, Michelsen Gerd, Wiek Arnim, *Sustainability Science, An Introduction*, Springer, 2016.

Hering, Achim, "Bc rueckreise 025 swartz bay 2ndary efflorescence alkali silica reactions." Retrieved from: https://commons.wikimedia.org/wiki/File:Bc_rueckreise_025_swartz_bay_2ndary_efflorescence_alkali_silica_reactions_1.png.

Hering, Achim, "This is one of a series of 46 pictures, which compares two bridges. The first bridge, (pictures 1–45) is part of the Canadian freeway QEW (Queen Elizabeth Way), crossing a river and "Oakwood Drive" in Niagara Falls, Ontario, Canada." Retrieved from: https://commons.wikimedia.org/wiki/File:Qew_bruecke_nf_beton_kaputt_25_von_46.jpg.

Hering, Achim, "This is one of a series of 46 pictures, which compares two bridges. The first bridge, The second bridge (picture 46) crosses a canal inside the City of Amsterdam." Retrieved from: https://commons.wikimedia.org/wiki/File:Qew_bruecke_nf_beton_kaputt_33_von_46.jpg.

Hering, Achim, "This is one of a series of 46 pictures, which compares two bridges. The first bridge, The second bridge (picture 46) crosses a canal inside the City of Amsterdam" Retrieved from: https://commons.wikimedia.org/wiki/File:Qew_bruecke_nf_beton_kaputt_34_von_46.jpg.

Iranian Institute for Research on Construction Industry, 9th topic of National Rules for Construction, "Concrete Structures", 2009.

Iranian National Management and Programming Organization, National Handbook of Concrete Structures, 2005.

Iranian Standard Organization, Standard Specification for Ready Mixed Concrete, ISIRI6044, 2015.

Isokoski, Jukka, "Constructing natural gas line in winter, Finland," 2007. Retrieved from: https://commons.wikimedia.org/wiki/File:Constructing_natural_gas_line_in_winter, _Finland.jpg#filehistory.

Janamian Kambiz, Aguiar Jose, *A Comprehensive Method for Concrete Mix Design*, Materials Research Forum LLC, 2020.

Mahmood Zadeh Amir, Iranpoor Jafar, Concrete Technology and Test (Farsi), Golhaye Mohammadi, 2007.

McKinely, Robyn, "Spraying the curing compound." Retrieved from: https://www.flickr.com/photos/elmiracollege/2902049697.

Mostofinejad Davood, *Concrete Technology and Mix Design (Farsi)*, Arkane Danesh, 2011.

Newman John, Choo Ban Seng, *Advanced Concrete Technology, Concrete Properties*, Elsevier, 2003.

Popovics Sandor, *Concrete Materials, Properties Specification and Testing*, NOYES Publications, 1992.

Ramachandran V.S, Beaudion James, *Handbook of Analytical Techniques in Concrete Science and Technology, Principles, Techniques and Applications*, William Andrew Publishing, 2001.

Ramezanianpoor Aliakbar, Arabi Negin, Cement and Concrete Test Methods (Farsi), Negarande Danesh, 2011.

Richardson M, *Fundamentals of Durable Reinforced Concrete*, SPON Press, 2004.

Richardson M, *Fundamentals of Durable Reinforced Concrete*, Spon Press, 2002.

Safaye Nikoo Hamed, *Introduction to Concrete Technology (Farsi)*, Heram Pub, 2008.

Shinkolobwe, "Surface of a concrete pillar of the building of the National Gallery of Canada at Ottawa presenting the typical crack pattern of the alkali-silica reaction (ASR)." Retrieved from: https://commons.wikimedia.org/wiki/File:ASR_concrete_pillar_National_Gallery_of_Canada_02.jpg.

Zandi Yousof, *Advanced Concrete Technology (Farsi)*, Forouzesh Pub, 2009.

Zandi Yousof, *Concrete Tests and Mix Design (Farsi)*, Forouzesh Pub, 2007.

2 Portland Cement

Portland cement is the most important constituent material for any kind of concrete. We know that this is a very dangerous material for the environment in the production process. But there is no way! If we would like to produce a high-quality concrete, we should use Portland cement.

For a concrete technologist, it is necessary to know about the Portland cement. From the production process to the chemical constituents, several types of Portland cement and the usage of this material in different types of concrete. The quality control process of Portland cement is also another important issue for concrete technologists.

In this chapter, we will start with the definition of the Portland cement, then we will talk about the raw materials for the production of cement and also the production process. In the next part, the subject is different types of Portland cement according to the ASTM standard. We will learn about other types of cement that we can use for other purposes instead of the production of normal concrete. The next issue is the other binders or mineral additives that we can use besides Portland cement for the production of high-quality concrete. Then we will talk about the quality control of Portland cement in the factory. Different tests that we should consider to evaluate the quality of cement to use in the production of concrete will be discussed. Finally, we will learn about the transportation and storage of cement for different uses.

2.1 WHAT IS THE PORTLAND CEMENT?

Portland cement (Figure 2.1) is a complex powder consisting of calcium silicates and calcium aluminates, which can react with water and become a hard material during

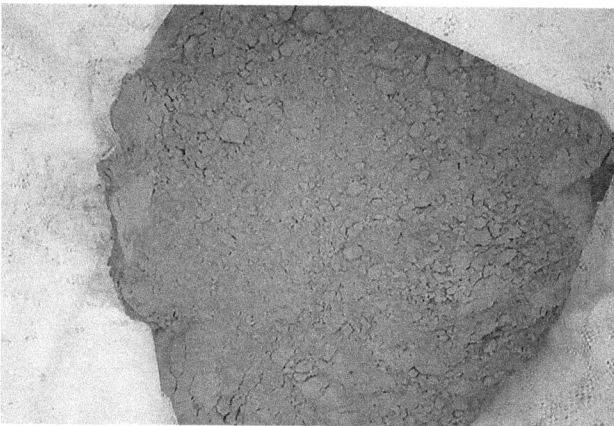

FIGURE 2.1   Portland cement. (Photograph by the author.)

DOI: 10.1201/9781003384243-2

time. The reaction of water and cement is called the hydration reaction, which is the reason for the hardening of cement-based mortars and concrete.

Portland cement is one of the most important construction materials all over the world. We can count the uses of this material in the construction industry as below:

- Production of different types of concrete: From light weight to heavy weight and with different kinds of mechanical properties.
- Production of shotcrete: We can use shotcrete for different purposes, like the stabilizing of tunnel walls.
- Production of mortar: We can use different types of mortar as the binder for any kind of finishing in construction projects.
- Production of lightweight mortars: This mortar can be used as the light-weight filler for roofs and also for the production of different types of light-weight bricks.
- Production of slurry: We can use slurry as the filler for different types of finishing or cracks and joints.

You can see that we have many different usages for the Portland cement in the construction industry. So, we have cement production factories in different countries of the world. Because the construction industry needs this material very much. The only problem is: "Portland cement production is very dangerous for the environment!"

As we would like to produce 1 ton of cement, we will release about 950 kg of carbon dioxide into the air (Figure 2.2). On the other hand, the use of resources like different kinds of fuels depending on the factory is very high for the production of Portland cement.

Because of these problems, countries all over the world try to modify the usage of cement in their country. Most of the time, they try to produce the cement according to their demand and the export of cement is not acceptable for many countries.

FIGURE 2.2 Air pollution by cement factories. ("Al Kufa Cement plant" by Carsten Wiehe.)

We should consider the minimum amount of cement in all types of concrete mix design. In fact, a good quality concrete with the minimum Portland cement content is the best choice for the production of concrete according to the sustainable development, which is a very important concept to protect resources and environment for the future generation.

2.2 CEMENT HISTORY

The history of cement refers to many years ago. We found so many ancient constructions made of materials similar to the concrete. Many of them are silicate and lime-based mortars. They are very strong mortars compared with the other ancient structures made of clay and brick.

The first example of a concrete concept goes to 300 BC for the Romans. They used volcano ash with lime and water to produce a very strong mortar which could be hard over time, exactly like the concrete. Volcano ash consists of silicate and aluminate. It can react with lime and water like cement to harden. So, Romans used this mortar for the construction of some important ancient structures like the Pantheon and Colosseum (Figures 2.3 and 2.4).

From that time, so many people tried to make composite mortars like today's concrete consisting of the fillers like sand and different types of binders like cement. All of the mortars harden during time with the reaction of the binder with water. So, we can call all of these mortars concrete and we can call all of the binders to produce these mortars, cement.

The evolution of cement with today's concept goes to the Joseph Aspdin, an English mason. He started the production of cement in 1824 and he named it the Portland cement because it was similar to the limestones of Portland island in the United Kingdom. This is the reason for the name of the Portland cement that we are using today.

FIGURE 2.3   The Pantheon, one of the most popular ancient buildings in Italy. ("Pantheon (Rome) – Front".)

FIGURE 2.4   The colosseum, one of the most popular ancient buildings in Italy. ("A 4×4 segment panorama of the Coliseum at dusk".)

As the Portland cement was a very good binder for the production of different types of mortars, most of the advanced countries at that time started to import the cement from the producers in the UK and then started to produce the Portland cement in their countries.

The first production of Portland cement in North America was in Pennsylvania, USA, in 1871. After that, in Canada also, they started to produce Portland cement in 1889. Now, we can say that in most of the countries all over the world we have at least one producer of the Portland cement. So, you can access many kinds of Portland cement with different specifications at any location in the world.

2.3   CEMENT RAW MATERIALS

There are different types of raw materials for the production of Portland cement. The most important subject is the availability of the raw material near the cement factory. The list of the raw materials for Portland cement is as follows:

- Silicates: you can find the silicates in marl, calcium silicate stones, clay, limestone (Figure 2.5), quartzite, or other types of stones with high amounts of silicate.
- Calcium: you can find calcium in calcite, limestone, clay, marble, marl, and other types of stones with high amounts of calcium.
- Aluminate: you can find aluminate in clay, bauxite, and limestone.
- Ferrite: you can find ferrite in clay or the different types of iron ore.
- Sulfate: you can use calcium sulfate in the production process of Portland cement.

FIGURE 2.5   Limestone as raw materials for the production of Portland cement. (Photograph by the author.)

As mentioned before, most of the cement producers use stones with different types of raw materials nearby their factory as the main raw material (Figure 2.6). But they should add the shortage of raw materials from the mines far from their factory.

2.4   CEMENT PRODUCTION

The main part of cement production is the production of cement clinker. Clinker is baked raw material of cement that we should grind to produce the final cement. Clinker is in the shape of gray grains with different sizes from about 10 mm or less to 20 mm or more (Figure 2.7).

There are different types of cement factories in the world. In fact, to operate a cement factory, we need to buy the instruments from the producers and assemble them

FIGURE 2.6   Mine of raw materials for a cement factory. (Photograph by the author.)

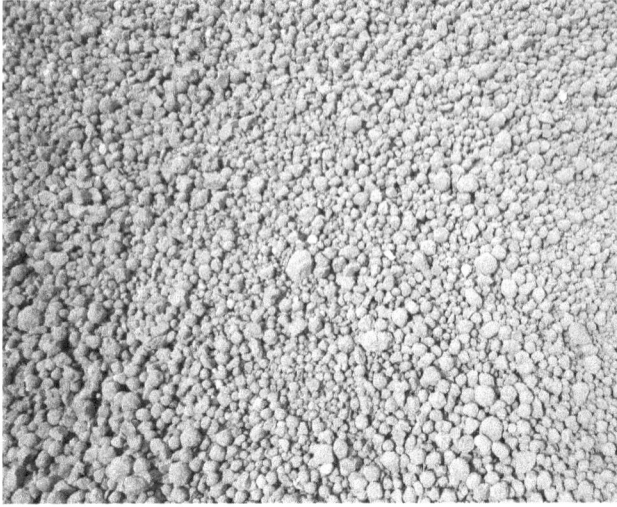

FIGURE 2.7   Portland cement clinker. (Photograph by the author.)

with the guidance of the producers and consults for this industry. So, the production process depends on the producer and the instruments of the factory (Figure 2.8).

But we can explain a rough process for the production of Portland cement step by step as follows:

- In the first step, raw materials are derived from the mine nearby the factory. They are in the form of mid-size to large stone parts.
- The stone parts should be crushed with a stone crusher to the defined particle size, less than 25 mm.

FIGURE 2.8   Production line in a cement factory. ("Factory of National Cement Share Company" by DFID- UK Department for International Development.)

FIGURE 2.9 The roller cylinder kiln of a cement factory. ("Cement kiln. Location: Gorazdze Cement Plant near Chorula (Poland)".)

- Then the raw materials should grind like a powder and also should mix together. In this part of the process, we should check the raw materials compositions to control the amount of any material that is necessary for the production of cement. Maybe we should add some other raw materials from other mines to our raw materials to balance the amount of each chemical for the best quality of cement.
- Ground raw materials should go to the cement kiln to bake. The kiln is a roller cylinder with a mild slope (Figure 2.9). Raw materials go to the kiln from the upper side of the kiln and the flame goes to the kiln from the other side. In this part of the process, cement will bake and we will have Portland cement clinker after the kiln.
- The final part of the process is grinding. We should grind Portland cement clinker after cooling to the ambient temperature with a little amount of calcium sulfate to produce the final Portland cement powder. The use of calcium sulfate is for the control of the immediate reaction of cement with water, which we will talk about later.

We can use two or more different kinds of grinders for cement. The bullet mill (Figure 2.10) and the roller mill (Figure 2.11). Particle size distribution for the cement ground with the roller mill is better, but the output of the bullet mills is usually better. We can use both of them in a cement factory.

2.5 COMPOSITIONS OF PORTLAND CEMENT

We talked about the production of Portland cement in the previous part. In this part, we are going to talk about the chemical compounds of Portland cement. There are four major compounds in any type of Portland cement as follows:

FIGURE 2.10   Picture of a bullet or ball mill. ("Top view of a cement ball mill".)

- Tricalcium silicate (C_3S): This is the main component of the Portland cement which is about 50% to 70% of it. The strength of concrete especially in the early ages is due to this compound. The hydration heat released from the reaction of C_3S and water is high and most of it will release between the age of 1 day and 7 days. C_3S reacts with water to produce calcium silicate hydrate (C-H-S) and $Ca(OH)_2$. The C-H-S is responsible for the strength and impermeability of concrete. So, it is the main product for the hydration reaction of the Portland cement.
- Dicalcium silicate (C_2S): This is the second important compound of the Portland cement which is about 20% to 30% of it. The strength of concrete for ages more than 7 days is due to the reaction of this compound. The hydration heat for C_2S reaction is not high because it will release over a long period of time. C_2S reacts with water to produce C-H-S and $Ca(OH)_2$. The only difference between C_3S and C_2S is the speed of the reaction which is slower in the case of C_2S.

FIGURE 2.11   Picture of a roller mill. This is for the raw materials, but similar ones can be used for cement. ("Hanson Cement, Clitheroe" by Alan Murray-Rust.)

- Tricalcium aluminate (C_3A): This compound is not important in the hydra-tion reaction. Because it will not produce the C-H-S. C_3A reacts with water exactly after the first contact of cement with water and produces the ettr-ingite which will cause the flash set of cement. So, we should control this flash set, because we would like to work with the concrete during time. To control this flash set, we should add a little amount of calcium sulfate to the Portland cement clinker during the grinding process. The amount of calcium sulfate could be between 2% and 3% and depends on the target set-ting time. If we increase the amount of calcium sulfate we can increase the setting time. Calcium sulfate will react with the C_3A and remove it from the reaction environment. So, the flash set will control.

  As mentioned before, C_3A is the main cause for the sensitivity of cement against sulfate attack. In fact, sulfate ion will attack the C_3A and the products of this chemical reaction will cause severe expansion. So, to control the sul-fate attack in concrete we should decrease the amount of C_3A in the cement.

- Tetra-calcium alumino-ferrite (C_4AF): This compound does not have any role in the strength and other mechanical properties of the Portland cement. But it will be produced in the production process of cement. The only effect of this compound is on the gray color of the cement.
- Free CaO: In the production of Portland cement it is possible to retain a little amount of free CaO in the final product. This free CaO will cause the expansion in the concrete or mortar. So, we should decrease it to the minimum amount to control this expansion. Standards did not give us the maximum amount for this free CaO. Instead of that we should control the expansion of cement with the autoclave expansion test that we will talk about it in the future.
- Total Alkalis: As mentioned before, to control the alkali aggregate reaction in concrete we can decrease the amount of total alkalis (Na_2O and K_2O) to the minimum. It is not possible all the time and for all of the cement manufacturers to control the amount of alkalis in their cement. Because sometimes, the type of raw materials do not allow them to produce low alkali cement. Instead of their effect on the alkali aggregate reaction, these alkalis have no positive or negative effect on the other properties of concrete and mortar.

2.6   TYPES OF PORTLAND CEMENT

Now that you learned about the compounds of the Portland cement, it is time to talk about different types of cement. There are different divisions for the types of Portland cement with different standards. Here, we are going to use the ASTM standards. According to the ASTM standard we have five main types of Portland cement as below:

- Portland cement type I: This type of cement is the ordinary Portland cement that we can use for several purposes instead of the places in which concrete will be in contact with the sulfate ion. In this case, this type of cement cannot resist the sulfate ion attack. We can produce this type of cement with different compressive strength types. We will talk about the test for cement compressive strength which is doing with a special mortar cast in 5 x 5 x 5 cm cubes. According to this standard test we can produce three strength types for this cement:

  Type I-325: This is the lowest strength type for type I cement. Most of the time, producers try to produce stronger ones. But if we have this kind of cement, we can use it as the mortar cement or for the production of low-strength concretes. You can see the compositions and properties of a sample I-325 cement in Tables 2.1 and 2.2.

TABLE 2.1
Chemical Composition of a Sample I-325 Cement

Chemicals	SiO_2	Al_2O_3	Fe_2O_3	CaO	MgO	Cl	So_3	Free Cao	Total Alkalis
Amount	23.5%	5.8%	3.1%	60.0%	3.1%	0.02%	2.0%	1.3%	0.75%

TABLE 2.2

Mechanical Properties of a Sample I-325 Cement

Parameter	Blaine (cm²/gr)	Initial Setting Time (min)	Final Setting Time (min)	3 Days Comp Strength (kg/cm²)	7 Days Comp Strength (kg/cm²)	28 Days Comp strength (kg/cm²)	Autoclave Expansion (%)
Amount	2950	120	180	140	280	350	0.08

Type I-425: This is the mid-range kind of type I Portland cement. Most of the cement producers can produce I-425 type of cement. We can use this cement in all kinds of concrete and mortar except for the sulfate exposure. You can see the compositions and properties of a sample I-425 cement in Tables 2.3 and 2.4.

Type I-525: This is the highest strength type of cement. Production of this cement is a little difficult. So, some of the producers cannot produce this type of cement. We can use this cement for all kinds of concrete and mortar. But most of the time, this cement is used for the production of high-strength concrete.

• Portland cement type II: This is the modified type I cement. In fact, the difference between this cement and type I cement is the better resistance of type II cement against sulfate attack.

As mentioned before, C_3A is the sensitive part of cement for the sulfate attack. A usual type I Portland cement has more than 8% of C_3A (sometimes up to 14%). But type II cement has between 5% and 8% of C_3A. So, this is a more resistant cement against sulfate attack. For the places with moderate exposure to the sulfate ion, we can use type II cement. For the places exposed to sulfate and chloride together, the best cement is type II. You can see the compositions and properties of a sample type II cement in Tables 2.5 and 2.6.

TABLE 2.3

Chemical Compositions of a Sample I-425 Cement

Chemicals	SiO_2	Al_2O_3	Fe_2O_3	CaO	MgO	Cl	SO_3	Free Cao	Total Alkalis
Amount	21.0%	5.0%	3.45%	64.0%	2.2%	0.02%	2.2%	1.3%	0.75%

TABLE 2.4

Mechanical Properties of a Sample I-425 Cement

Parameter	Blaine (cm²/gr)	Initial Setting Time (min)	Final Setting Time (min)	3 Days Comp Strength (kg/cm²)	7 Days Comp Strength (kg/cm²)	28 Days Comp Strength (kg/cm²)	Autoclave Expansion (%)
Amount	3100	110	160	180	390	490	0.08

TABLE 2.5
Chemical Compositions of a Sample II Cement

Chemicals	SiO_2	Al_2O_3	Fe_2O_3	CaO	MgO	Cl	So_3	Free Cao	Total Alkalis
Amount	21.1%	4.9%	4.0%	64.2%	2.2%	0.02%	2.2%	1.3%	0.75%

TABLE 2.6
Mechanical Properties of a Sample II Cement

Parameter	Blain (cm²/g)	Initial Setting Time (min)	Final Setting Time (min)	3 Days Comp Strength (kg/cm²)	7 Days Comp Strength (kg/cm²)	28 Days Comp Strength (kg/cm²)	Autoclave Expansion (%)
Amount	3000	120	190	150	330	410	0.08

- Portland cement type III: This type of cement is the accelerated type. The amount of C_3A and also C_3S in this type of cement is high. We can use this type of cement in cold climate conditions and in cold seasons for more heat of hydration and control of concrete early age freezing.

  Today we have strong accelerator chemical admixtures to use in concrete. We can modify the type and amount of acceleration that we would like for concrete. So, the production of type III cement is reduced in cement factories. In fact, the demand for this type of cement is reduced because of the accelerator chemical admixtures.

- Portland cement type IV: unlike type III, this is the retarded type of Portland cement. The amount of C_2S in this type of cement is higher than the others. So, strength increase will be very slow in this type.

  As we have very good retarding chemical admixtures to control and increase the setting time and strength grow of concrete, the production of this kind of cement is reduced nowadays.

- Portland cement type V: This type of cement is the sulfate-resistant cement. It contains less than 5% of C_3A. So, the resistance of this cement against sulfate attack is higher than type II. For the concrete elements which are in contact with a high amount of sulfate ion, we should use type V Portland cement. You can see the compositions and properties of a sample type V cement in Tables 2.7 and 2.8.

TABLE 2.7
Chemical Compositions of a Sample V Cement

Chemicals	SiO_2	Al_2O_3	Fe_2O_3	CaO	MgO	Cl	So_3	Free Cao	Total Alkalis
Amount	21.7%	4.2%	5.2%	63.9%	2.1%	0.02%	1.9%	1.3%	0.75%

TABLE 2.8
Mechanical Properties of a Sample V Cement

Parameter	Blain (cm²/g)	Initial Setting Time (min)	Final Setting Time (min)	3 Days Comp Strength (kg/cm²)	7 Days Comp Strength (kg/cm²)	28 Days Comp Strength (kg/cm²)	Autoclave Expansion (%)
Amount	3000	160	210	130	250	330	0.08

For more information about the minimum and maximum amounts of compositions in different types of Portland cement, you can see Tables 2.9 and 2.10.

2.7 OTHER TYPES OF CEMENT

There are other types of cement instead of five ASTM types mentioned before. We can use them for special cases. In this section, we are going to talk about them and their usage. We can name other types of cement as follows:

- Portland cement type IA: According to the ASTM standard, we have type IA cement. It is exactly type I cement with air-entraining effect. We mentioned that for special cases where the concrete element is exposed to freezing and thawing, we need to entrain air bubbles into the concrete. For this

TABLE 2.9
Minimum and Maximum Amounts of Compositions in Different Types of Portland Cement

	SiO_2 (%)	Al_2O_3 (%)	Fe_2O_3 (%)	CaO (%)	MgO (%)	SO_3 (%)
Type I	18–23	4–7	1.5–5	60–67	0.5–4	1.5–5
Type II	20–24	3–6	2–5	60–67	0.5–5	2–4
Type III	18–22	2–6	1–5	60–67	0.5–5	2–5
Type IV	21–23	3–6	3–6	62–64	1–4	1.5–3
Type V	20–24	2–5	3–6	62–66	0.5–4	1.5–4

TABLE 2.10
Minimum and Maximum Amounts of Compositions in Different Types of Portland Cement

	C_3S (%)	C_2S (%)	C_3A (%)	C_4AF (%)	Blaine (cm²/g)
Type I	40–65	10–35	6–14	5–13	2900–3600
Type II	35–65	6–35	2–7	7–14	2900–3700
Type III	45–70	5–25	5–15	4–14	3200–4200
Type IV	35–45	25–40	3–5	11–18	2800–3300
Type V	45–65	10–30	1–5	10–20	2800–3500

case, we can use type IA cement. Today, we have air-entraining admixtures with high quality and low dosage to entrain air into the concrete. So, the use of type IA cement is very low and cement manufacturers don't like to produce this kind of cement.

- Portland cement type IIA: Like before, this is exactly type II cement with air-entraining effect. This type of cement also is not a regular cement for production and use in the industry exactly like type IA.
- Portland cement type IIIA: This is also exactly type III cement with air-entraining effect. This type of cement also is not a regular cement for production and use in the industry.
- White cement: This type of cement is for architectural usages when we would like to produce a concrete or mortar with light color. To produce the white cement, we should control the amount of C_4AF in the final product. As you know, this is responsible for the gray color of Portland cement. So, if we reduce the amount of ferrite and magnesium oxides in the raw materials of cement, we can produce white cement. Most of the time, white cement is like type I Portland cement in the case of other compositions instead of C_4AF.
- Color cement: This type of cement is only for architectural usage. To produce this type of cement, we should only add some pigments to the ground white cement. We can produce different colors of concrete or mortar with color cements. But the more cheap technique for the production of colored concrete or mortar is the usage of pigments separately from the white cement. Also, we can use these pigments with normal gray cement for the production of some colors like black, brown, dark green, dark blue, or red. So, with the use of pigments and gray or white cement, we can produce many different kinds of colored concrete. So, the production of color cements was reduced too much in the cement factories.
- Masonry cement: This is a kind of cement that is the mix of Portland cement with powdered limestone and other additives to produce a special cement for mortar. For this kind of cement, the strength and mechanical properties are not much important. Instead of that, the properties like price, controlled setting time, good workability, and sometimes the durability is very important. The production process and additives for this kind of cement will cause lower carbon dioxide release during the production process.
- Expanding cement: This is a special kind of cement with expanding properties. Unlike Portland cement, this kind of cement will expand after hydration. So, we can use it for the special purposes for which we need a concrete without shrinkage. Although this is not a common cement, you can find it in many countries like European countries or the North America.
- High blain cement: This is a Portland cement with higher blain than normal. As mentioned before, the normal blain of cement should be more than $2800 \, cm^2/g$ and it is most of the time between 3000 and $3200 \, cm^2/g$. If we grind this cement more than normal, then we will have a cement

with the blain more than 3200 and up to 4200 cm²/g. This cement has special properties like high initial strength, high final strength, and short setting time. Sometimes and for some special projects, we need these properties. So, we can use this kind of cement. The production of high blain cement is possible for all of the Portland cement manufacturers in the world. The most important point is good particle size distribution curve for high blain cement that should be considered. In fact, the particle size distribution shape of high blain cement should be the same as normal blain cement. The only difference should be only finer total distribution of particles.

- Calcium aluminate cement: This is a kind of cement that has completely different properties and constituent material than the Portland cement. The main material is calcium aluminate which can hydrate with water and give us special properties like very high initial strength, good resistance to some chemical attacks, and good resistance to the high temperature.
- Geopolymer cement: This is a kind of natural cement that is completely different from the Portland cement. This is a kind of aluminosilicate system with an alkaline solution. These materials will react and harden like the Portland cement at room temperature. As we can use some materials like metakaolin or fly ash for the production of this kind of cement or concrete, we can say that this is more environment-friendly than the Portland cement. So, researches on this cement are growing for increasing the usage of this material instead of Portland cement.

2.8 HYDRATION REACTION

As mentioned in the first chapter of this book, the hydration reaction is an exothermic chemical reaction between the most important composition of cement and water. The most important part of the reaction is as below:

$$(C_3S, C_2S) + H_2O \rightarrow C\text{-}H\text{-}S + Ca(OH)_2$$

As you can see, this is a reaction between C_3S and C_2S with water. It will produce C-H-S and $Ca(OH)_2$ during the time. The start of the reaction is from the first contact of cement particles with water and it can continue for time to 90 days and even more. But the most important time is 28 days after the production of concrete. You know that the most important chemical composition for the mechanical properties and durability of concrete is the C-H-S. But the $Ca(OH)_2$ is not a useful material itself. You can see in the next part that, we can use calcium hydroxide in the reaction with the mineral additives to increase the quality of concrete.

Although the other components of cement will react with the water, the products of the reaction between C_3A and C_4AF and water is not as important as the C_3S and C_2S products for the properties of concrete.

You can see the hydration reaction products for different compositions of the Portland cement in Table 2.11.

TABLE 2.11
Hydration Reaction of the Portland Cement

Cement Composition	Reaction
C_3S	Tricalcium silicate + Water = Calcium silicate hydrate + Calcium hydroxide
C_2S	Dicalcium silicate + Water = Calcium silicate hydrate + Calcium hydroxide
C_3A	Tricalcium aluminate + Gypsum + Water = Ettringite
	Tricalcium aluminate + Ettringite + Water = Calcium mono sulfo aluminate
	Tricalcium aluminate + Calcium hydroxide + Water = Tricalcium aluminate hydrate
C_4AF	Tetracalcium alumino ferrite + Calcium hydroxide + Water = Calcium alumino ferrite hydrate

2.9 MINERAL ADDITIVES (SUPPLEMENTARY CEMENTITIOUS MATERIALS)

The subject of mineral additives in concrete technology is very important. So, we will talk about this subject in Chapter 3. But for now, you should know briefly about these materials.

The most important component in these materials is the active SiO_2. As mentioned before, the hydration reaction between C_3S and C_2S with water will produce C-H-S and calcium hydroxide. The C-H-S is responsible for the strength and quality of concrete. But there is no special effect for the $Ca(OH)_2$ itself.

We can use calcium hydroxide with mineral additives. The active SiO_2 can react with the calcium hydroxide and produce more C-H-S. So, we can consume the $Ca(OH)_2$ by using mineral additives.

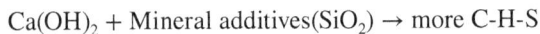

$$Ca(OH)_2 + Mineral\ additives(SiO_2) \rightarrow more\ C\text{-}H\text{-}S$$

The most important mineral additives are as below:

- Silica fume: This is the by-product of ferro-silica alloy factories. We can derive it from the releasing chimney of these factories.

  This is the most powerful pozzolan material with more than 90% of active SiO_2.
- Fly ash: This is the by-product of electricity power plants whose fuel is the collier. So, if we have this kind of power plants in a country, then we can have the fly ash. The amount of active SiO_2 in the fly ash could be different depending on the type of collier and the power plant system. But it is about 50% and more.
- Ground granulated blast furnace slag (GGBS): This is the by-product of steel factories in the process of deriving iron from the minerals. If we cool down the slag rapidly, we will have a good quality GGBS that we can use in concrete. The amount of SiO_2 in the GGBS is different depending on the production process of the steel factory. But most of the time, it is less than 40%.

- Natural pozzolans: These are some materials in the nature with the pozzolanic activity. The amount of SiO_2 and activity of these natural pozzolans are different depending on the mineralogy of the material. But, we can find good quality natural pozzolans in different parts of the world.

2.10 BLENDED CEMENT

We can use Portland cement and mineral additives separately in the concrete mix. But some of the cement producers mix these additives with the Portland cement to produce blended cement. If the production process of the blended cements will be accurate enough, using these cements is much better and simple than the separate use of the Portland cement and mineral additives. But unfortunately, in some countries, the production process and the quality of blended cements are not good. So, most of the time, it is better to use the Portland cement and mineral additives separately in the concrete mix.

There are different considerations for the good quality of any blended cement as below:

- The quality of mineral additives itself should be high. It depends on the amount of active SiO_2 in the mineral additive.
- The quality of the Portland cement clinker should be high enough to produce a good quality blended cement.
- The grinding process of the Portland cement clinker and the mineral additives should be separate because the hardness of these materials is different from each other. This is especially the case of the GGBS and some of the natural pozzolans which need grinding. In the case of silica fume and fly ash, they are in the powder form. So, there is no need for grinding.
- The fineness of mineral additives should be more than the fineness of the Portland cement. Because the activity of the mineral additives is less than the Portland cement.
- The mixing of ground Portland cement and mineral additive powder should be uniform. So, the amount of Portland cement and the mineral additive should be the same in any portion of the mix.

There are six types of blended cement in ASTM standard. You can see these cements in Table 2.12.

TABLE 2.12
Blended Cements in ASTM Standard

Type of Cement	Explanation
Type IS	Portland cement mixed with the slag
Type IP	Portland cement mixed with Pozzolan
Type P	Pozzolan cement when there is no consideration for the early age strength
Type I (PM)	Modified pozzolan cement
Type I (SM)	Modified slag cement
Type S	Slag cement to use with pure Portland cement in concrete

TABLE 2.13
Different Cement Types in EN197

Type of Cement	Explanation	Number of Types
CEM I	Pure Portland cement	1 type
CEM II	Different blended cement	19 types
CEM III	Slag cement	3 types
CEM IV	Pozzolan cement	2 types
CEM V	Composite cement	2 types

In European standard, there are different types of blended cement. You can see the different types of cement in EN197, European standard in Table 2.13.

As you can see in the table, the CEM I is the only pure Portland cement in European standard with about 95% to 100% of Portland cement clinker. For the other types you can see Figure 2.12 which is from the EN197, European standard.

We should consider the use of blended cement in all countries of the world. Because to produce sustainable concrete, it is very important to use these kinds of cements in the production process.

Furthermore, it is very important for the cement producers to produce high-quality blended cements for the concrete industry. By doing this, they will have below benefits:

- Environment-friendly production process.
- More efficient and beneficial products.
- Several types of products, So, they can sell their cement to all kinds of projects in any kind of environment.
- Help for the building of durable concrete structures.
- Using waste materials from other industries.

2.11 QUALITY CONTROL OF CEMENT

Like any other material to be used in an industrial production process, we should control the quality of cement before using it in the concrete production. The only problem with cement is the hardness and difficulties in the quality control process. We can name below points as the difficulties for the quality control of cement:

- Expensive test equipment for cement quality control. So, for a project or ready mixed plant it is difficult to buy them.
- Specified technician lab for cement quality control. There is a need for physical and chemical lab technician and also concrete lab technician. As you know, in a construction project or a ready mixed plant, most of the time, we have a concrete technician lab.
- Number of cement specimens in a concrete plant will be too much because the amount of cement trucks that transport the cement for a project is too much and the laboratory should take a specimen from each truck.

Cement Type	Designation	Notation	Clinker K	G.G.B.S. S	Silica fume D	Pozzolana — Natural P	Pozzolana — Industrial Q	Fly ashes — Silic. V	Fly ashes — Calcar W	Burnt Shale T	Limestone L	Limestone LL	Minor Additional constit.
I	Portland Cement	I	95-100	-	-	-	-	-	-	-	-	-	0-5
	Portland Slag Cement	II / A-S	80-94	6-20	-	-	-	-	-	-	-	-	0-5
		II / B-S	65-79	21-35	-	-	-	-	-	-	-	-	0-5
	Portland Silica Fume Cement	II / A-D	90-94	-	6-10	-	-	-	-	-	-	-	0-5
	Portland Pozzolana Cement	II / A-P	80-94	-	-	6-20	-	-	-	-	-	-	0-5
		II / B-P	65-79	-	-	21-35	-	-	-	-	-	-	0-5
		II / A-Q	80-94	-	-	-	6-20	-	-	-	-	-	0-5
		II / B-Q	65-79	-	-	-	21-35	-	-	-	-	-	0-5
II	Portland Fly Ash Cement	II / A-V	80-94	-	-	-	-	6-20	-	-	-	-	0-5
		II / B-V	65-79	-	-	-	-	21-35	-	-	-	-	0-5
		II / A-W	80-94	-	-	-	-	-	6-20	-	-	-	0-5
		II / B-W	65-79	-	-	-	-	-	21-35	-	-	-	0-5
	Portland Burnt Shale Cement	II / A-T	80-94	-	-	-	-	-	-	6-20	-	-	0-5
		II / B-T	65-79	-	-	-	-	-	-	21-35	-	-	0-5
	Portland Limestone Cement	II / A-L	80-94	-	-	-	-	-	-	-	6-20	-	0-5
		II / B-L	65-79	-	-	-	-	-	-	-	21-35	-	0-5
		II / A-LL	80-94	-	-	-	-	-	-	-	-	6-20	0-5
		II / B-LL	65-79	-	-	-	-	-	-	-	-	21-35	0-5
	Portland Composite Cement	II / A-M	80-94	<-------- 6-20 -------->									0-5
		II / B-M	65-79	<-------- 21-35 -------->									0-5
III	Blastfurnace Cement	III / A	35-64	35-65	-	-	-	-	-	-	-	-	0-5
		III / B	20-34	66-80	-	-	-	-	-	-	-	-	0-5
		III / C	5-19	81-95	-	-	-	-	-	-	-	-	0-5
IV	Pozzolanic Cement	IV / A	65-89	-	<-------- 11-35 -------->				-	-	-	-	0-5
		IV / B	45-64	-	<-------- 36-55 -------->				-	-	-	-	0-5
V	Composite Cement	V / A	40-64	18-30	<-------- 18-30 -------->				-	-	-	-	0-5
		V / B	20-39	31-50	<-------- 31-50 -------->				-	-	-	-	0-5

FIGURE 2.12 All types of cement in European standard EN197. (Photo retrieved from EN197.)

- Some of the quality control test results will be reachable after some days or weeks. After this period of time, a concrete plant cannot store the cements. They should consume all of the cement that they received.

So, what should we do about the difficulties mentioned above?

Unfortunately, most of the construction projects and ready mixed plants do not test the properties of cement in their laboratory. They should trust the quality control of cement factories and get the results from them.

Sometimes, we can check the results by testing them in another laboratory that can do the cement quality control tests. In fact, the projects can give the test results from the QC of cement factory in a defined period of time. For example, each week or month or season. Then to check these results, they can get some samples from a random truck and get them to another laboratory. This will be the best process that we can do to ensure the quality of cement as the most important constituent material of concrete.

In this part of the chapter, we are going to talk about the most important tests for the quality control of cement.

2.11.1 FINENESS OF CEMENT

As mentioned before, the last process for the production of Portland cement is the grinding of clinker with calcium sulfate.

One of the most important properties of Portland cement is the fineness of the particles which has a very high effect on the quality of cement. We can name below points as the effects of cement fineness on the cement and concrete properties:

- A finer cement means more activity for cement.
- A finer cement means more heat of hydration in the hydration reaction process.
- A finer cement means higher early age strength and less growth in the strength of concrete between the age of 7 and 28 days.
- A finer cement will need more water for the hydration reaction.
- A finer cement will consume more plasticizer and super-plasticizer for the same workability.
- A finer cement will cause less slump retention effect in the transportation process of concrete.

As you can see above, it is very important to know the fineness of Portland cement in the projects and the uniformity of the fineness is very important in the quality control of the concrete. Because if the fineness of cement will differ from one part to another one, it will have a drastic effect on the properties of concrete. From early age compressive strength to the amount of water and the final compressive strength. From the dosage of plasticizer and super-plasticizer to the slump retention effect in the transportation process of concrete.

The most important test for the evaluation of fineness is the blain test or ASTM C204 (Figure 2.13). In this test, we can evaluate the area of cement particles in a unit mass. So, the finer cement will have more area in a unit mass. The unit for the blain

FIGURE 2.13   Blain test equipment. (Photograph by the author.)

test is cm²/g. So, if we do the blain test for a common cement, we will have the result for about 2800 to 3300 cm²/g.

2.11.2   PARTICLE SIZE DISTRIBUTION

In the last part of the chapter, we mentioned fineness as a very important property of cement. In this part, we would like to talk about another very important property of cement which depends on the grinding process like fineness.

Particle size distribution is much important than the blain of cement because it is possible to have two cements with the same blain and different particle size distribution. In this case, the properties of these cements in the concrete production process will be different. So, the effect of particle size distribution on the properties of concrete is more important than the blain.

FIGURE 2.14   Sieve for cement particle size testing. (Photograph by the author.)

The best size for cement particles is between 5 and 45 μm. The particles finer than 5 μm have very high activity and can cause problems like drastically decreasing the setting time, slump loose, and very high amount of hydration heat especially at early ages. On the other hand, the particles coarser than 45 μm have very low activity. Their reaction to the production of C-H-S is very low. So, we should restrict the particles to coarser than 45 μm.

We have a test technique that is not accurate enough to find the percentage of very coarse particles in cement. We can use 45, 75, or 90 μm sieve and wash the cement on it. Then we can calculate the amount of cement on the sieve and it will show us the nonactive or low active part of cement. The standard test method for wet washing of cement is ASTM C786 or ASTM C430 (Figure 2.14).

But the most accurate test for the particle size distribution of cement is the test with the laser diffraction method. The instrument for this test is very expensive and accurate. This kind of instrument is used for the particle size distribution of other kinds of powders. But the researchers are working on the best methods to use this instrument, especially for the cement particles. For example, the standard test method for the particle size distribution of metal powders is ASTM B822. You can see a picture of the instrument for the laser diffraction method test in Figure 2.15 and a sample particle size distribution curve for cement in Figure 2.16.

FIGURE 2.15 Particle size distribution analyzer. ("Particle size distribution analyzer 990" by CILAS).

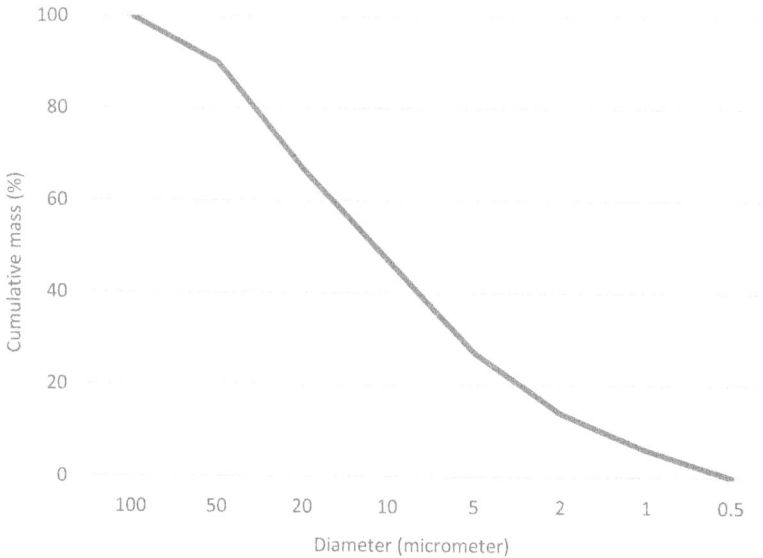

FIGURE 2.16 A typical particle size distribution curve for cement. Graph created by the author.

2.11.3 Cement Health

One of the most important specifications of Portland cement is the stability of volume in concrete or mortar. If the amount of some materials like free CaO in cement will be high, the cement mortar or concrete would like to expand. It can cause deep cracks in the surface and inside a concrete element. So, we can call a cement with controlled expansion a healthy cement.

To control cement health or expansion, we can use the autoclave expansion test according to ASTM C151. You can see the exact process for this test in the standard text. But briefly, we can explain the test as follows:

FIGURE 2.17 Cement prism mold. (Photograph by the author.)

We should make a cement mortar with special sand, cement, and water. Then we should mold it and make special prisms. To control the expansion of these prisms, we should put them in the autoclave and check their dimensions. If the dimension change will be in the defined area, we can use the cement for the production of concrete. You can see a picture of cement prism molds in Figure 2.17 and a picture of autoclave for the expansion test in Figure 2.18.

2.11.4 SETTING TIME OF CEMENT

When we make a concrete, we can see that in three phases:

- Phase 1: This is the fresh concrete phase. In this phase concrete can flow to a mold, we can shape it and make different types of structural elements, we can work with concrete, we can pump the concrete to the upper places of the structure, we can trowel the concrete surface, and we can mix concrete in a truck mixer or with other types of mixing elements.

  As you can see, the hydration reaction did not start in this phase or it is in the starting state.

  In the last point of this phase, the cement in concrete is going to the initial setting point. So, we can call the time between the first contact of cement and water to the last of fresh concrete phase the initial setting time. For a type I cement without any admixture or additive and in the normal temperature conditions (About 20°) it is usually about 70 to 110 minutes.

- Phase 2: the second phase of concrete is more or less the same as jelly phase. We can mix the concrete in this phase, but not as easy as before. If we walk on the surface of concrete element, our shoe trace will stand on the surface. So, we can say that the concrete is not hard enough to load and it is not soft enough to work.

  If we try to mix the concrete in this phase, we are cracking the bonds related to the hydration reaction. So, the concrete will be weaker than usual.

  In the last point of this phase, the cement in the concrete is going to the final setting point. So, we can call the time between the first contact of

FIGURE 2.18   Picture of autoclave. ("Instituto Butantan, São Paulo, Brazil.Autoclave" by Mike Peel.)

cement and water to the last of jelly concrete phase the final setting time. For a type I cement without any admixture or additive and in the normal temperature condition, it is usually about 150 to 200 minutes.

- Phase 3: This is the hardened concrete phase. In fact, the start of this phase is the final setting time. Concrete in this phase starts to be hard enough for loading. We can measure the compressive strength of concrete in this phase. It will start in the first hours of the hardened phase and will grow over time. The most important time for the compressive strength is 28 days after the final setting time.

We should know the initial setting time of cement to set the working time of concrete made with the defined cement. We should know the final setting time for the mold release of some concrete elements and precast concrete segments.

We have different types of setting time tests according to the different standards in the world. But the most important and usable test is the setting time test with Vicat needle according to the ASTM C 191.

In this test, first we should make a standard cement paste with normal consistency according to the ASTM C187. In fact, we should determine the amount of water we

FIGURE 2.19   Vicat test instrument. (Photograph by the author.)

need for the normal consistency of cement according to this standard. Then we should use this paste in the cone of the Vicat test instrument (Figure 2.19). After different times, we should release the 1 mm needle of Vicat test instrument as you can see in the figure to fall down on the surface of the cement paste and we should measure the amount of needle penetration into the cement paste. When the 1 mm needle penetration will be 25 mm, we can call that time the initial setting time, and when the needle sink cannot be visible to the paste, it will be the final setting time for the cement.

2.11.5   CEMENT COMPRESSIVE STRENGTH

If you would like to make a good concrete with defined compressive strength, and by special attention to the economy, you should know the minimum compressive strength of cement that you would like to use. So, this is a very important test for the concrete mix design accuracy.

Another important fact is the standard deviation for the compressive strength results for a defined cement. If the standard deviation will be very large, it shows that the quality control or production accuracy of cement is not good enough to guarantee the same quality for the cement during the time. It is possible for a cement producer to give us a cement with a very high compressive strength on some days and a very low compressive strength on other days. This cement will not be a good choice. Because first, you should use the minimum compressive strength of cement for your concrete mix design and it will not be economical for concrete production. Second, if you use this cement, your concrete quality will be different over time and it is not good for your resume. So, if you can find a cement with a mean value of the compressive strength but little standard deviation it would be the best choice for you.

As most of the time, we cannot measure the compressive strength of cement in a concrete laboratory in the projects or ready mixed concrete factories, we should give the monthly results from the cement laboratory.

The standard test method for the compressive strength of cement is ASTM C109. To measure the compressive strength of cement, we should use a special mortar with a defined well-graded silica sand, cement, and water mold in 5 x 5 x 5 cm cubes (Figure 2.20).

The mixing water was achieved according to the normal consistency test ASTM C187. The mixing of the mortar should be very good to achieve a steady mortar. Then the molded specimens should maintain a perfect curing condition in the curing room with controlled temperature and moisture. Then we can measure the compressive strength of these cubes with a special test machine (Figure 2.21).

Most of the time, we will use 1 day, 3 days, 7 days, 11 days, 28 days, 42 days, and 90 days compressive strength. But, the most important one is the 28 days compressive strength that we will use in the calculations of concrete mix design.

After testing the cement's compressive strength, we can compare the results with the standard specification for that kind of cement. For example, for type I Portland cement we have three types:

- Type I-525: This is a cement type I with the 28 days compressive strength of more than 525 kg/cm^2 for all of the tests. This is a high-strength cement with high blain and is suitable for the production of high-strength concrete and/or high initial strength concrete.
- Type I-425: This is a cement type I with the 28 days compressive strength of more than 425 kg/cm^2 for all of the tests. This is a normal strength cement and can be used for the production of all kinds of concrete instead of the places where we should use type II or V cement.

FIGURE 2.20 Molds for cement compressive strength test. (Photograph by the author.)

FIGURE 2.21 Test machine for the cement compressive strength test. (Photograph by the author.)

- Type I-325: This is a cement type I with the 28 days compressive strength of more than 325 kg/cm^2 for all of the tests. This is a low-strength cement and is not suitable for the production of concrete. It is better to use it for the production of special mortars.

2.11.6 SPECIFIC GRAVITY OF CEMENT

We have two kinds of specific gravity for cement and any other powder material as below:

- Particle-specific gravity: This is the specific gravity of cement particles without the air voids inside the cement powder. In this book, when we talk about the cement-specific gravity, we mean the particle-specific gravity.
- Bulk-specific gravity: This is the specific gravity of the bulk powder with the air voids inside that. We need this kind of specific gravity for the calculation

of the silo capacity. But as you can imagine, this is not a constant digit for the powders. It depends on the pressure on the surface of the cement. For example, if you have a huge silo for the storage of bulk cement, the bulk-specific gravity of cement in the lowest level of silo will be much higher than the specific gravity of cement in the upper level because the pressure of higher levels on the surface of lower levels will cause the cement powder to compress more and more. Also, if you store cement for a long time in a silo, it will compress during the time and the bulk-specific gravity of cement and the capacity of the silo will be higher during the time.

As mentioned before, we will use cement particle-specific gravity in the calculations of concrete mix design. So, it is important for us to know that.

The standard test method for the specific gravity of cement is the ASTM C188. In this method, we will use the Le Chatelier flask to measure the specific gravity of cement particles (Figure 2.22).

This is a very simple test that we can do in all cement and concrete laboratories. But you have to know that, for all types of pure Portland cement, the specific gravity is about 3.15 to 3.16 kg/L. it depends on the cement clinker. So, if the clinker did not change, there is no need to repeat the specific gravity test for the cement.

If you use blended cement, you should test the specific gravity to achieve the exact result because it depends on the type of blended material and the percent of use. By using this test for all parts of the cement, you can give an idea about the quality of the blended cement, especially about the percentage of the blended material in the cement.

2.11.7 CEMENT HEAT OF HYDRATION

As mentioned before, the hydration reaction between the cement particles and water is an exothermic reaction. But the amount of heat released is different for different types of cement. If we would like to name the types of cement from highest releasing of heat to the lowest one, it will be as follows:

- Type III
- Type I
- Type II
- Type V
- Type IV

But for each type of cement for example Type I, the heating release could be different according to the production conditions like the blain, particle size distribution, amount of C_3A in the clinker, and the amount of calcium sulfate in the grinding process.

On the other hand, if you use blended cement, the heat of hydration will be different according to the type of blended material and the percent of that. So, like the specific gravity test, you can examine the quality consistency of the blended cement by measuring the heat of hydration.

The standard test method for the heat of hydration is ASTM C186. There is a special instrument that you can use to measure the heat of hydration for cement.

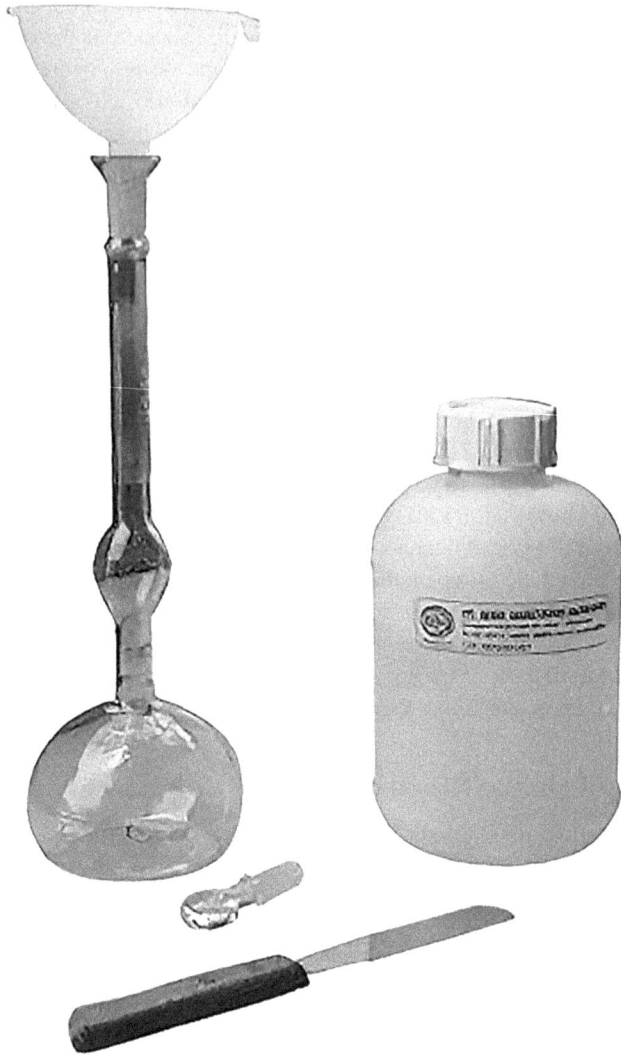

FIGURE 2.22   The Le Chatelier flask for cement-specific gravity test. ("Specific gravity test".)

2.12   ANALYSIS OF CEMENT

To estimate several properties of the Portland cement, there is a need for the chemical and physical analysis of cement. You can give a special sheet from the cement producer which is the analysis of cement and you can check different properties of cement from that. You can see three analysis sheets for three types of cement from a producer in Figures 2.23–2.25.

As you can see in the figures above, we can understand most of the properties of cement from the analysis sheet. For example, for the amount of C_3A, we have the following:

Chemical analysis of cement Type I			
Composition	Requirements according to ASTM C150	Requirements according to EN 197	Test Results
LOI (%)	Max 3.0	Max 5.0	1.29
SiO$_2$ (%)	----	----	21.05
Al$_2$O$_3$ (%)	----	----	5.31
Fe$_2$O$_3$ (%)	----	----	4.08
CaO (%)	----	----	63.15
MgO (%)	Max 6.0	----	2.1
SO$_3$ (%)	Max 3.5	Max 3.5	2.1
Na$_2$O (%)	----	----	0.21
K$_2$O (%)	----	----	0.63
Cl (%)	----	----	0.02
IR (%)	Max 1.5	Max 5.0	0.22
Free CaO (%)	----	----	1.98
LSF (%)	----	----	90.89
SIM (%)	----	----	2.24
ALM (%)	----	----	1.3
C$_3$S (%)	----	----	49.52
C$_2$S (%)	----	----	23.5
C$_3$A (%)	----	----	7.17
C$_4$AF (%)	----	----	12.4
Total Alkalies (%)	----	----	0.62
Physical analysis of cement Type I			
Composition	Requirements according to ASTM C150	Requirements according to EN 197	Test Results
Fineness (cm^2/gr)	Min 2800	----	3650
Initial setting time (min)	Min 45	Min 60	120
Final setting time (min)	Max 360	Max 360	190
2 days comp strength (kg/cm^2)	Min 100	Min 100	215
3 days comp strength (kg/cm^2)	----	----	270
7 days comp strength (kg/cm^2)	----	----	401
28 days comp strength (kg/cm^2)	Min 425	Min 425	522
Autoclave expansion (%)	----	----	0.03

FIGURE 2.23 Analysis sheet for a type I cement. (Photograph created by the author.)

- For type I cement, there is no restriction.
- For type II cement, we have maximum amount of 8% and for this cement, we have 6.98%.
- For type V cement, we have the maximum amount of 5% and for this cement, we have 4.08%.

And for the amount of autoclave expansion, we have the maximum standard amount of 0.8 for all of the three types and we have 0.03 for the sample type I cement, 0.01 for the sample type II cement and 0.04 for the sample type V cement.

Chemical analysis of cement Type II			
Composition	Requirements according to ASTM C150	Requirements according to EN 197	Test Results
LOI (%)	Max 3.0	----	1.3
SiO$_2$ (%)	Min 20.0	----	21.05
Al$_2$O$_3$ (%)	Max 6.0	----	5.31
Fe$_2$O$_3$ (%)	Max 6.0	----	4.19
CaO (%)	----	----	63.15
MgO (%)	Max 6.0	----	2.19
SO$_3$ (%)	Max 3.0	----	1.92
Na$_2$O (%)	----	----	0.21
K$_2$O (%)	----	----	0.59
Cl (%)	----	----	0.02
IR (%)	Max 1.5	----	0.26
Free CaO (%)	----	----	1.99
LSF (%)	----	----	90.99
SIM (%)	----	----	2.22
ALM (%)	----	----	1.27
C$_3$S (%)	----	----	49.9
C$_2$S (%)	----	----	23.17
C$_3$A (%)	Max 8.8	----	6.98
C$_4$AF (%)	----	----	12.7
Total Alkalies (%)	----	----	0.60
Physical analysis of cement Type II			
Composition	Requirements according to ASTM C150	Requirements according to EN 197	Test Results
Fineness (cm^2/gr)	Min 2800	----	3500
Initial setting time (min)	Min 45	----	160
Final setting time (min)	Max 360	----	220
2 days comp strength (kg/cm^2)	----	----	185
3 days comp strength (kg/cm^2)	Min100	----	240
7 days comp strength (kg/cm^2)	Min 175	----	360
28 days comp strength (kg/cm^2)	Min 315	----	480
Autoclave expansion (%)	----	----	0.02

FIGURE 2.24 Analysis sheet for a type II cement. (Photograph created by the author.)

For the 28 days compressive strength, we have:

- For type I cement, the minimum standard amount is 425 kg/cm^2 and sample cement has the compressive strength of 532 kg/cm^2.
- For type II cement, the minimum standard amount is 315 kg/cm^2 and sample cement has the compressive strength of 505 kg/cm^2.
- For type V cement, the minimum standard amount is 270 kg/cm^2 and sample cement has the compressive strength of 486 kg/cm^2.

To calculate the amount of C$_3$S, C$_2$S, C$_3$A, and C$_4$AF we need the amount of different chemicals which you can see in the upper rows of the tables.

Chemical analysis of cement Type V			
Composition	Requirements according to ASTM C150	Requirements according to EN 197	Test Results
LOI (%)	Max 3.0	Max 5.0	1.45
SiO$_2$ (%)	----	----	21.56
Al$_2$O$_3$ (%)	----	----	4.61
Fe$_2$O$_3$ (%)	----	----	4.81
CaO (%)	----	----	62.92
MgO (%)	Max 6.0	----	1.94
SO$_3$ (%)	Max 2.3	Max 3.0	1.8
Na$_2$O (%)	----	----	0.16
K$_2$O (%)	----	----	0.65
Cl (%)	----	----	0.01
IR (%)	Max 1.5	Max 5.0	0.24
Free CaO (%)	----	----	1.12
LSF (%)	----	----	89.45
SIM (%)	----	----	2.29
ALM (%)	----	----	0.96
C$_3$S (%)	----	----	49.25
C$_2$S (%)	----	----	25.12
C$_3$A (%)	Max 5.0	Max 5.0	4.08
C$_4$AF (%)	----	----	14.62
Total Alkalies (%)	----	----	0.59
Physical analysis of cement Type V			
Composition	Requirements according to ASTM C150	Requirements according to EN 197	Test Results
Fineness (cm^2/gr)	Min 2800	----	3350
Initial setting time (min)	Min 45	----	175
Final setting time (min)	Max 360	----	230
2 days comp strength (kg/cm^2)	----	----	160
3 days comp strength (kg/cm^2)	85	----	195
7 days comp strength (kg/cm^2)	Min 150	----	314
28 days comp strength (kg/cm^2)	Min 270	----	456
Autoclave expansion (%)	----	----	0.04

FIGURE 2.25   Analysis sheet for a type V cement. (Photograph created by the author.)

So, this is very important to achieve the exact amounts of any chemicals in the cement. We should use the ASTM C114 test method for this reason.

The best technique for the chemical analysis of cement is the XRD or X-ray diffraction method. This method studies the crystal structure to identify the crystalline phases present in a material and thereby reveal chemical composition information (Figure 2.26).

2.12.1   CEMENT ANALYSIS FORMULAS

There are some important features in the cement quality control that you should check any time that you would like to use a cement for concrete production. We will describe some of the most important ones here. Before starting this part, you should know the abbreviations that we are going to use in the formulation as you can see in Table 2.14.

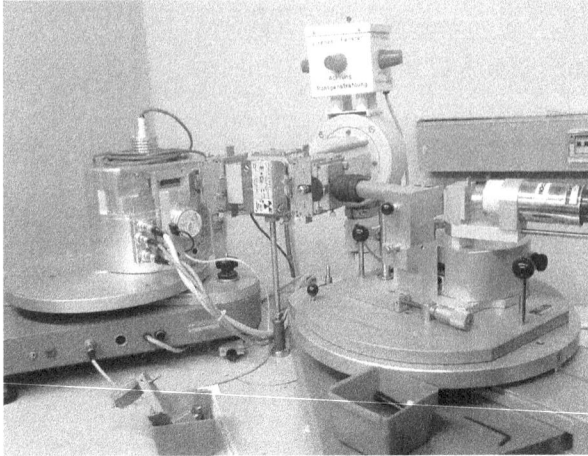

FIGURE 2.26   XRD test instrument. ("X-Ray Diffractometer" by GFDL.)

TABLE 2.14
Abbreviations for Cement Formulas

Abbreviation	Exact Chemical
C	CaO
S	SiO_2
M	MgO
A	Al_2O_3
F	Fe_2O_3

- Lime Saturation Factor (LSF): This factor shows us the performance of clinker baking in the kiln and it should be more than 90 but it cannot be more than 98.

$$\text{L.S.F. } (MgO < 2) = 100 \ (C + 0.75M)/(2.85S + 1.18A + 0.65F)$$

$$\text{L.S.F. } (MgO > 2) = 100 \ (C + 1.5)/(2.85S + 1.18A + 0.65F)$$

- Alkali Equivalent (AE): This shows the amount of alkalis in the cement. Most of the time, it should be less than 0.6. But as mentioned before to control alkali aggregate reaction we should use a low alkali cement with the AE less than 0.3.

$$AE = Na_2O + 0.659 \ K_2O$$

- Alumina Ratio (AR): this shows the amount of alumina in the cement which is very important for the process of baking clinker. This should be between 1.3 and 2.5 and the best and ideal number is 1.38. If the AR is less than 1.3, the cement will have low early strength and low heat of hydration, and if the AR is more than 2.5, it shows that the cement will have high early strength and high heat of hydration.

$$AR = A/F$$

- Bouge formulas: We can calculate the amount of cement compositions by the below formulations:

If $AR > 0.64$, we have:

$$C_3S = 4.071C - 6.6\ S - 6.718\ A - 1.43\ F - 2.852\ SO_3$$

$$C_2S = 2.867\ S - 0.754\ C_3S$$

$$C_3A = 2.650\ A - 1.692\ F$$

$$C_4AF = 3.04\ F$$

2.13   CEMENT TRANSPORTATION

After finalizing the cement production, we need to transport the cement to the place of use. This is very important to transport the cement without any distractions to protect the quality. For example, cement should prevent from the moisture during transportation. According to the above mentioned, we have two types of cement in the market:

- Bulk cement: This is the cement to use in the concrete industry and for large scale of any production of concrete or mortar. This kind of cement can store in the siloes in cement factory and can transport with the cement bunkers (Figure 2.27).

   Loading process of a bunker is from the top vent in the cement factory and the discharge of it is by the pressure of air from inside the Bunker which causes

FIGURE 2.27   A picture of cement bunker. (Photograph by the author.)

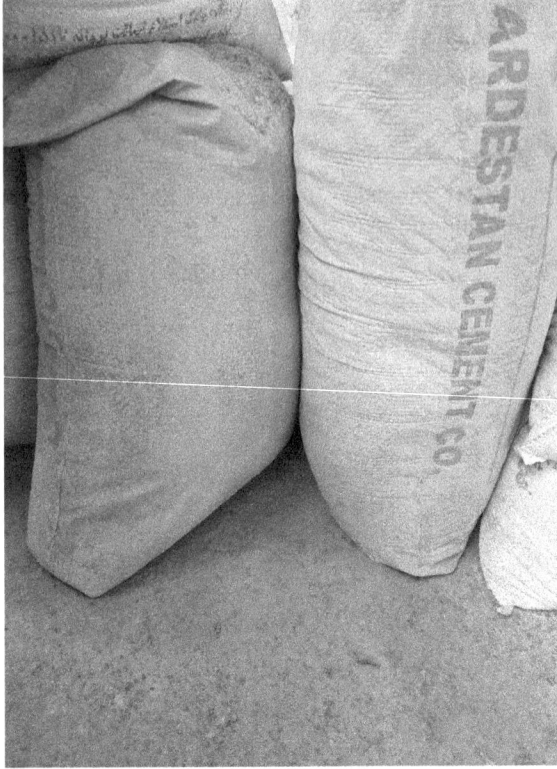

FIGURE 2.28   Cement pockets stored on a pallet. (Photograph by the author.)

the cement to go out of the tank with a good discharging capacity. For example, we can discharge 20 tons of cement from a bunker in less than 15 minutes.

- Pocket cement: This is a cement to use in the production of mortar or concrete in a smaller scale. The weight of these pockets is different around the world. But it can differ from 25 to 50 kg per bag. Also, the bags material could be different. They can be made from paper (Figure 2.28) or some types of plastic yarn.

2.14   CEMENT STORAGE

Cement is very sensitive to the moisture. So, we should consider prevention from any kind of moisture for the storage of cement.

For the storage of pocket cement, as you can see in Figure 2.28, you can put the cement on suitable pallets. So, the ground moisture or flowing water cannot filtrate into the cement. On the other hand, you should put these pallets inside an indoor area because rain can transit from the bags of cement.

By doing above mentioned, you can store pocket cement for at least 6 months from the production date. It doesn't mean that after 6 months, you cannot use the cement but it means that you should test it before using it.

FIGURE 2.29 Cement siloes for the storage of bulk cement. ("Cement silos of Günter Papenburg Beton company" by Christian Schroder.)

For the storage of bulk cement, you should use special cement siloes (Figure 2.29).

These siloes are specially designed for the storage of bulk cement or any other powder material. They can prevent the cement from the ground and air moisture and rainwater. You can load these siloes from the cement bunkers very easily by the air pressure and you can unload cement to the bucket of the batching plant by using special spiral pipes.

REFERENCES

Aitcin P.C., High Performance Concrete, E&FN SPON, 2004.

American Society for Testing and Materials, Standard Practice for Making and Curing Concrete Test Specimens in the Laboratory, ASTM C192-00.

American Society for Testing and Materials, Standard Practice for Capping Cylindrical Concrete Specimens, ASTM C617-98.

American Society for Testing and Materials, Standard Specification for Ready-Mixed Concrete, ASTM C94-00.

American Society for Testing and Materials, Standard Specification for Portland Cement, ASTM C150-00.

American Society for Testing and Materials, Standard Specification for Flow Table for use in Test of Hydraulic Cement, ASTM C230-98.

American Society for Testing and Materials, Standard Test Method for Compressive Strength of Hydraulic Cement Mortars, ASTM C109-99.

American Society for Testing and Materials, Standard Test Method for Chemical Analysis of Hydraulic Cement, ASTM C114-00.

American Society for Testing and Materials, Standard Test Method for Sieve Analysis of Fine and Coarse Aggregates, ASTM C136-01.

American Society for Testing and Materials, Standard Test Method for Autoclave Expansion of Portland Cement, ASTM C151-00.

American Society for Testing and Materials, Standard Test Method for Heat of Hydration of Hydraulic Cement, ASTM C186-98.

American Society for Testing and Materials, Standard Test Method for Density of Hydraulic Cement, ASTM C188-95.

American Society for Testing and Materials, Standard Test Method for Time of Setting of Hydraulic Cement by Vicat Needle, ASTM C191-99.

American Society for Testing and Materials, Standard Test Method for Flow of Hydraulic Cement Mortar, ASTM C1437-99.

American Society for Testing and Materials, Standard Test Method for Compressive Strength of Hydraulic Cement Mortars, ASTM C109-99.

CILAS, "Particle size distribution analyzer 990 particle size analysis CILAS." Retrieved from: https://commons.wikimedia.org/wiki/File:Particle_size_distribtion_analyzer_990_-_Particle_size_analysis_-_CILAS.jpg.

Connor Jerome J, Faraji Susan, *Fundamentals of Structural Engineering*, Springer, 2016.

DFIDUK Department for International Development, "Factory of National Cement Share Company." Retrieved from: https://commons.wikimedia.org/wiki/File:Factory_of_National_Cement_Share_Company.jpg.

Diliff, "A 4×4 segment panorama of the Coliseum at dusk." Retrieved from: https://commons.wikimedia.org/wiki/File:Colosseum_in_Rome,_Italy_-_April_2007.jpg.

European Standard Organization, Cement Composition, Specifications and Conformity Criteria for Common Cements, EN197-1: 2000.

European Standard Organization, Concrete-Part 1: Specification, Performance, Production and Conformity, EN206-1, 2000.

European Standard Organization, Methods of Testing Cement, EN196 Series.

GFDL, "X Ray Diffractometer." Retrieved from: https://commons.wikimedia.org/wiki/File:X_Ray_Diffractometer.JPG.

Hauschild Michael, Rosenbaum Ralph K, Olsen Sting Irving, *Life Cycle Assessment, Theory and Practice*, Springer, 2018.

Heinrichs Harald, Martens Pim, Michelsen Gerd, Wiek Arnim, *Sustainability Science, An Introduction*, Springer, 2016.

Iranian Institute for Research on Construction Industry, 9th topic of National Rules for Construction, "Concrete Structures", 2009.

Iranian National Management and Programming Organization, National Handbook of Concrete Structures, 2005.

Iranian Standard Organization, Concrete Specification of Constituent Materials, Production and Compliance of Concrete, ISIRI2284-2, 2009.

Iranian Standard Organization, Standard Specification for Ready Mixed Concrete, ISIRI6044, 2015.

Ivbuiliev, "Top view of a cement ball mill." Retrieved from: https://commons.wikimedia.org/wiki/File:Cement_mill_-_top_view.jpg.

Janamian Kambiz, Aguiar Jose, *A Comprehensive Method for Concrete Mix Design*, Materials Research Forum LLC, 2020.

Jb957, "Cement kiln. Location: Gorazdze Cement Plant near Chorula (Poland)." Retrieved from: https://commons.wikimedia.org/wiki/File:Cement_kiln_in_Gorazdze_Cement_plant.JPG.

Lamond F.Joseph, Pielert H.James, *Significance of Tests and Properties of Concrete and Concrete Making Materials*, ASTM International, 2006.

Mahmood Zadeh Amir, Iranpoor Jafar, Concrete Technology and Test (Farsi), Golhaye Mohammadi, 2007.

Mostofinejad Davood, Concrete Technology and Mix Design (Farsi), Arkane Danesh, 2011.

Murray-Rust, Alan, "Hanson Cement, Clitheroe, it grinds the previously crushed raw material into fine powder for feeding into the kiln." Retrieved from: https://www.geograph.org.uk/photo/4480152.

Nawy G.Edward, *Concrete Construction Engineering Handbook*, CRC Press, 2008. NAYANA PB, "specific gravity test" Retrieved from: https://commons.wikimedia.org/wiki/File:SpecificGrafityofcement.gif.

Newman John, Choo Ban Seng, *Advanced Concrete Technology, Concrete Properties*, Elsevier, 2003.

Peel, Mike (www.mikepeel.net), "Instituto Butantan, São Paulo, Brazil. Autoclave." Retrieved from: https://commons.wikimedia.org/wiki/File:Instituto_Butantan_2016_081_-_Autoclave.jpg.

Popovics Sandor, *Concrete Materials, Properties Specification and Testing*, NOYES Publications, 1992.

Rabax63, "Pantheon (Rome) – Front." Retrieved from: https://commons.wikimedia.org/wiki/File:Pantheon_Rom_1_cropped.jpg.

Ramachandran V.S, Beaudion James, *Handbook of Analytical Techniques in Concrete Science and Technology, Principles, Techniques and Applications*, William Andrew Publishing, 2001.

Ramachandran, Paroli, Beaudion, Delgado, *Handbook of Thermal Analysis of Construction Materials*, NOYES Publications, 2002.

Ramezanianpoor Aliakbar, Arabi Negin, Cement and Concrete Test Methods (Farsi), Negarande Danesh, 2011.

Safaye Nikoo Hamed, *Introduction to Concrete Technology (Farsi)*, Heram Pub, 2008.

Schroder, Christian, "Cement silos of Günter Papenburg Beton company located at Lohweg in Misburg-Sued quarter of Hannover, Germany." Retrieved from: https://commons.wikimedia.org/wiki/File:Cement_silo_Papenburg_Lohweg_Misburg-Sued_Hannover_Germany.jpg.

Shekarchizade Mohammad, Liber Nicolas Ali, Dehghan Solmaz, Poorzarrabi Ali, Concrete Admixtures Technology and Usages (Farsi), Elm & Adab, 2012.

Wiehe, Carsten "Al Kufa Cement plant." Retrieved from: https://commons.wikimedia.org/wiki/File:Al_Kufa_Cement_plant_-_panoramio.jpg.

Zandi Yousof, *Advanced Concrete Technology (Farsi)*, Forouzesh Pub, 2009.

Zandi Yousof, *Concrete Tests and Mix Design (Farsi)*, Forouzesh Pub, 2007.

3 Mineral Additives (Supplementary Cementitious Materials)

Mineral additives or the supplementary cementitious materials are mineral powders or by-product of other industries that we can use besides cement to increase the performance of concrete. They can improve the mechanical properties and help us to produce impermeable concrete which can prevent the attack of the environmental chemicals on the concrete or the steel bars.

On the other hand, as most of these materials are by-products of other industries, using them can help to prevent environmental pollution by these materials as waste. So, using these additives is going to be necessary in the production of concrete all over the world in many countries.

The third reason for using these materials is the economic benefits. As these materials are by-products and/or mineral materials, the production process for them is only the grinding process. So, these are very cheaper than the Portland cement and as we should replace some amount of cement with these mineral additives, we can say that they will make the concrete more economic.

According to the above mentioned, most of the concrete technologists suggest using supplementary cementitious materials for all types of concrete. So, these are very important raw materials for concrete production.

In this chapter, we will discuss the supplementary cementitious materials, their types, and their effects on the properties of fresh and hardened concrete. Also, we will talk about the use of these materials for different needs in concrete production.

3.1 WHAT ARE THE SUPPLEMENTARY CEMENTITIOUS MATERIALS?

Supplementary cementitious materials are different types of powders that contain active SiO_2 which can react with water besides cement hydration reaction and produce more calcium silicate hydrate (C-H-S) which is the main cause of concrete mechanical properties and performance.

So, the most important material in the supplementary cementitious materials is the active SiO_2 and the pozzolanic activity and power of these materials refer to the amount of SiO_2.

According to the above mentioned, about the active SiO_2, we can name the most important supplementary cementitious materials from the higher activity to the lowest activity as below:

DOI: 10.1201/9781003384243-3

- Silica fume (in some places of the world they name that as micro-silica but it is different than powdered silica)
- Fly ash
- Metakaolin and natural pozzolans
- Ground granulated blast furnace slag (GGBS)

You can find the above materials in most places of the world. But you can find some of them better than the others in some countries. Also, it is possible that you cannot find some of these materials in some countries. For example, there are no coal-fired power plants in some countries like most of the countries in the middle east because they have many mines of fossil fuels in their countries. So, they are not going to fire the coal for electricity production and you cannot find fly ash in these countries.

Another difference between these materials is their prices in different parts of the world. Silica fume is very expensive in all parts of the world, and in many countries, it is more expensive than cement. But for the other supplementary cementitious materials, you can see different prices in different parts of the world which depends on the suppliers in a country. For example, the mean price of fly ash in China is cheaper than in Europe because there are more coal-fired power plants in China than in Europe (Figure 3.1).

In Figure 3.1 the color black refers to the operating coal-fired power plants, and the other colors refer to the other types of power plants. You can find the places of the world where there are more coal-fired power plants and you can find cheap price fly ash.

Without any attention to the price of these materials, you can find many benefits of using them in concrete. But you should find the best choice depending on the place you are living and the defined properties of your concrete.

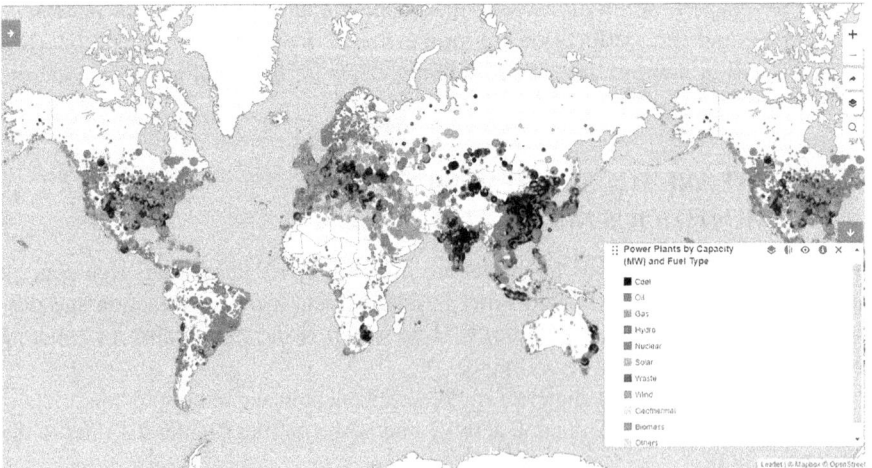

FIGURE 3.1 Coal-fired power plants in the world. ("Global power plants by generation sources" by Global Energy Observatory, Google, KTH Royal Institute of Technology in Stockholm, Enipedia, World Resources Institute.)

3.2 REACTION WITH CEMENT AND WATER

It is very important that the mineral additives itself doesn't react with water, but they react with water besides cement, and we can explain the reaction as below:

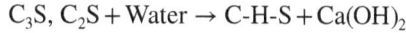

$$C_3S, C_2S + Water \rightarrow C\text{-}H\text{-}S + Ca(OH)_2$$

$$Ca(OH)_2 + Active\ SiO_2 \rightarrow C\text{-}H\text{-}S$$

You can see from the equations above that, first C_3S and C_2S react with water and make the C-H-S, which is the main cause of concrete mechanical properties and performance, and $Ca(OH)_2$, which has not any effect on the performance of concrete itself. After more time the active SiO_2 in the supplementary cementitious materials reacts with the $Ca(OH)_2$ and makes more C-H-S. So, the mechanical properties and performance of concrete will increase.

The speed of reaction depends on the amount of active SiO_2 in the supplementary cementitious material and the reaction environment. For example, you can accelerate the reaction by using plasticizers and super-plasticizers like polycarboxylate ether. But if you do not use any super-plasticizer the effect of mineral additives will decrease drastically. Using these materials will increase the water demand for concrete as they are powder materials and also the nature of these materials or the fineness of some of them like silica fume will cause more water demand. So, if you don't use any super-plasticizer, the water you should use will increase and the water-to-binder ratio of concrete will increase which can cause the decrease of mechanical properties and this reduction of mechanical properties will terminate the effect of supplementary cementitious materials.

Although we speed up the reaction of mineral additives in concrete, we can see the effect of the minimum in the age of 11 days and the best time to see the effect of these materials is 42 days and it will continue to 90 days. So, you can make two theoretical concrete as below:

- Concrete No. 1: A concrete with a defined 28 days compressive strength of 40 MPa, but without the usage of any supplementary cementitious material.
- Concrete No. 2: A concrete with the same 28 days compressive strength of the concrete No. 1 but with the usage of a supplementary cementitious material (e.g., GGBS).

  You can see the behavior of these concretes in Figure 3.2.

From the above figure, you can see below explanations:

- As we have the targeted compressive strength of 40 MPa, we should design a concrete with higher achieved compressive strength (here 45 MPa).
- As we have the same targeted compressive strength, we should design two concretes for the same achievement of compressive strength in the 28 days (here 45 MPa). We will explain the mixed design of concrete with the usage of supplementary cementitious materials in the future.
- At 7 days, the compressive strength of concrete No. 1 is about 20% more than the concrete No. 2.

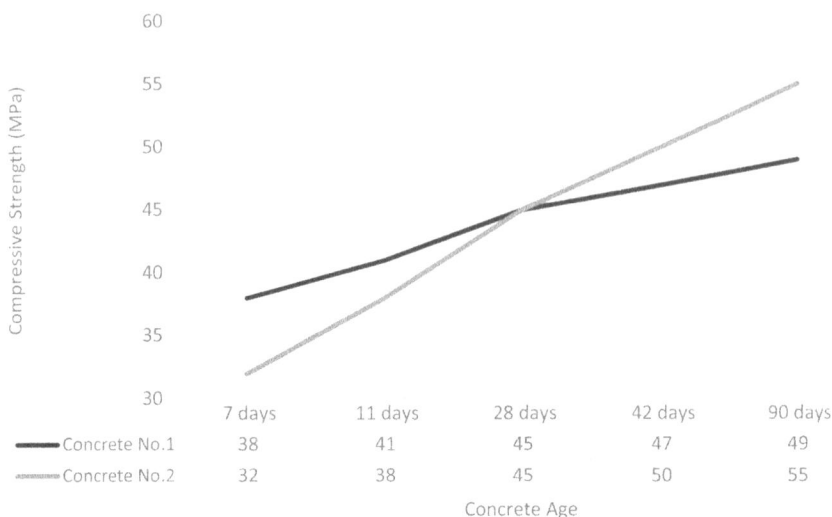

	7 days	11 days	28 days	42 days	90 days
Concrete No.1	38	41	45	47	49
Concrete No.2	32	38	45	50	55

FIGURE 3.2 The effect of supplementary cementitious materials in concrete. Graph created by the author.

- At 11 days, the compressive strength of concrete No. 1 is about 10% more than the concrete No. 2. You can see the first effects of the GGBS at this age by decreasing the gap of two concrete compressive strengths.
- At 28 days, you can see the same compressive strength for two concretes.
- At 42 days, the compressive strength of concrete No. 2 is about 7% more than the concrete No. 1. You can see the victory of concrete No. 2 at this age.
- At 90 days, the compressive strength of concrete No. 2 is about 12% more than the concrete No. 1.

So, when you use the supplementary cementitious materials, you will increase the properties of concrete in the ages more than 28 days.

Now, it is time to explain more about each of the supplementary cementitious materials in detail.

3.3 SILICA FUME

Silica fume is a very fine powder that is the by-product of ferro-alloy industries. It is in fact the dust derived from the funnels of these factories.

Silica fume contains more than 90% of active SiO_2. On the other hand, it is a very fine powder (more than 50 times of the Portland cement). The blain of silica fume could be between 150000 and 300000 cm^2/g. The structure of this pozzolan is spherical. As the activation of Silica fume is very high compared with the other supplementary cementitious materials, we can call that a super-pozzolan. You can see the chemical analysis of silica fume in Table 3.1.

TABLE 3.1

Chemical Analysis of Silica Fume

Material	SiO_2	C	Fe_2O_3	Al_2O_3	CaO	MgO	Na_2O	K_2O	P_2O_5	SO_3	Cl
Percent	94.43	0.3	0.87	1.32	0.49	0.97	0.31	1.01	0.16	0.1	0.04

FIGURE 3.3 A picture of silica fume powder. (Photograph by the author.)

As you can see in the table, the amount of active SiO_2 in this silica fume is 94.43%.

3.3.1 SILICA FUME SPECIFICATION

We can name the specification of silica fume as below:

- Appearance: It is a light gray and very fine powder. Its color is lighter than the Portland cement (Figure 3.3).
- Specific gravity: It is lighter than cement. Its specific gravity is about 2.2 kg/L.
- Blain: As mentioned before, it is a very fine powder. Its blain is different for several producers in the different parts of the world. But we can say that it is between 150000 and 300000 cm²/g.
- Amount of use in concrete: We have different suggestions for the percentage of use for this material. But you have to use more than 5% by weight of total binder (Portland cement plus supplementary cementitious material) and less than 12%.

3.3.2 Effect of Silica Fume on the Properties of Fresh Concrete

You can see the effect of silica fume on the properties of fresh concrete as below:

- Effect on the water demand: As silica fume is a very fine powder and it contains a high amount of silica, it absorbs too much water. So, water demand in concrete will increase, when we use silica fume.
- Concrete workability: As we would like to talk about the workability, there are too many factors that we should consider. So, for the silica fume because of more water demand, it can reduce the workability. But as silica fume is a very fine powder it can help the concrete to pump better and the texture of concrete will improve.
- Segregation and bleeding: As the silica fume is a very fine powder, it can absorb the excess water. So, it can reduce the segregation and bleeding of concrete.
- Entrapped air: The amount of entrapped air in concrete will decrease when we use silica fume because it is a fine powder and reduces the fraction between the constituent materials in concrete.
- Heat of hydration: When we use silica fume in concrete, we should replace some amount of cement with this material. So, we can reduce the heat of hydration.
- Setting time: As we use a little amount of silica fume in concrete (most of the time less than 10%). So, it has not any considerable effect on the setting time of concrete.
- Pumpability: We have silica fume as a very fine powder. So, it will increase the amount of cement paste and it has a positive effect on the viscosity and pumpability of concrete.

3.3.3 Effect of Silica Fume on the Properties of Hardened Concrete

You can see the effect of silica fume on the properties of hardened concrete as below:

- Compressive strength: Silica fume acts as a super-pozzolan. It produces more C-H-S in concrete. So, it can increase the compressive strength of concrete, especially in the ages more than 11 days.
- Freeze-thaw resistance: We can produce an impermeable concrete by using of the silica fume. Moisture cannot go inside an impermeable concrete element. So, the destroying effects of freeze-thaw cycles can be controlled by using the silica fume.
- Shrinkage and crack: There are no straight effects from the silica fume on the shrinkage of concrete. But as it needs much amount of water, concrete made with it will need much curing than normal. So, silica fume will increase the probability of cracking in concrete.
- Permeability: As mentioned before, we can produce more C-H-S in concrete when we use silica fume. On the other hand, the fineness of silica fume can decrease the number of pores in the concrete texture. So, it can reduce the permeability of concrete.

- Alkali aggregate reaction: We use the silica fume as a percent of the total binder in concrete. So, by using it, we are reducing the amount of cement alkalis for the alkali aggregate reaction and we can control this reaction.
- Sulfate and chloride resistance: As we can reduce the permeability of concrete by using the silica fume, we can control the moisture and aggressive ions like sulfate and chloride inside the concrete. So, silica fume can improve the resistance of concrete against sulfate and chloride attack.

3.3.4 Use of Silica Gel or Silica Slurry

If we would like to name a deficiency for the silica fume, it is the carcinogenic effect of this material on humans. Scientists talked about different reasons which can cause silica fume a carcinogen material. For example, the super-fineness of this powder will cause the particles breathed to stand in the interior layers of human lung and cannot go out by cough or mucus. So, using silica fume in concrete has some restrictions for the laborers who are going to work with it.

To prevent this effect, we can use silica gel or silica slurry that you can see the definitions as below:

- Silica slurry (Figure 3.4): This is a combination of the silica fume powder and water that they mixed very well together with a good stirrer. The amount of silica fume in the slurry could be about 20% to 30% by weight. Sometimes, we can use a little amount of super-plasticizer (specially polycarboxylate ether) to ensure the good dispersing of silica fume in the water. But the amount of super-plasticizer is not too much that we should consider it in the concrete mix design.

  As the production process of the silica slurry is in a factory with a defined protection process, the safety of this material is much more than the silica fume powder.
- Silica gel: This is a combination of the silica fume powder, water, and a special amount of super-plasticizer (most of the time polycarboxylate ether base) which are mixed very well together with a good stirrer. The amount of silica fume in the silica gel is about 20% to 40% by weight and the amount of the super-plasticizer could be between 0.5% and 1.5% by weight. So, we should consider this amount of super-plasticizer in the concrete mix design.

Like the silica slurry, as the production process for this material is in a factory with a defined protection process, the safety of this material is much more than the silica fume powder.

The other important reason to use silica slurry and gel instead of silica fume powder is the importance of dispersion for the silica fume in the total texture of concrete with the same dosage. Researchers showed that the dispersing of silica gel and slurry is much better in the construction of concrete than the silica fume powder.

Do not forget to use all kinds of powder additives with the suitable super-plasticizers in concrete because the super-plasticizers can help better and homogeneous dispersion of these materials in concrete.

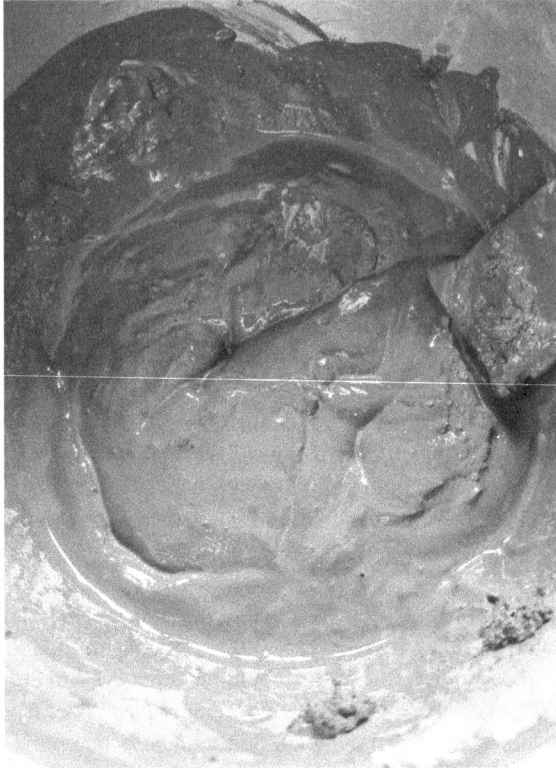

FIGURE 3.4   Silica slurry ready to use in concrete. (Photograph by the author.)

3.4   FLY ASH

Fly ash is a very fine powder which is the by-product of electricity power plants that use coal as fuel. It derives from the funnels of these power plants (Figure 3.5).

Fly ash contains between 30% and 50% of active SiO_2. On the other hand, its blain is about 4000–4500 cm²/g (between 30% and 50% more than normal Portland cement). The activation of the fly ash is less than the silica fume but more than the other supplementary cementitious materials.

We have two types of fly ash which depend on the type of coal burned in the power plant:

- Fly ash type F: It is a low calcium content fly ash.
- Fly ash type C: It is a moderate or high calcium content fly ash.

You can see the chemical analysis of a fly ash type F in Table 3.2 and a fly ash type C in Table 3.3.

As you can see in the tables above, the amount of active SiO_2 in the fly ash is different in types F and C. As type F fly ash has more SiO_2, it is a very good pozzolan. But for type C we may have a little cementing activity. It means that type C fly ash

FIGURE 3.5 A coal fire electricity power plant. ("Electricity power plant".)

TABLE 3.2
Chemical Analysis of Fly Ash Type F

Material	SiO_2	Al_2O_3	Fe_2O_3	CaO	SO_3	Na_2O	K_2O
Percent	54	24	11.5	5	1.5	1.5	2.5

TABLE 3.3
Chemical Analysis of Fly Ash Type C

Material	SiO_2	Al_2O_3	Fe_2O_3	CaO	SO_3	Na_2O	K_2O
Percent	37	20	7	24	4.5	6	1.5

can activate with water itself because it has high amount of CaO. But the cementing activity is not too much. So, we use this type of fly ash in concrete production for its pozzolanic activity.

3.4.1 Fly Ash Specification

We can name the specification of the fly ash as below:

- Appearance: It is a gray to light yellow powder. Its fineness is a little more than the Portland cement. You can see a picture of the fly ash in Figure 3.6.
- Specific gravity: It is lighter than the cement. Its specific gravity is between 1.9 and 2.6 kg/L.

FIGURE 3.6   Fly ash. (Photograph by the author).

- Blain: As mentioned before, it is about 30% to 50% finer than the normal Portland cement. Its blain is different for several producers in the different parts of the world. But it is between 4000 and 4500 cm²/g.
- Amount of use in concrete: We have different suggestions for the percent of the use of this material. You have to use it between 10% and 35% by weight of total binder (Portland cement plus supplementary cementitious material).

3.4.2   EFFECT OF FLY ASH ON THE PROPERTIES OF FRESH CONCRETE

You can see the effect of fly ash on the properties of fresh concrete as below:

- Effect on the water demand: As The fly ash has different blain for different producers, we cannot say anything about the water demand. If the blain is high (about 4500 cm²/g), it can increase the amount of water, but if the blain is lower, like 4000 cm²/g, it has no considerable effect on the water demand of concrete.
- Concrete workability: As the fly ash is a fine powder and it will increase the amount of total binder in concrete, it can help the concrete to pump better and the texture of concrete will improve. So, we can say that the fly ash has positive effect on the workability of concrete.
- Segregation and bleeding: Fly ash is a fine powder that will increase the total binder. So, it can absorb the excess water and reduce the segregation and bleeding of concrete.
- Entrapped air: The amount of entrapped air in concrete will decrease when we use fly ash because it is a fine powder and reduces the fraction between the constituent materials in concrete.
- Heat of hydration: When we use fly ash in concrete, we should replace some amount of cement with this material. So, we can reduce the heat of hydration.

- Setting time: When we use high amount of fly ash in concrete, we are replacing the pure Portland cement with that. So, as the activity of fly ash is less than the Portland cement and it will start the reaction in the ages above 11 days, using the fly ash can retard the setting time of concrete.
- Pumpability: We have the fly ash as a fine powder. So, it will increase the amount of binder paste and it has a positive effect on the viscosity and pumpability of concrete.

3.4.3 EFFECT OF FLY ASH ON THE PROPERTIES OF HARDENED CONCRETE

You can see the effect of fly ash on the properties of hardened concrete as below:

- Compressive strength: Fly ash acts as a pozzolan material. It produces more C-H-S in concrete. So, it can increase the compressive strength of concrete, especially in the ages more than 11 days.
- Freeze-thaw resistance: We can produce an impermeable concrete by using the fly ash. Moisture cannot go inside an impermeable concrete. So, the destroying effects of freeze-thaw cycles can be controlled by using the fly ash.
- Shrinkage and crack: There are no straight effects from the fly ash on the shrinkage of concrete and there is no effect on the probability of cracking of concrete.
- Permeability: As mentioned before, we can produce more C-H-S in concrete when we use the fly ash. So, it can reduce the permeability of concrete.
- Alkali aggregate reaction: We use fly ash as a percent of total binder in concrete. So, by using it, we are reducing the amount of cement alkalis for the alkali aggregate reaction and we can control this reaction.
- Sulfate and chloride resistance: As we can reduce the permeability of concrete by using fly ash, we can control the moisture and penetration of sulfate and chloride ions inside concrete. So, we can improve the sulfate and chloride resistance of concrete.

3.5 GROUND GRANULATED BLAST FURNACE SLAG (GGBS)

Ground granulated blast furnace slag (GGBS) is the by-product of steel production industries (Figure 3.7). It is the remained material of the melted steel when it comes out of the furnace, which is cooled down rapidly after coming out. The cooling process is very important for the activity of SiO_2. In fact, the GGBS for concrete produces has two stages:

- Stage 1: The production process of steel and cooling process of slag in the steel factory.
- Stage 2: The grinding process in a cement factory which should be separated by the cement grinding because of the different hardness of cement clinker and slag.

FIGURE 3.7   Steel complex factory. ("Foolad Mobarakeh Steel Mill. Isfahan." by Hasan Majidi.)

GGBS contains about 30% of active SiO_2. As we should grind the slag for the production of the GGBS, we can adjust the fineness of the final product. As the activity of the GGBS is much less than the Portland cement, we should grind it more than the cement clinker. Most of the time, the good blain for GGBS is about $4000 cm^2/g$. The activity of GGBS is less than the silica fume and fly ash. But it is a very good material for the production of a durable concrete. On the other hand, as the production of slag from the production of steel is very much in the world, it is a good idea to use it in concrete production for environmental protection. You can see the chemical analysis of a GGBS in Table 3.4.

As you can see in the table above, the amount of SiO_2 in this GGBS is 35%. With a good cooling process, we can activate 99% of this SiO_2. Nevertheless, it is not a good slag for use in concrete production.

3.5.1   GGBS Specification

We can name the specification of the GGBS as below:

- Appearance: It is a gray to light yellow powder. We can adjust the fineness more than the Portland cement to adjust its activity. You can see a picture of the slag before grinding in Figure 3.8.

TABLE 3.4
Chemical Analysis of GGBS

Material	SiO_2	Al_2O_3	Fe_2O_3	CaO	SO_3	Na_2O	K_2O
Percent	35	13	1	41	9	0.5	0.5

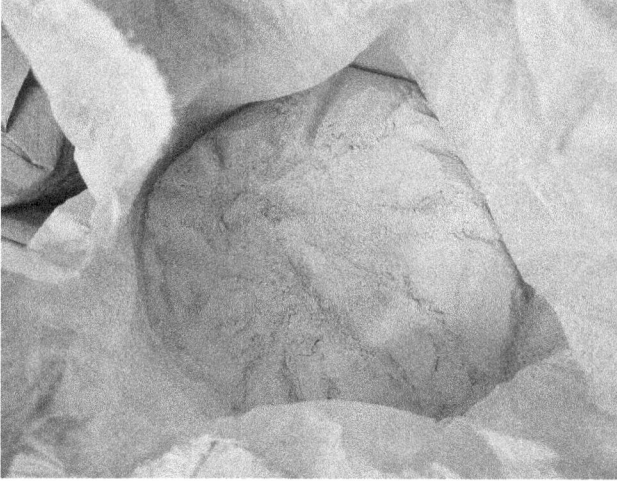

FIGURE 3.8   Picture of slag. (Photograph by the author.)

- Specific gravity: Most of the time, it is lighter than the cement. Its specific gravity is about 2.9 kg/L.
- Blain: As mentioned before, we can adjust the blain in the cement factory. Most of the time, the GGBS blain is about 4000 cm²/g. It should be softer than the Portland cement for its lower activity.
- Amount of use in concrete: There are different suggestions for the percent of use for this material. For example, in Europe, you can use it up to 95% of the total binder. But here we suggest to use it between 20% and 50% by weight of the total binder (Portland cement plus supplementary cementitious material).

3.5.2   EFFECT OF GGBS ON THE PROPERTIES OF FRESH CONCRETE

You can see the effect of GGBS on the properties of fresh concrete as below:

- Effect on the water demand: As the GGBS has different blain for different producers, we cannot say anything about the water demand. If the blain is so high (about 4500 cm²/g), it can increase the water demand, but if the blain is lower like 4000 cm²/g, it has no considerable effect on the water demand of concrete.
- Concrete workability: As the GGBS is a fine powder and it will increase the amount of total binder in concrete, it can help the concrete to pump better and the texture of the concrete will improve. So, we can say that the GGBS has a positive effect on the workability of concrete.
- Segregation and bleeding: GGBS is a fine powder that will increase the total binder of concrete. So, it can absorb the excess water and reduce the segregation and bleeding.

- Entrapped air: The amount of entrapped air in concrete will decrease when we use the GGBS because it is a fine powder and reduces the fraction between the constituent materials in concrete.
- Heat of hydration: When we use GGBS in concrete, we should replace some amount of cement with this material. So, we can reduce the heat of hydration.
- Setting time: When we use a high amount of the GGBS in concrete, we are replacing the pure Portland cement with that. So, as the activity of GGBS is less than the Portland cement and it will start the reaction in the ages above 11 days, using the GGBS can retard the setting time of concrete.
- Pumpability: We have the GGBS as a fine powder. So, it will increase the amount of binder paste and it has a positive effect on the viscosity and pumpability of concrete.

3.5.3 Effect of GGBS on the Properties of Hardened Concrete

You can see the effect of GGBS on the properties of hardened concrete as below:

- Compressive strength: GGBS acts as a pozzolan material. It produces more C-H-S in concrete. So, it can increase the compressive strength of concrete, especially in the ages more than 11 days. But the effect of GGBS on the compressive strength of concrete is less than the fly ash and silica fume.
- Freeze-thaw resistance: Using the GGBS can help to produce an impermeable concrete. Moisture cannot go inside an impermeable concrete. So, the destroying effects of freeze-thaw cycles can control by using the GGBS.
- Shrinkage and crack: There is no straight effect from the GGBS on the shrinkage and probability of cracking in concrete.
- Permeability: As mentioned before, we can produce more C-H-S in concrete when we use the GGBS. So, it can reduce the permeability of concrete.
- Alkali aggregate reaction: We use the GGBS as a percent of the total binder in concrete. So, by using it, we are reducing the amount of cement alkalis for the alkali aggregate reaction and we can control this reaction.
- Sulfate and chloride resistance: As we can reduce the permeability of concrete by using the GGBS, we can control the moisture and sulfate/chloride ions inside the concrete. So, we can improve the resistance of concrete against chloride and sulfate attacks.

3.6 NATURAL POZZOLANS

There are too many materials in the nature which have pozzolanic activity. We can call them natural pozzolans. For example, we have metakaolin as the anhydrous calcined form of a special kind of clay called Kaolinite which we use most of the time for the production of the porcelain.

Another example is the different types of volcanic ashes in different parts of the world which have different pozzolanic activities. In fact, you can find these types of natural pozzolans in places where we have a volcanic mountain (Figure 3.9).

FIGURE 3.9   A volcanic mountain. ("Popocatepetl [Nahuatl for "Smoking Mountain"]".)

Like the other types of pozzolans, the main material which is very important in natural pozzolans is the active SiO_2. So, the activity of different types of natural pozzolans is different. But most of the time, the activity of natural pozzolans is the same as fly ash and is more than the GGBS. You should make trials with the natural pozzolans when you would like to use them in the production of any industrial concrete to ensure their activity and effects on the properties of concrete.

Chemical analysis of different types of natural pozzolans is different according to its type and the place of origin. But as mentioned before, the most important material is the active SiO_2, which is more than 50% for almost all types of the natural pozzolans. For example, the amount of SiO_2 in metakaolin is between 50% and 60%.

3.6.1   NATURAL POZZOLANS SPECIFICATION

We can name the specification of the natural pozzolans as below:

- Appearance: The appearance of different types of natural pozzolans is different from each other. For example, metakaolin is a light yellow powder and volcanic ash is light gray powder. You can see a picture of the metakaolin under the microscope in Figure 3.10.
- Specific gravity: All types of natural pozzolans are lighter than the cement. Their specific gravity is about 2.5 kg/L.
- Blain: Natural pozzolans are very fine powders. They are much finer than the Portland cement. But the blain is different for different types of them. For example, the blain of metakaolin is more than $180000 \, cm^2/g$ which is about 50 times more than the Portland cement.
- Amount of use in concrete: There are different suggestions for the percent of use for these materials. It depends on the type of natural pozzolan, its activity, and the reasons of use. You should make trials before using them in

FIGURE 3.10   A picture of metakaolin. (Photograph by the author.)

the production of industrial concrete. Here we suggest to use them between 10% and 30% by weight of the total binder (Portland cement plus supplementary cementitious material).

3.6.2   EFFECT OF NATURAL POZZOLANS ON THE PROPERTIES OF FRESH CONCRETE

You can see the effect of natural pozzolans on the properties of fresh concrete as below:

- Effect on the water demand: As natural pozzolans have different blain and different properties according to the type and place of origin, their effect on the water demand could be different. But as they are very fine powders and if we use them more than 10% of the total binder in concrete, they can increase water demand.
- Concrete workability: As the natural pozzolans are fine powders and they will increase the amount of total binder in concrete, they can help the concrete to pump better and the texture of the concrete will improve. So, we can say that the natural pozzolans have a positive effect on the workability of concrete. On the other hand, you should consider the water demand and its effect on the workability of concrete.

- Segregation and bleeding: Natural pozzolans are fine powders that will increase the total binder in concrete. So, they can absorb the excess water and reduce the segregation and bleeding of concrete.
- Entrapped air: The amount of entrapped air in concrete will decrease when we use natural pozzolans because they are fine powders and can reduce the fraction between the constituent materials in concrete.
- Heat of hydration: When we use natural pozzolans in concrete, we should replace some amount of cement with these materials. So, we can reduce the heat of hydration.
- Setting time: When we use a high amount of natural pozzolans (more than 10% by weight of total binder) in concrete, we are replacing the pure Portland cement with that. So, the activity of any kind of natural pozzolans is less than the Portland cement and they will start the reaction in the ages above 11 days, and using them can retard the setting time of concrete.
- Pumpability: We have natural pozzolans as very fine powders. So, they will increase the amount of binder paste and they have a positive effect on the viscosity and pumpability of concrete.

3.6.3 EFFECT OF NATURAL POZZOLANS ON THE PROPERTIES OF HARDENED CONCRETE

You can see the effect of natural pozzolans on the properties of hardened concrete as below:

- Compressive strength: Natural pozzolans produce more C-H-S in concrete. So, it can increase the compressive strength of concrete, especially in the ages more than 11 days. The effect of natural pozzolans on the compressive strength of concrete is similar to the fly ash.
- Freeze-thaw resistance: Using natural pozzolans can help to produce an impermeable concrete. Moisture cannot go inside an impermeable concrete. So, the destroying effects of freeze-thaw cycles can control by using natural pozzolans.
- Shrinkage and crack: There is no straight effect from the natural pozzolans on the shrinkage of concrete. But if we use a high amount of them in concrete, they can increase the water demand and increase the probability of cracks in concrete.
- Permeability: As mentioned before, we can produce more C-H-S in concrete when we use natural pozzolans. So, it can reduce the permeability of concrete.
- Alkali aggregate reaction: We use natural pozzolans as a percent of the total binder in concrete. So, by using them, we are reducing the amount of cement alkalis for the alkali aggregate reaction and we can control this reaction.
- Sulfate and chloride resistance: As we can reduce the permeability of concrete by using natural pozzolans, we can control the moisture and chloride/sulfate ions inside the concrete. So, we can improve the resistance against the chloride and sulfate attack.

3.7   COMPARISON BETWEEN DIFFERENT SUPPLEMENTARY CEMENTITIOUS MATERIALS

We mentioned all of the properties of supplementary cementitious materials in the parts before and we discuss the reasons for each property in detail. In this part, you can see all of the properties in tables to compare these materials with each other for the best decision-making about using these additives in concrete. You can see the comparison between the specification of the supplementary cementitious materials in Table 3.5.

You can see the different suggestions for the amount of use of supplementary cementitious materials in Table 3.6.

For the amount of use, it is very important to know the target properties of concrete to decide about the use of any supplementary cementitious material and also the amount of use for each one. You can also use a mixture of two of these additives for the production of a high-quality concrete. For example, use of silica fume and GGBS together is very common for the production of concrete which is in contact with sulfate and chloride attack.

You can see a comparison between the effect of supplementary cementitious materials on the properties of fresh concrete in Table 3.7.

You can see a comparison between the effect of supplementary cementitious materials on the properties of hardened concrete in Table 3.8.

TABLE 3.5
Comparison Between the Specification of Supplementary Cementitious Materials

Type of Additive	Appearance	Specific Gravity (kg/L)	Blain (cm²/g)
Silica fume	Light gray powder	About 2.2	150000–300000
Fly ash	Gray to light yellow powder	1.9–2.6	4000–5000
GGBS	Gray to light yellow powder	About 2.9	About 4000
Natural Pozzolans	Light yellow or light gray powder	About 2.5	About 180000

TABLE 3.6
Different Suggestions for the Amount of Use of Supplementary Cementitious Materials

Type of Additive	Suggestion of EN197 (%)	Suggestion of ACI 211.1 (%)	Suggestion of This Book (%)
Silica fume	6–10	5–15	5–12
Fly ash	6–35	15–35	10–35
GGBS	20–95	25–70	20–50
Natural Pozzolans	6–35	10–20	10–30

TABLE 3.7

Comparison Between the Effect of Supplementary Cementitious Materials on the Properties of Fresh Concrete

Type of Additive	Water Demand	Workability	Segregation and Bleeding	Entrapped Air	Hydration Heat	Setting Time	Pumpability
Silica fume	Increase	Decrease slump but improve texture	Decrease	Decrease	Decrease	No considerable effect	Improve
Fly ash	No considerable effect	Improve texture	Decrease	Decrease	Decrease	Can retard	Improve
GGBS	No considerable effect	Improve texture	Decrease	Decrease	Decrease	Can retard	Improve
Natural Pozzolans	Can increase	Improve texture	Decrease	Decrease	Decrease	Can retard	Improve

TABLE 3.8

Comparison Between the Effect of Supplementary Cementitious Materials on the Properties of Hardened Concrete

Type of Additive	Compressive Strength	Freeze-Thaw Resistance	Probability of Cracking	Permeability	Alkali-Aggregate Reaction	Sulfate-Chloride Resistance
Silica fume	Increase	Can help	Increase	Decrease	Can control	Can help
Fly ash	Increase	Can help	No effect	Decrease	Can control	Can help
GGBS	Increase	Can help	No effect	Decrease	Can control	Can help
Natural Pozzolans	Increase	Can help	Can increase	Decrease	Can control	Can help

3.8 USE OF MINERAL ADDITIVES IN CONCRETE PRODUCTION

To use mineral additives in concrete production you should consider below points:

- First, you should decide which of the supplementary cementitious materials is the best choice for your concrete according to the target properties of concrete and the access to that material.
- You should decide about the amount of use of the supplementary cementitious material according to the target properties of concrete.
- You should design a concrete mix by using the mineral additives that we will talk about it in the future.

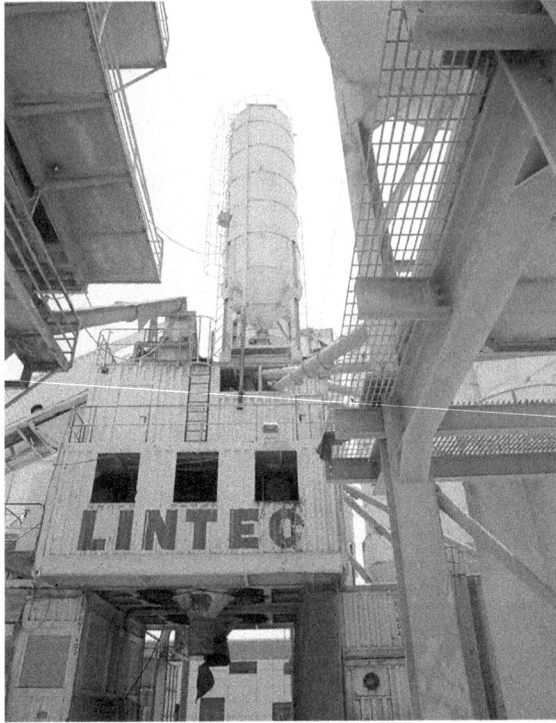

FIGURE 3.11   A special silo for silica fume direct use in concrete. (Photograph by the author.)

- It is very important to use mineral additives easily to avoid any mistakes. For example, you should use them like the Portland cement from the siloes of the batching plant (Figure 3.11).
- It is very important to disperse the mineral additives in concrete homogenously. So, you should use them with good super-plasticizers and a suitable dosage of them. The best one is the polycarboxylate base that we will talk about them in the future.
- For better dispersing of mineral additives in concrete texture, it is better to produce concrete with a slump more than 180 mm. You will make this concrete with a super-plasticizer. So, the dispersing process will be perfect.

REFERENCES

Aitcin P.C., High Performance Concrete, E&FN SPON, 2004.
American Society for Testing and Materials, Standard Practice for Making and Curing Concrete Test Specimens in the Laboratory, ASTM C192-00.
American Society for Testing and Materials, Standard Specification for Ready-Mixed Concrete, ASTM C94-00.
American Society for Testing and Materials, Standard Specification for Portland Cement, ASTM C150-00.

American Society for Testing and Materials, Standard Specification for Flow Table for use in Test of Hydraulic Cement, ASTM C230-98.

American Society for Testing and Materials, Standard Specification for Chemical Admixtures for Concrete, ASTM C494-99.

American Society for Testing and Materials, Standard Specification for Coal fly Ash and Raw or Calcined Natural Pozzolan for Use as a Mineral Admixture in Concrete, ASTM C618-00.

American Society for Testing and Materials, Standard Specification for Use of Silica Fume as a Mineral Admixture in Hydraulic Cement Concrete, Mortar and Grout, ASTM C1240-00.

American Society for Testing and Materials, Standard Test Method for Density, Absorption and Voids in Hardened Concrete, ASTM C642-97.

American Society for Testing and Materials, Standard Test Method for Flow of Hydraulic Cement Mortar, ASTM C1437-99.

American Society for Testing and Materials, Standard Test Method for Compressive Strength of Hydraulic Cement Mortars, ASTM C109-99.

Bertolini L, Elsener B, Pedeferri P, Polder R, *Corrosion of Steel in Concrete, Prevention, Diagnosis, Repair*, WILEY-VCH, 2004.

-Chilangabacho, "Popocatepetl (Nahuatl for "Smoking Mountain") is one of the few active volcanoes in Mexico and in the Americas." Retrieved from: https://commons.wikimedia.org/wiki/File:The_volcano, _Popocatepetl_(Nahuatl_for_%22Smoking_Mountain%22,_on_the_east_side_of_the_Valley_of_Mexico, _errupts_on_Dec.2018.jpg.

-"Electricity power plant." Retrieved from: https://pxhere.com/en/photo/559959.

European Standard Organization, Admixtures for Concrete Mortar and Grout, EN934 Series.

European Standard Organization, Admixtures for Concrete, Mortar and Grout Test Methods, EN480 Series.

European Standard Organization, Cement Composition, Specifications and Conformity Criteria for Common Cements, EN197-1: 2000.

European Standard Organization, Concrete-Part 1: Specification, Performance, Production and Conformity, EN206-1, 2000.

European Standard Organization, Methods of Testing Cement, EN196 Series.

European Standard Organization, Testing Fresh Concrete, EN12450 Series.

European Standard Organization, Testing Hardened Concrete, EN12390 Series.

Gjorv Odd E, *Durability Design of Concrete Structures*, Taylor & Francis, 2009.

Global Energy Observatory, Google, KTH Royal Institute of Technology in Stockholm, Enipedia, World Resources Institute. 2019. Global Power Plant Database v1.2.0. Published on Resource Watch (http://resourcewatch.org/) and Google Earth Engine (https://earthengine.google.com/). Accessed through Resource Watch, (date). www.resourcewatch.org. "Global power plants by generation sources" Retrieved from: https://commons.wikimedia.org/wiki/File:Global_power_plants_by_generation_sources.png.

Hauschild Michael, Rosenbaum Ralph K, Olsen Sting Irving, *Life Cycle Assessment, Theory and Practice*, Springer, 2018.

Heinrichs Harald, Martens Pim, Michelsen Gerd, Wiek Arnim, *Sustainability Science, An Introduction*, Springer, 2016.

Iranian Institute for Research on Construction Industry, 9th topic of National Rules for Construction, "Concrete Structures", 2009.

Iranian Institute for Research on Construction Industry, National Concrete Mix Design Method, 2015.

Iranian National Management and Programming Organization, National Handbook of Concrete Structures, 2005.

Iranian Standard Organization, Concrete Admixtures, Specification, ISIRI2930, 2011.

Iranian Standard Organization, Concrete Specification of Constituent Materials, Production and Compliance of Concrete, ISIRI2284-2, 2009.

Iranian Standard Organization, Standard Specification for Ready Mixed Concrete, ISIRI6044, 2015.

Janamian Kambiz, Aguiar Jose, A Comprehensive Method for Concrete Mix Design, Materials Research Forum LLC, 2020.

Lamond F.Joseph, Pielert H.James, Significance of Tests and Properties of Concrete and Concrete Making Materials, ASTM International, 2006.

Majidi, Hasan, "Foolad Mobarakeh Steel Mill. Isfahan." Retrieved from: https://commons.wikimedia.org/wiki/File:Foolad_Mobarakeh7.jpg.

Nawy G.Edward, *Concrete Construction Engineering Handbook*, CRC Press, 2008.

Popovics Sandor, *Concrete Materials, Properties Specification and Testing*, NOYES Publications, 1992.

Ramachandran V.S, Beaudion James, *Handbook of Analytical Techniques in Concrete Science and Technology, Principles, Techniques and Applications*, William Andrew Publishing, 2001.

Ramachandran V.S, *Concrete Admixtures Handbook, Properties, Science and Technology*, NOYES Publications, 1995.

Ramachandran, Paroli, Beaudion, Delgado, *Handbook of Thermal Analysis of Construction Materials*, NOYES Publications, 2002.

Ramezanianpoor Aliakbar, Arabi Negin, Cement and Concrete Test Methods (Farsi), Negarande Danesh, 2011.

Richardson M, *Fundamentals of Durable Reinforced Concrete*, SPON Press, 2004.

Richardson M, *Fundamentals of Durable Reinforced Concrete*, Spon Press, 2002.

Safaye Nikoo Hamed, Introduction to Concrete Technology (Farsi), Heram Pub, 2008.

Shekarchizade Mohammad, Liber Nicolas Ali, Dehghan Solmaz, Poorzarrabi Ali, Concrete Admixtures Technology and Usages (Farsi), Elm & Adab, 2012.

Zandi Yousof, Advanced Concrete Technology (Farsi), Forouzesh Pub, 2009.

Zandi Yousof, Concrete Tests and Mix Design (Farsi), Forouzesh Pub, 2007.

4 Aggregates

Aggregates are the skeleton of concrete. They are the biggest constituent material of concrete from the volume and weight point of view. More than 60% of the volume and more than 65% by weight of any kind of concrete is the aggregates. So, the properties and specification of them are very important for the production of high-quality and performance concrete.

Aggregates effect is on the properties of fresh and hardened concrete. Some of the most important effects of aggregates on the properties of fresh concrete are as below:

- The texture and softness of concrete depend on the texture of aggregates.
- Pumpability of concrete depends on the texture of aggregates.
- Workability and malleability of concrete depend on the texture of aggregates.
- The final temperature of concrete depends on the temperature of the aggregates.
- The water demand of concrete depends on the texture and kind of aggregates.
- The effectiveness of chemical admixtures, especially the super-plasticizers, depends on the kind and texture of aggregates.

Some of the most important effects of aggregates on the properties of hardened concrete are as below:

- For normal strength concrete, the compressive strength of concrete depends on the shape of aggregates.
- For high-strength concrete, the compressive strength of concrete depends on the hardness of aggregates.
- Uniform gradation of aggregates can help to control the permeability of concrete and also it can increase the compressive strength.
- Some of the added residue materials in aggregates can cause decreasing of the compressive strength of concrete.
- The texture of aggregates can vary the final surface of concrete elements.

You can see that the aggregates are very important for the properties of fresh and hardened concrete. So, we should test them before use in the production of concrete and we should check them with standards to ensure their quality.

In this chapter, we are going to talk about the properties of aggregates and the tests for quality control before using them in the production of concrete. We will talk about the coarse and fine aggregates and the natural and crushed ones. We will talk about the sieve analysis test, specific gravity, and water absorption test of aggregates which are the most important tests for the concrete mix design. We will talk about the alkali aggregate reaction which is one of the most important defects of some aggregate used in concrete.

DOI: 10.1201/9781003384243-4

4.1 TYPES OF NATURAL STONES

There are several types of natural stones in nature. We cannot use all of them as the aggregate for concrete because the stones which are suitable for use in concrete as aggregate should have the below specifications:

- The stone should have enough strength that we can produce a durable concrete with higher strength with it.
- The mines of the stones should be available in the places where we would like to make concrete.
- The price of some types of stones is too high because they are used as the decorative stones. So, they are not suitable for the production of concrete.
- The chemical behavior of the stone should be controlled because some of the materials in some types of stones will react with the cement constituents and cause alkali aggregate reaction which will be very destructive for concrete in the future.
- As we should crush the stones and grade them for use in concrete, the hardness of the stone should not be too high which can cause the depreciation of the aggregate production instruments.

To choose a suitable type of stone for the aggregate production, you should know several types of stones in nature. We have three main types of stone or rocks in nature as bellow:

- Igneous rocks: They are formed from the solidification of molten materials which came from the volcanoes. They are divided into two types as below:
- Intrusive igneous rocks: They are crystallized below the surface of the earth. Some examples of this type of rock are Granite and Gabbro (Figure 4.1).
- Extrusive igneous rocks: They are crystallized after coming out of the volcanoes on the surface of the earth. Some examples of this type of rock are Basalt and Andesite (Figure 4.2).

FIGURE 4.1 Picture of granite rock in left and gabbro rock in right. ("Granite" by James St. John), ("Gabbro" by James St. John.)

FIGURE 4.2   Picture of basalt rock in left and andesite rock in the right. ("Basalt" by James St. John), ("Andesite" by James St. John.)

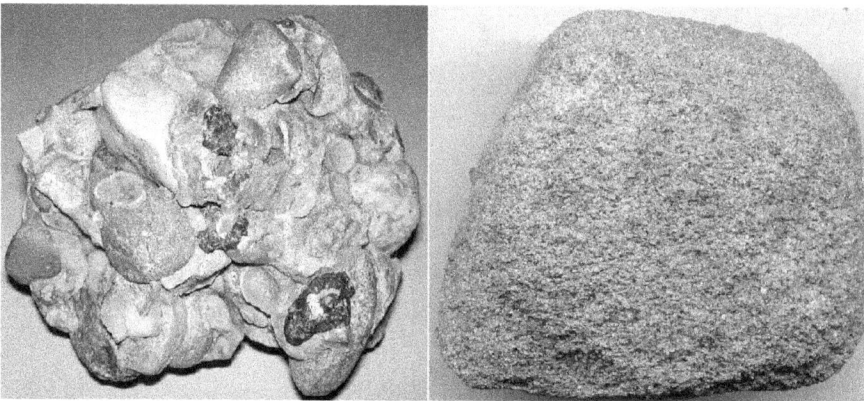

FIGURE 4.3   Picture of conglomerate rock in the left, sandstone in the right. ("Travertine-cemented conglomerate" by James St. John), ("Sandstone" by James St. John.)

- Sedimentary rocks: They are formed by the accumulation of sediments over time. These rocks are divided into three types as below:
- Clastic sedimentary rocks: They are formed from the accumulation of mechanical weathering debris. Some examples of this type of rock are Conglomerate, sandstone, and shale (Figure 4.3).

Chemical sedimentary rocks: They are formed when dissolved materials precipitate from the solution. Some examples of this type of rock are dolomite and limestone (Figure 4.4).

- Organic sedimentary rocks: They are formed from the accumulation of plant or animal debris. Some examples of this type of rocks are Diatomite and Chalk (Figure 4.5).
- Metamorphic rocks: They are modified by heat, pressure, and chemical processes from the other types of rocks. They are divided into two types as below:

FIGURE 4.4   Picture of dolomite rock in the left and limestone rock in the right. ("Dolomite rock and calcite" by Junpei Satoh), ("Oolitic limestone from the Mississippian of Indiana, USA" by James St. John.)

FIGURE 4.5   Picture of diatomite rock in the left and chalk rock in the right. ("Diatomite" by James St. John), ("Chalk from the Cretaceous of Britain" by James St. John.)

- Foliated metamorphic rocks: They have layers that are produced because of the exposure to direct heat or pressure. Some examples of this type of rock are Gneiss and Schist (Figure 4.6).
- Non-foliated metamorphic rocks: they don't have a layered appearance. Some examples of this type of rock are Marble and Quartzite (Figure 4.7).

According to the above descriptions, you can see which kind of natural rocks in your area should be suitable for the production of concrete. These natural stones can be derived from the mines and brought to the aggregate production plants. Then they will be crushed and graded to different sizes and it is possible to wash them for some reasons like the separation of soil and other residue materials. Finally, they will be ready to transport into the concrete production plant for the production of concrete. You can see pictures of the aggregate production plants in Figures 4.8 to 4.10.

FIGURE 4.6   Picture of Gneiss rock in the left and Schist rock in the right. ("Gneiss" by James St. John), (Source "Biotite Schist" by James St. John.)

FIGURE 4.7   Picture of marble rock in the left and quartzite rock in the right. ("Marble" by James St. John), ("Sioux Quartzite" by James St. John.)

For the production of finer aggregates, we can use the left-over sands in the river-beds. We call it natural sand. We can send them to the aggregate production plant and then crush, grade, and wash them for the production of natural sand. Also, we can use the coarser sizes as the gravel for the production of concrete. We will talk about the natural and crushed aggregates in this chapter.

4.2   AGGREGATES AND THE DENSITY OF CONCRETE

The most important constituent material has effect on the density of concrete in the aggregates. By using different types of aggregates, we can produce concrete with different densities. On the other hand, the density of concrete should be very important for us. So, we are going to talk about the different types of concrete according to the factor of density. You can see the density of some of the most important types of

FIGURE 4.8   A mobile stone crusher for the production of aggregates. ("Stone crusher" by Richard Webb.)

FIGURE 4.9   A sand washing plant. ("Feed section for Evowash sand washing plant" by Peter Craven.)

rocks in Table 4.1. These ranges are because of the difference in the origins of rocks and the probable impurities.

Instead of the aggregates, the amount of air in concrete is the other important factor effecting the density of concrete. When we use a higher amount of air, we will have a lighter concrete. But unfortunately, it can cause drastically reducing of compressive strength and other mechanical properties.

The effect of other constituent materials on the density of concrete is very low. So, to evaluate or predict the density of concrete, we should pay special attention to the aggregates and the amount of air.

FIGURE 4.10 An aggregate production plant. (Photograph by the author.)

TABLE 4.1
Density of Different Types of Rocks

Type of Rock	Granite	Sandstone	Quartzite	Limestone	Dolomite	Marble	Gneiss
Density (kg/L)	2.6–2.8	2.2–2.8	2.6–2.8	1.8–2.8	2.6–2.9	2.5–2.7	2.7–2.8

According to the above mentioned, we have three types of concrete in the case of the differentiation of density:

- Normal weight concrete: When we use any type of normal aggregates made with the types of rocks mentioned in Section 4.1, we will have a normal weight concrete. The density of this type of concrete which is the most common type (more than 90% of concrete in the world) is between 2350 and 2450 kg/m^3. You can see a picture of a normal type aggregate in Figure 4.11.

We can use this type of concrete in any kind of structure like below examples:

- Columns, walls, roofs, and foundations of urban structures
- Columns, decks, and foundations of the bridges
- Water tank, storages, and pools
- Pavements
- Offshore structures
- Light weight concrete: This is a special type of concrete with lower density than the normal concrete. We can make light weight concrete with two techniques:
- Make light weight concrete with air entraining and special gradation of aggregates: We can make a very light concrete by increasing the amount of air bubbles inside the concrete. With this technique we can make a concrete with the density between 300 and 1800 kg/m^3.

FIGURE 4.11   Normal type aggregate. (Photograph by the author.)

TABLE 4.2
Examples of Light Weight Aggregates

Type of Light Weight Aggregate	Explanation
Leca	This is the light weight expanded clay aggregate made by heating the clay to around 1200°C in a kiln
Pumice	It is a very light grained volcanic rock
Perlite	This is a kind of volcanic glass which can expand when heated and make a light weight aggregate

- Make light weight concrete by using light weight aggregates: We can use light weight aggregates as total amount of aggregates in concrete or part of it. We can make a light weight concrete with the density between 600 and 1800 kg/m³ by using the light weight aggregates.

We have several types of light weight aggregates. You can see some types of it in Table 4.2. You can see pictures of Leca, Pumice, and Perlite in Figures 4.12 to 4.14.

We have two types of light weight concrete:

- Structural light weight concrete: This is a concrete with good or enough mechanical properties so that we can use it as the structural element. Most of the time, this type of light weight concrete will be used for light weight aggregates. The density of this concrete is usually more than 800 kg/m³ and less than 2000 kg/m³ and the compressive strength should be more than 25 MPa.

  Making a structure with the light weight concrete can reduce the dead and earthquake load and we can make a structure with good resistance against the earthquake.

FIGURE 4.12 Picture of light expanded clay (Leca) aggregates. ("LECA Aggregates Granules".)

FIGURE 4.13 Picture of pumice aggregates. ("Rhyodacite pumice & volcanic ash" by James St. John.)

- Non-structural light weight concrete: This is a kind of light weight concrete that we can use it for the production of blocks or other types of void fillers in the construction industry. This type of concrete is very light (between 300 and 800 kg/m^3) but has not enough mechanical properties to use them in the structural elements (Figure 4.15).
- Heavy weight concrete: This is a special type of concrete that is heavier than the normal weight concrete. It is made by using special heavy weight aggregates. You can see some types of heavy weight aggregates in Table 4.3.

Heavy weight concrete is used for shielding the radiation in the areas like X-ray photography or in the central part of the nuclear reactors to control the radiation out of the allowed area (Figure 4.16).

You can see pictures of Magnetite, Hematite, and iron ore in Figures 4.17 to 4.19. We can make a heavy weight concrete with the density between 3000 and 6000 kg/m^3 by using the above aggregates as the total aggregate in concrete or a part of it.

FIGURE 4.14 Picture of perlite aggregates. (Photograph by the author.)

FIGURE 4.15 A type of roof block made with non-structural light weight concrete. (Photograph by the author.)

TABLE 4.3
Examples of Heavy Weight Aggregates

Type of Aggregate	Density (kg/L)	Explanation
Magnetite	5.1–5.2	This is a type of iron oxide mineral
Hematite	More than 5.2	This is a type of ferric oxide with high iron content
Iron aggregates	More than 7.2	Different size iron or steel particles

FIGURE 4.16 An X-ray medical photographer in left and nuclear reactor in right. They should protect with the heavy weight concrete. ("X-ray table" by Broken Sphere), ("Nuclear electricity power plant".)

FIGURE 4.17 Magnetite rock which can be used for heavy weight aggregate production. ("Magnetite rock" by James St. John.)

FIGURE 4.18 Hematite rock which can be used for heavy weight aggregate production. ("Hematite rock" by James St. John.)

FIGURE 4.19 Iron ore used as heavy weight aggregates. ("Specularite" by James St. John.)

4.3 AGGREGATE SIZE

To produce a compact and dense structure concrete, we should use different sizes of aggregates from the bigger size to the finer ones. So, we should check the compatibility of aggregates in the case of their size and shape for concrete that we will talk about it in the future.

In ASTM standard, we have different sizes of sieves, which we should use to check the aggregate size by using sieve analysis. We will talk about the sieve analysis test of aggregates in the next part of this chapter. Now you can see the sieve sizes in Table 4.4.

TABLE 4.4

ASTM Sieve Sizes for Aggregates

Sieve Size (mm)	25	19	12.5	9.5	4.75	2.36	1.18	0.6	0.3	0.15	0.075
Sieve No	–	–	–	–	4	8	16	30	50	100	200

We have different sieve sizes in (mm) and different sieve numbers from 4.75 mm (No. 4) to 0.075 mm (No. 200). According to these sizes, we can separate two kinds of aggregates from the size point of view:

- Coarse aggregates
- Fine aggregates

4.3.1 COARSE AGGREGATES

Coarse aggregates are defined as aggregates coarser than 4.75 mm. So, we can say the border between the fine and coarse aggregates is 4.75 mm. The limit of coarse aggregate is very broad. We can have an aggregate with a size of 25 mm or coarser or we can have an aggregate with a size of 9.5 mm or a little coarser all as coarse aggregates. So, in different parts of the world, we can see different aggregate sizes as the coarse aggregate. Here we mention the most common types:

- Size 5–12: The aggregates with a size of between 4.75 and 12.5 mm. The nominal size of this kind of aggregate is about the size of a pea (Figure 4.20).
- Size 12–19: The aggregate with a size of between 12.5 and 19 mm. The nominal size of this kind of aggregate is about the size of an almond.
- Size 12–25: The aggregate with a size of between 12.5 and 25 mm. In this case also, the nominal size is about the size of an almond. The only difference is the coarser size which is 25 mm in this type. We can call 12–19 and 12–25 aggregates as gravel (Figure 4.21).

To produce a well-designed concrete, most of the time, we should use 5–12 coarse aggregates with one of the 12–19 or 12–25 ones. We cannot mix 12–19 and 12–25 for the production of one concrete. Sometimes, we may have a kind of coarse aggregate with a size of between 4.75 and 19 or 25 mm. In this case, there is no need to use 5–12 aggregates separately.

4.3.2 MAXIMUM SIZE OF COARSE AGGREGATE

One of the most important topics for concrete mix design is the choose of maximum size of coarse aggregate for the desired concrete. We can choose 25, 19, or 12.5 mm as the maximum size of coarse aggregates. If we choose 12.5 mm as the maximum size, we should use only 5–12 aggregates as coarse. If we use 19 mm as the maximum size, we should use a mix of 5–12 and 12–19 or only 12–19 as the coarse aggregates and if we use 25 mm as the maximum size, we should use a mix of 5–12 and 12–25 or only 12–25 as the coarse aggregates.

FIGURE 4.20   Picture of 5–12 aggregates. (Photograph by the author.)

FIGURE 4.21   Picture of 12–25 aggregates. (Photograph by the author.)

To choose the maximum size of coarse aggregate in concrete, we should consider several factors as below:

- We should consider the type of concrete in the case of needed compressive strength. If we need more compressive strength, we should use a smaller aggregate size.
- We should consider the slump and rheology of concrete. To produce a high slump concrete if we use a smaller aggregate size, we need more water or super-plasticizer.

- We should consider the type of structural elements in the case of any rebar congestion or using concrete in the thin elements or in huge structures. For huge structures, it is better to use larger aggregate size and for thin or congested structures it is better to use finer aggregates.
- We should consider the availability of coarse aggregates. Sometimes, it is possible that we are convicted to use a type and size of coarse aggregates. In this case, we should adjust some of the specifications of concrete with the aggregates.

In the case of structural elements ACI 211-1 recommends as below:

"The maximum size of coarse aggregate should not exceed one fifth of free distance between forms and one third of slab diameter and three fourth of distance between rebars in the structural element."

In the case of structural type with new technology and concrete compressive strength, we recommend Table 4.5 to choose the maximum size of coarse aggregate.

4.3.3 Fine Aggregates

Fine aggregates are defined as aggregates finer than 4.75 mm. So, all of the particles passed by sieve No. 4 are defined as fine aggregates. In most parts of the world, they call fine aggregates as sand. Particle sizes for fine aggregates are not as broad as the coarse aggregates. But we have four types of fine aggregates:

- Sand 0–2.36: This is the aggregate with a size between 0 and 2.36 mm. This is a very soft sand. We can use it for the production of concrete. But we should consider the total aggregates sieve analysis to get the best result. Most of the time, we will use this type of sand for the production of mortar.
- Sand 0–4.75: This is the aggregate with a size between 0 and 4.75 mm. This is the best type of sand for the production of concrete. But sometimes, we cannot find a 0–4.75 sand in some regions of the world because the production process will be harder than the other ones.

TABLE 4.5

Recommendation for the Maximum Size of Coarse Aggregates According to the Type of Structure and Concrete Compressive Strength

Concrete Compressive Strength (MPa)	Structural Element	Recommended Max Size of Coarse Aggregate (mm)
Less than 30	Foundations	25
Less than 30	Floors, columns, walls	19
30–45	Foundations, floors	19
30–45	Columns, walls	12.5
45–70	All types of elements	12.5
More than 70	All types of elements	9.5

- Sand 0–9.5: This is the aggregate with a size between 0 and 9.5 mm. In this type of sand, we have some percent of coarse aggregate that we should consider in the mix design (Figure 4.22).
- Dune sand: This is the aggregate with a size between 0 and 0.6 mm. You can find this type of sand in the deserts of some regions of the world like the Middle East. We can use this type of sand for the production of concrete. But most of the times, we use it mixed with other types of sand (Figure 4.23).

FIGURE 4.22   Picture of 0–8 sand. (Photograph by the author.)

FIGURE 4.23   Picture of dune sand. (Photograph by the author.)

4.4 PRODUCTION OF AGGREGATES

As mentioned before, aggregates are made from different types of rocks in nature. But the production process for each kind of aggregate should be different. In this case, we have two types of aggregates:

* Natural aggregates
* Crushed aggregates

4.4.1 NATURAL AGGREGATES

This is a type of aggregate that is derived from the riverbeds. So, the shape of them is circular (Figure 4.24). The production process for this type of aggregate is the separation of different sizes with special riddles. Sometimes, for better quality, we may use a sand maker machine to crush some of the aggregates for the production of natural sand. It can help us to have better sizing and also, we can have some crushed aggregates mixed with the natural which can help to give better compressive strength in concrete. Finally, we should wash these aggregates for the separation of clay and other harmful materials. You can see a picture of a sand washer in Figure 4.9 and a sand maker machine in Figure 4.25.

The positive effects of using natural aggregates in concrete are because of the circular shape of them, which can cause better movement of aggregates inside fresh concrete. These effects are as below:

* Improvement of the rheology and workability of concrete.
* Improvement of concrete pumpability.
* Prevention of segregation and bleeding in concrete.
* The ability to produce concrete with a higher initial slump.

The most important negative effect of using natural aggregates in concrete which is because of its circular shape is the reduction of concrete compressive strength

FIGURE 4.24 A deposit of natural sand in a ready mixed plant. (Photograph by the author.)

FIGURE 4.25   A sand maker machine. (Photograph by the author.)

because of the weaker stress transition area on the surface of aggregates. On the other hand, the natural resources in riverbeds for the production of natural aggregates are limited and using these resources is very dangerous for the environment because it can cause very bad erosion effects in the environment.

4.4.2   CRUSHED AGGREGATES

This is a kind of aggregate that is derived from mountain rocks (Figure 4.26). In fact, for the production of crushed aggregates, we should use a crushing machine to crush and separate the rocks derived from the mountain. Then we can wash them to separate the harmful materials. Most of the times, there is no much harmful material in these aggregates. So, there is no need to wash them. On the other hand, the production of concrete with crushed aggregates needs more fillers. As washing can remove some of the useful fillers, it is better to use crushed aggregates without washing. You can see a picture of a crusher in Figure 4.8.

FIGURE 4.26 A deposit of crushed gravel in a ready mixed plant. (Photograph by the author.)

The positive effects of using crushed aggregates in the production of concrete are as below:

- Improvement of concrete compressive strength because of the crushed shape of aggregates which can cause better interlocking inside the concrete.
- Better quality control of aggregates because of the control in the whole production process.
- Lower effect on the environment because of more resources in nature.
- The ability to produce high-strength concrete for special purposes.

The most important negative effect of using crushed aggregates in the production of concrete is the reduction of the workability and pumpability because of the crushed shape and interlocking inside concrete.

To decide about the use of natural or crushed aggregates in the production of concrete you should consider below points:

- The access to the minerals of crushed or natural aggregates is very important for decision-making.
- For lower compressive strength, we can use natural aggregates and for higher compressive strength it is better to use crushed or mixed aggregates.
- The workability and pumpability of concrete are very important. If you use crushed aggregates with low amount of fillers (passed by sieve No. 100), the pumping of concrete will be very difficult.
- Using crushed coarse aggregate is the best decision for most types of concrete. We can use crushed fine aggregates if it has enough amount of filler or we can use some fillers like dune sand or stone powder beside that. Nevertheless, it is better to use natural fine aggregates.

4.5   AGGREGATES TEST AND QUALITY CONTROL

Like other concrete constituent materials, you should test the aggregates before using them in the production of concrete. We have special standards for the aggregates to check before use. Sometimes, it is possible for us to reach only one or two types of aggregates in some places of the world. In this case, there is no way instead of using these aggregates in the production of concrete even if they don't have enough quality. But we should consider some other modifications in the concrete mix design to control the deficiency of aggregates.

For example, in some parts of the world, there is only crushed sand with very low amount of fillers. If we use them in the production of concrete, the workability and pumpability will decrease drastically. We should consider some modifications in concrete mix design to use these aggregates. We can use added fillers, like dune sand or stone powder. This will cause using more amount of water or super-plasticizer. We can use viscosity modifier admixtures.[1] These admixtures can adjust the viscosity and pumpability of concrete. All of the above mentioned consist of more cost in concrete production. So, you should compare the cost and quality using local aggregates with low quality and the above considerations or using other aggregates from a place far from the project that it will cost more for transportation.

In this part, we are going to talk about the most important quality control tests for the aggregates. By doing these tests, you can ensure the quality of aggregates. On the other hand, for concrete mix design, you will need data from these tests like sieve analysis and the specific gravity of aggregates. You can see some of the most important tests of aggregates according to the ASTM standard in Table 4.6.

4.5.1   ABRASION RESISTANCE

Resistance of aggregates against the abrasion is a very important indicator for the quality. It can show the strength of the rocks used for the production of aggregates.

The standard test method for the abrasion resistance of aggregates is ASTM C131. In this test, we will use the Los Angeles test machine to measure the abrasion resistance of aggregates (Figure 4.27).

Los Angeles test machine consists of a rotating cylinder. Inside the cylinder, there are some steel balls. We should wash the coarse aggregates and let them dry in the oven. After that, we should measure the exact weight of aggregates, then put them inside the cylinder and let the cylinder rotate for a defined period of time. Then we should put the aggregates out of the cylinder and measure the weight again after washing and drying. Finally, we can calculate the amount of healthy aggregates to indicate the abrasion resistance.

4.5.2   SIEVE ANALYSIS TEST

Sieve analysis test is the most important quality control test for aggregates. As you can see in Table 4.6, the standard test method for this test is ASTM C136. Before the description of the test method, you can see different sieve sizes in ASTM standard in Table 4.4.

FIGURE 4.27 Los Angeles test machine. (Photograph by the author.)

TABLE 4.6

The Most Important Tests for Aggregates According to the ASTM Standard

ASTM C131	Abrasion resistance of aggregates by Los Angeles machine
ASTM C136	Sieve analysis test for coarse and fine aggregates
ASTM C127	Specific gravity and absorption of coarse aggregates
ASTM C128	Specific gravity and absorption of fine aggregates
ASTM C117	Minerals finer than $75\,\mu m$ in aggregates by washing
ASTM C40	Organic impurities in fine aggregates
ASTM C227	Potential alkali reactivity of cement aggregate combination
ASTM C289	Potential alkali silica reactivity of aggregates
ASTM C586	Potential alkali reactivity of carbonate rocks

FIGURE 4.28   Sieves for the coarse aggregates and the amount remained after shaking. (Photograph by the author.)

For the sieve analysis of coarse aggregates, you should put the sieves from 25 to 2.36 mm from coarser to the finer one or you can select according to the size of coarse aggregates (Figure 4.28). Then you should put a container below the last sieve and put the coarse aggregates on the top of the coarser sieve. Finally, you should put the door on the top of the sieves collection and put it in a shaker machine (Figure 4.29). After enough shaking, you should weigh the amount of aggregates remaining on the top of each sieve and passed by the last sieve and write it in a table like in Tables 4.7 and 4.8.

After calculation of the passed percent from each sieve like in the above tables, you should draw a figure which you can compare the last column of these tables with the standard amounts (Figures 4.30 and 4.31). So, you can see if these aggregates can be according to the standard amounts or not. The standard amounts are special for different parts of the world. You can see standard amounts for coarse aggregates according to local concrete regulations in Table 4.9.

The procedure for testing the fine aggregates is the same but you should use sieves from 4.75 to 0.15 mm (Figure 4.32). You can use sieve No. 200, but the percent passed by this sieve is not accurate enough that you can trust it. To achieve the exact amount passed by sieve No. 200, you should use ASTM C117 test.

Like the coarse aggregates, you should compare the percent passed by each sieve with the standard amounts. You can see the standard amount for fine aggregates according to the local concrete regulations in Table 4.10.

For a better understanding of sieve analysis test of fine aggregates, you can see an example in Table 4.11 and you can see the chart in Figure 4.33.

As you can see in Table 4.11, there is an additional column in the table which is the cumulative percent remaining on sieves. This column is used for the calculation of the fineness module of the sand. You should divide the total amount of this column by 100 to achieve the fineness module. For the sand in Table 4.11 fineness module is 3.47. For the meaning of different ranges of sand fineness modules, you can see Table 4.12.

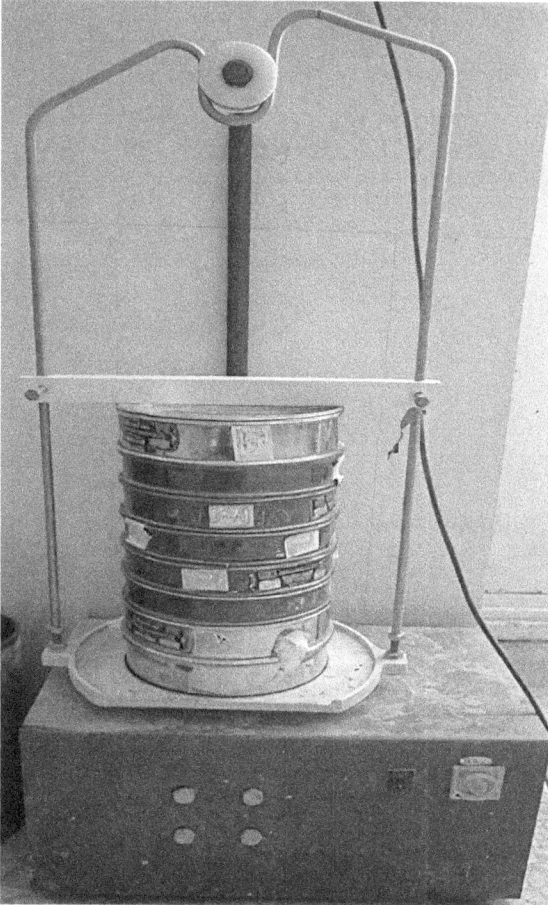

FIGURE 4.29   Shaker machine. (Photograph by the author.)

As you can see from the descriptions in the above table, you can use any kind of sand for the production of concrete. But the quality of sand is more important than the coarse aggregates. In fact, you can compensate for the defects of the coarse aggregates easier than the sand. Another important factor for the sand quality is the amount of filler passed by sieve No. 200 that we will talk about it in the future.

4.5.3   DENSITY AND WATER ABSORPTION OF AGGREGATES

One of the other important properties of aggregates for concrete mix design is the density and water absorption of them. So, it is very important to test and calculate these properties for any kind of aggregates that we are going to use for concrete production.

TABLE 4.7
Sieve Analysis Table for a 12–25 Gravel

Sieve Size (mm)	Weight of Aggregates Remained on Sieve (g)	Weigh of Aggregates Passed by Sieve (g)	Percent of Aggregates Remained on Sieve (%)	Percent of Aggregates Passed by Sieve (%)
25	0	1671	0.0	100
19	308	1363	18.4	81.6
12.5	971	392	58.2	23.4
9.5	353	39	21.1	2.3
4.75	39	0	2.3	0
Total	1671	–	100	–

TABLE 4.8
Sieve Analysis Table for a 5–12 Gravel

Sieve Size (mm)	Weight of Aggregates Remained on Sieve (g)	Weigh of Aggregates Passed by Sieve (g)	Percent of Aggregates Remained on Sieve (%)	Percent of Aggregates Passed by Sieve (%)
19	0	1524	0.0	100
12.5	27	1497	1.8	98.2
9.5	456	1041	29.9	68.3
4.75	896	145	58.8	9.5
2.36	145	0	9.5	0
Total	1524	-	100	-

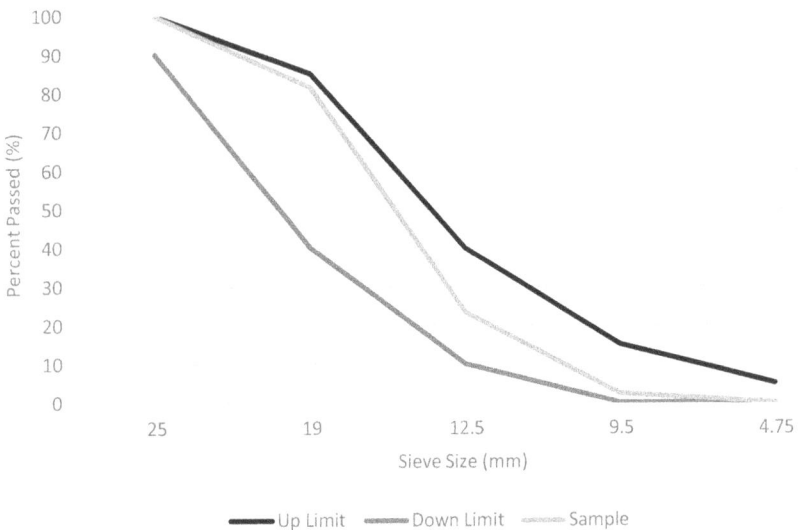

FIGURE 4.30 Sieve analysis chart for 12–25 gravel. (Graph created by the author.)

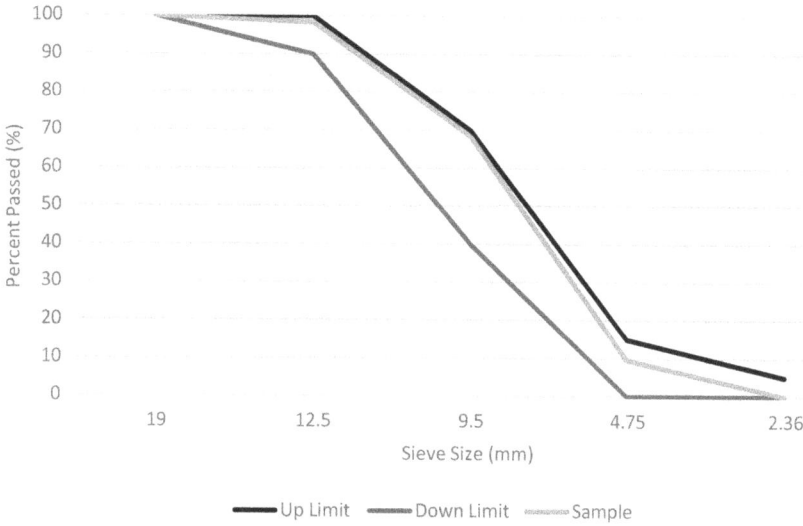

FIGURE 4.31 Sieve analysis chart for 5–12 gravel. (Graph created by the author.)

TABLE 4.9
Standard Amounts of Percent Passed for the Coarse Aggregates According to the Local Concrete Regulations

Coarse Aggregate Type (mm)	Passed by 25 mm (%)	Passed by 19 mm (%)	Passed by 12.5 mm (%)	Passed by 9.5 mm (%)	Passed by 4.75 mm (%)	Passed by 2.36 mm (%)
12-25	90–100	40–85	10–40	0–15	0–5	–
11-19	100	90–100	20–55	0–15	0–5	–
5-12	–	100	90–100	40–70	0–15	0–5

On the other hand, using aggregates with high water absorption is a limit for using them in the production of concrete because it will cause more water demand and more water demand means lower quality for concrete. So, the water absorption of aggregates is one of the quality control factors for aggregates. For normal aggregates, the water absorption should not exceed 3% by weight of them.

Most of the time, the aggregates with very low density are not good quality aggregates, especially in the case of compressive strength. So, by checking the aggregate density, we can give an idea about their quality and compressive strength. For Limestone and silica aggregates, the density is between 2.5 and 2.9 kg/L.

Before talking about the test methods for density and water absorption, you should know about the aggregate moisture conditions. Any aggregate in the project or batching plant could be in four conditions:

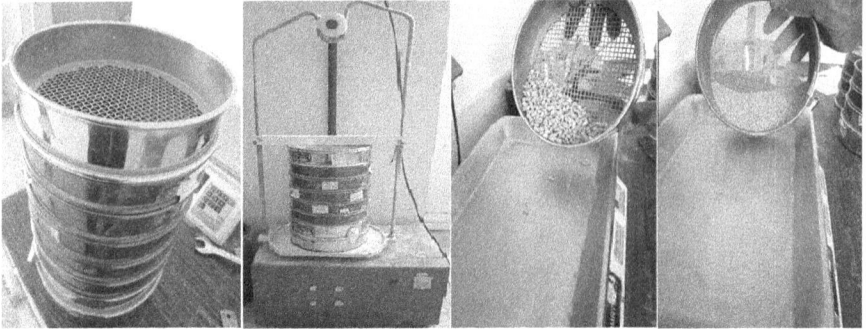

FIGURE 4.32   Sieve analysis test for fine aggregates. (Photograph by the author.)

TABLE 4.10

Standard Amounts of Percent Passed for the Fine Aggregates According to the Local Concrete Regulations

Fine Aggregate	Passed by 4.75 mm (%)	Passed by 2.36 mm (%)	Passed by 1.18 mm (%)	Passed by 0.6 mm (%)	Passed by 0.3 mm (%)	Passed by 0.15 mm (%)
Sand	89–100	60–100	30–90	15–54	5–40	0–15

TABLE 4.11

Sieve Analysis Table for Sand

Sieve Size (mm)	Weight of Aggregates Remained on Sieve (g)	Weigh of Aggregates Passed by Sieve (g)	Percent of Aggregates Remained on Sieve (%)	Percent of Aggregates Passed by Sieve (%)	Cumulative Percent Remained on Sieve (%)
4.75	144	1418	9.2	90.8	9.2
2.36	392	1026	25.1	65.7	34.3
1.18	356	670	22.8	42.9	57.1
0.6	206	464	13.2	29.7	70.3
0.3	181	283	11.6	18.1	81.9
0.15	188	95	12	6.1	93.9
Total	1562	–	100	–	346.7

- Oven dry: It means that there is no moisture inside the aggregates. The exact definition for an oven dry aggregate is the aggregates that are dried in the oven. But many types of crush aggregates especially coarse ones are from mountain rocks. So, if they store in a dry condition prevented from rain, we can say they are dry. In fact, we can neglect the amount of little moisture inside the aggregates. You should calculate the amount of saturation limit of aggregates. This is the amount of water you should add to the concrete water in mix design when you use oven-dried aggregates.

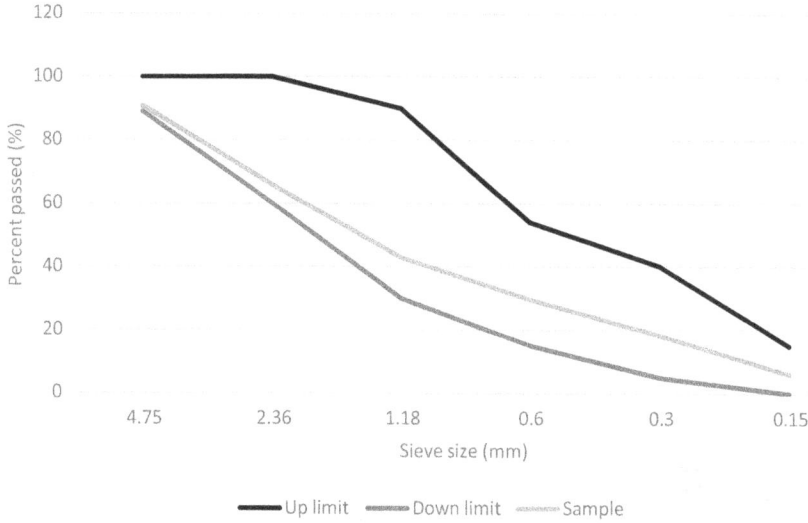

FIGURE 4.33   Sieve analysis chart for sand. (Graph created by the author.)

TABLE 4.12
Different Ranges of Sand Fineness Module

Fineness Module Range	Description
Less than 3	It means that we have a very fine sand. This type of sand will be good for concrete production. But we should consider the amount of water demand.
Between 3 and 3.5	It means that we have a normal sand which is very good for the production of concrete
Between 3.5 and 4	It means that we have a coarse sand that we can use it for the production of concrete. But we should consider the pumpability of concrete.
More than 4	It means that we have a very coarse sand. It is better to not using of this sand for the production of concrete. If there is no way we should consider the pumpability and also segregation and bleeding.

- Dried in the air: It means that the aggregates had some amount of moisture but they dried in the air. So, there is some amount of moisture in the aggregates but less than the saturation limit. You should calculate the amount of saturation limit of aggregates. Then you can find the amount of water shortage in the aggregates and add it to the total water of concrete in the mix design.
- Saturated surface dry (SSD): It means that the amount of moisture in the aggregate is exactly the same as the saturation limit. This condition is very important in concrete technology. We should assume the aggregate moisture condition for many tests and also for concrete mix design as the SSD. But we can say that this condition is only a hypothetical definition. The aggregates could not be in this condition in real projects and batching plants.

You should calculate concrete mix design water by assuming that the aggregates are in this condition. Then according to the real conditions, you should add or subtract the amount of shortage or excess water.

- Wet condition: It means that the aggregates have moisture more than the saturation limit. Most of the washed sands are in this condition in the projects and batching plants. You should calculate the amount of total moisture and the saturation limit. Then you can calculate the amount of excess moisture in the aggregates. After finalizing the concrete mix design, you should subtract this amount of excess moisture from the water of the concrete.

To calculate the density and water absorption of coarse aggregates you should use ASTM C127 and for fine aggregates, you should use ASTM C128.

For fine and coarse aggregates first, you should make the aggregates into the SSD condition. To do that, you should put them in water for more than 24 hours. For coarse aggregates, you should dry the surface with a suitable napkin. For fine aggregates, you should dry the surface with warm air and check the aggregates condition by using a special cone (Figure 4.34). You should put the sand inside the cone like the slump test and pull it up. If the sand stands like the cone shape it is not in the SSD condition. If the sand falls down, you can say that it is in the SSD condition. But you should take care that the sand condition should not be drier than the SSD.

After that, you should use a pycnometer for the measurement of the aggregate volume (Figure 4.35) and use an apparatus for the measurement of the weight (Figure 4.36). So, you will have the density of aggregates.

FIGURE 4.34   The instrument for checking the sand SSD condition. (Photograph by the author.)

FIGURE 4.35   Pycnometer. (Photograph by the author.)

For the saturation limit of the aggregates, you should weigh the aggregates in SSD condition. Then dry them for more than 24 hours in the oven and weigh them in dry condition. Then you can calculate the amount of moisture for the saturation point and calculate the percent of the moisture for the saturation point.

For the moisture content of aggregates, you should weigh the aggregates. Then you should dry them in the oven and weigh them again. Now, you can calculate the amount of moisture and its percent in the aggregates. For a better understanding you can see Table 4.13 for calculations:

4.5.4   FILLERS AND PASSING BY SIEVE NO. 200

We call passing by sieve No. 100 fillers (Figure 4.37). Fillers are a very important part of fine aggregates. Fillers, cement, and water make a very important part of concrete that we name fine mortar. This part of concrete is very important, especially for fresh concrete properties.

FIGURE 4.36   Special table and apparatus for weighting the aggregates. (Photograph by the author.)

TABLE 4.13

Calculations for the Density, Water Absorption and Moisture Content for Aggregates

Weight of aggregates in SSD condition: 268.6 g   Volume of aggregates in SSD condition: 100 mL

Weight of dried aggregates: 265.1 g   Weight of aggregates in normal condition: 272.2 g

Density = 268.6 / 100 = 2.686 kg/L

Water absorption = ((268.6 – 265.1) / 265.1) × 100 = 1.3%

Moisture content = ((272.2 – 265.1) / 265.1) × 100 = 2.7%

Excess water = 2.7 – 1.3 = 1.4%

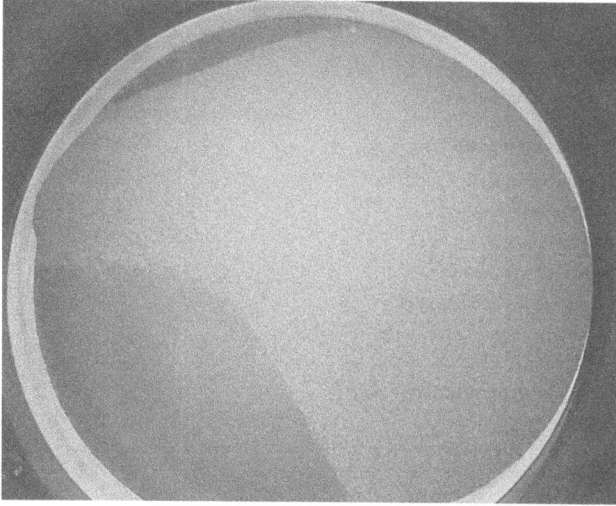

FIGURE 4.37 Fillers of a natural sand. (Photograph by the author.)

We can name some of the most important specifications of concrete related to the fillers as below:

- Pumpability of concrete: To produce a pumpable concrete, we need enough fillers for good viscosity. Good viscosity will achieve by enough amount of fillers in the sand.
- Malleability of concrete: If the amount of fillers in the sand will be enough, we can make a good surface for concrete and with lack of it, making a soft surface will be very hard.
- Prevention of segregation and bleeding: If you have a coarse grading aggregate, the water would like to come on the surface of concrete especially when you are using a high range water reducer chemical admixture. So, enough amount of fillers can help the water to remain inside the concrete and you can prevent segregation and bleeding.
- Possibility for the production of high slump concrete: When you would like to produce a high slump concrete, the probability of segregation and bleeding will increase. So, you should use a higher amount of cement, other binders, and filler to prevent that.

You can measure the amount of fillers in fine aggregates by sieve analysis test. It is possible to have some amount of fillers in the coarse aggregates as the stone powder. So, you should check the passed by sieve No. 100 for coarse aggregates to calculate the total amount of fillers in your concrete.

There are different suggestions for the amount of fillers in the text and lectures. Here you can find some useful suggestions in Table 4.14.

Although fillers are very good for concrete quality, passing by sieve No. 200 can be dangerous for some specifications of concrete as below:

TABLE 4.14
Suggestions for the Amounts of Fillers

Type of Concrete and Aggregates	Suggested Amount of Fillers (kg in 1 m³ of Concrete)
Normal slump concrete with natural sand	80–100
Normal slump concrete with crushed sand	100–120
High slump concrete with natural sand	100–120
High slump concrete with crushed sand	120–150
Self-compacting concrete with fine grading	140–180
Normal slump concrete: concrete with slump between 120 and 180 mm	
High slump concrete: concrete with slump more than 180 mm	
Self-compacting concrete: No slump concrete with flow circle more than 600 mm	

- High amount of passing by sieve No. 200 will increase water demand and more water can decrease the mechanical properties, especially the compressive strength of concrete.
- Water absorption of particles passed by sieve No. 200 is very high. So, they can decrease the workability of concrete over time and using concrete after a period of time will be hard.
- Super-plasticizers can adsorb to the particles passed by sieve No. 200. So, high amount of these very fine particles in concrete can increase the dosage of super-plasticizer which is non-economical.
- High amount of passing by sieve No. 200 can increase the amount of entrapped air in concrete. So, it can cause the amount of air in concrete out of control which is very dangerous for the quality of concrete.

You can use sieve No. 200 in sieve analysis test of aggregates (Figure 4.38). But you cannot measure the exact amount of passing by sieve No. 200 with this test because these very fine particles are very adhesive. So, you cannot pass them through the sieve by shaking. To measure the exact amount, you should use the ASTM C117 test.

In this test first, you should dry the aggregates in the oven and weigh them. By using two sizes of sieve, 1.18 mm in the up, 75 microns (No. 200) in the bottom, dishwashing liquid, and water you can wash and pass all of the particles passing by sieve No. 200. Finally, you should dry the remained aggregates in the oven and weight them again. By subtracting two weights, you can measure the exact amount of passing by sieve No. 200 in the aggregates.

FIGURE 4.38 Passing of sieve No. 200 for a natural sand. (Photograph by the author.)

4.5.5 HARMFUL MATERIALS IN AGGREGATES

As the aggregates come from the nature, it is possible to have harmful material for concrete. So, we should understand the effects of these harmful materials on concrete quality and properties. Here you can find each of the harmful materials that you may find in the aggregates and their effect on the quality and properties of fresh or hardened concrete:

- Alkali impurities: They can retard the setting time and also, they can decrease the final compressive strength of concrete.
- Clay: It can increase water demand and decrease the compressive strength. On the other hand, it can cause rapid slump loss in fresh concrete.
- Chloride ion: It can cause rebar corrosion in the concrete. Also, some salts which they content chloride can accelerate the setting time of concrete.
- Sulfate ion: It can cause corrosion in the concrete and like the chloride ion some salts contain the sulfate ion which can retard or accelerate the setting time.
- Wood or other not related particles: They can decrease the compressive strength of concrete. On the other hand, they can decrease the quality of the concrete surface especially when the final surface of the concrete is very important.

According to the above mentioned, we should test the aggregates for these harmful materials before using them in concrete production. Most of the time, for each mine of aggregates, if we check them at the first time of usage, it could be enough. Because if there was not a source for these harmful materials in a mine, the aggregates should be safe for future usage.

NOTE

1. VMA.

REFERENCES

Aitcin P.C, High Performance Concrete, E&FN SPON, 2004.

American Society for Testing and Materials, Standard Practice for Making and Curing Concrete Test Specimens in the Laboratory, ASTM C192-00.

American Society for Testing and Materials, Standard Specification for Concrete Aggregates, ASTM C33-01.

American Society for Testing and Materials, Standard Specification for Ready-Mixed Concrete, ASTM C94-00.

American Society for Testing and Materials, Standard Specification for Light-weight Aggregates for Structural Concrete, ASTM C330-00.

American Society for Testing and Materials, Standard Test Method for Organic Impurities in Fine Aggregates for Concrete, ASTM C40-99.

American Society for Testing and Materials, Standard Test Method for Surface Moisture in Fine Aggregates, ASTM C70-94.

American Society for Testing and Materials, Standard Test Method for Materials Finer than 75μm in Aggregates by Washing, ASTM C117-95.

American Society for Testing and Materials, Standard Test Method for Specific Gravity and Absorption of Coarse Aggregates, ASTM C127-88.

American Society for Testing and Materials, Standard Test Method for Specific Gravity and Absorption of Fine Aggregates, ASTM C128-97.

American Society for Testing and Materials, Standard Test Method for Sieve Analysis of Fine and Coarse Aggregates, ASTM C136-01.

American Society for Testing and Materials, Standard Test Method for Air Content of Freshly Mixed Concrete by the Volumetric Method, ASTM C173-01.

American Society for Testing and Materials, Standard Test Method for Potential Alkali Reactivity of Cement-Aggregate Combination, ASTM C227-97.

American Society for Testing and Materials, Standard Test Method for Potential Alkali Silica Reactivity of Aggregates (Chemical Method), ASTM C289-94.

American Society for Testing and Materials, Standard Test Method for Density, Absorption and Voids in Hardened Concrete, ASTM C642-97.

American Society for Testing and Materials, Standard Test Method for Length Change of Concrete Due to Alkali-Carbonate Rock Reaction, ASTM C1105-95.

American Society for Testing and Materials, Standard Test Method for Potential Alkali Reactivity of Aggregates (Mortar Bar Method), ASTM C1260-94.

Broken Sphere, "X-ray table." Retrieved from: https://commons.wikimedia.org/wiki/File:X-ray_table.JPG.

Connor Jerome J, Faraji Susan, *Fundamentals of Structural Engineering*, Springer, 2016.

Craven, Peter, "Feed section for Evowash sand washing plant." Retrieved from: https://commons.wikimedia.org/wiki/File:Feed_section_for_Evowash_sand_washing_plant_(5591191605).jpg.

Ervanne Heini, Hakanen Martti, *Analysis of Cement Super-plasticizer and Grinding Aids A Literature Survey*, Posiva Oy, 2007.

European Standard Organization, Concrete-Part 1: Specification, Performance, Production and Conformity, EN206-1, 2000.

European Standard Organization, Testing Fresh Concrete, EN12450 Series.

European Standard Organization, Testing Hardened Concrete, EN12390 Series.

European Standard Organization, Tests for General Properties of Aggregates, EN932 Series.

European Standard Organization, Tests for Geometrical Properties of Aggregates, EN933 Series.

Hauschild Michael, Rosenbaum Ralph K, Olsen Sting Irving, *Life Cycle Assessment, Theory and Practice*, Springer, 2018.

Heinrichs Harald, Martens Pim, Michelsen Gerd, Wiek Arnim, *Sustainability Science, An Introduction*, Springer, 2016.

Iranian Institute for Research on Construction Industry, 9th topic of National Rules for Construction, "Concrete Structures", 2009.

Iranian Institute for Research on Construction Industry, National Concrete Mix Design Method, 2015.

Iranian National Management and Programming Organization, National Handbook of Concrete Structures, 2005.

Iranian Standard Organization, Concrete Admixtures, Specification, ISIRI2930, 2011.

Iranian Standard Organization, Concrete Specification of Constituent Materials, Production and Compliance of Concrete, ISIRI2284-2, 2009.

Iranian Standard Organization, Standard Specification for Ready Mixed Concrete, ISIRI6044, 2015.

Janamian Kambiz, Aguiar Jose, A Comprehensive Method for Concrete Mix Design, Materials Research Forum LLC, 2020.

Lamond F.Joseph, Pielert H.James, *Significance of Tests and Properties of Concrete and Concrete Making Materials*, ASTM International, 2006.

Mahmood Zadeh Amir, Iranpoor Jafar, Concrete Technology and Test (Farsi), Golhaye Mohammadi, 2007.

Mexca, "LECA Aggregates Granules." Retrieved from: https://commons.wikimedia.org/wiki/File:LECA_-_Aggregates_Granules.jpg.

Mostofinejad Davood, Concrete Technology and Mix Design (Farsi), Arkane Danesh, 2011.

Nawy G.Edward, *Concrete Construction Engineering Handbook*, CRC Press, 2008.

Newman John, Choo Ban Seng, *Advanced Concrete Technology, Concrete Properties*, Elsevier, 2003.

Nuclear electricity power plant photo. Retrieved from: https://pxhere.com/en/photo/540951.

Ramachandran V.S, Beaudion James, *Handbook of Analytical Techniques in Concrete Science and Technology, Principles, Techniques and Applications*, William Andrew Publishing, 2001.

Ramezanianpoor Aliakbar, Arabi Negin, Cement and Concrete Test Methods (Farsi), Negarande Danesh, 2011.

Richardson M, *Fundamentals of Durable Reinforced Concrete*, Spon Press, 2002.

Safaye Nikoo Hamed, *Introduction to Concrete Technology (Farsi)*, Heram Pub, 2008.

Satoh, Junpei, "Dolomite rock and calcite." Retrieved from: https://commons.wikimedia.org/wiki/File:Dolomite_rock_and_calsite.jpg.

Shekarchizade Mohammad, Liber Nicolas Ali, Dehghan Solmaz, Poorzarrabi Ali, Concrete Admixtures Technology and Usages (Farsi), Elm & Adab, 2012.

St. John, James, "Andesite." Retrieved from: https://www.flickr.com/photos/jsjgeology/8455600595.

St. John, James, "Basalt." Retrieved from: https://commons.wikimedia.org/wiki/File:Basalt_5 (48674275908).jpg.

St. John, James, "Biotite Schist." Retrieved from: https://www.flickr.com/photos/jsjgeology/16896874736.

St. John, James, "Chalk from the Cretaceous of Britain." Retrieved from: https://commons.wikimedia.org/wiki/File:Chalk_(%22Upper_Chalk%22_Formation,_Upper_Cretaceous;_White_Cliffs_of_Dover,_England,_southern_Britain).jpg.

St. John, James, "Diatomite." Retrieved from: https://www.flickr.com/photos/jsjgeology/16656348400.

St. John, James, "Gabbro." Retrieved from: https://www.flickr.com/photos/jsjgeology/16541055767.

St. John, James, "Gneiss." Retrieved from: https://www.flickr.com/photos/jsjgeology/16310656713.

St. John, James, "Granite." Retrieved from: https://commons.wikimedia.org/wiki/File:Granite_47_(49201189712).jpg.

St. John, James, "Hematite rock (Biwabik Iron-Formation, Paleoproterozoic, ~1.878 Ga; Thunderbird Mine, Mesabi Iron Range, Minnesota, USA) 1." Retrieved from: https://www.flickr.com/photos/jsjgeology/34666757885.

St. John, James, "Magnetite rock (Biwabik Iron-Formation, Paleoproterozoic, ~1.878 Ga; Thunderbird Mine, Mesabi Iron Range, Minnesota, USA) 1." Retrieved from: https://www.flickr.com/photos/jsjgeology/34505350602.

St. John, James, "Marble (Yule Marble, Middle Miocene, 12 Ma; Marble, northern Gunnison County, western Colorado, USA) 2." Retrieved from: https://www.flickr.com/photos/jsjgeology/16887651452.

St. John, James, "Oolitic limestone from the Mississippian of Indiana, USA." Retrieved from: https://commons.wikimedia.org/wiki/File:Oolitic_limestone_(Salem_Limestone,_Middle_Mississippian;_Bedford,_Indiana,_USA)_2.jpg.

St. John, James, "Rhyodacite pumice & volcanic ash in the Holocene of Orgeon, USA." Retrieved from: https://commons.wikimedia.org/wiki/File:Rhyodacite_pumice_fall_deposit_(Holocene;_south_of_Grouse_Hill,_Crater_Lake_Caldera,_Oregon,_USA)_3.jpg.

St. John, James, "Sandstone" Retrieved from: https://www.flickr.com/photos/jsjgeology/16170306843.

St. John, James, "Sioux Quartzite (Paleoproterozoic, 1.65 to 1.70 Ga; Transcontinental Arch, USA)" Retrieved from: https://commons.wikimedia.org/wiki/File:Sioux_Quartzite_(Paleoproterozoic,_1.65_to_1.70_Ga;_Transcontinental_Arch,_USA)_17.jpg.

St. John, James, "Specularite (high-grade iron ore) (Soudan Iron-Formation, Neoarchean, ~2.722 Ga; Soudan Mine, Soudan, Minnesota, USA) 9." Retrieved from: https://www.flickr.com/photos/jsjgeology/19026398982.

St. John, James, "Travertine-cemented conglomerate." Retrieved from: https://commons.wikimedia.org/wiki/File:Travertine-cemented_conglomerate_10.jpg.

Webb, Richard, "Stone crusher." Retrieved from: https://commons.wikimedia.org/wiki/File:Stone_crusher_-_geograph.org.uk_-_436184.jpg.

Zandi Yousof, Advanced Concrete Technology (Farsi), Forouzesh Pub, 2009.

Zandi Yousof, Concrete Tests and Mix Design (Farsi), Forouzesh Pub, 2007.

5 Chemical Admixtures

Using of concrete admixtures in concrete production is inevitable nowadays. We can say that all types of concrete need the chemical admixtures for the improvement of quality.

Although the amount of use for chemical admixtures compared with the other constituent materials is neglectable (most of the times less than 10 kg in 1 m^3 of concrete) but their effect on concrete quality and performance is huge. So, the selection of chemical admixtures and the manufacturer is very important for many projects in the world.

According to the above mentioned, it is very important for engineers to give enough information about the chemical admixtures for concrete. In many texts on concrete technology, the chapter on chemical admixtures is very brief. But here we discuss these admixtures more, because in the 21st century usage of these admixtures, especially plasticizers and super-plasticizers, is more important than the other concrete constituent materials.

5.1 THE REASON FOR USING CONCRETE ADMIXTURES

Using of concrete admixtures goes to many years ago. When the engineers understood that they cannot improve the properties of concrete only with cement, aggregates, and water, they tried to find other materials to use in concrete to achieve better performance.

The need to change some of the specifications of concrete is very important in many projects. So, some of the projects in the world could be built only with special types of concrete. Some of the most important properties of concrete which are not possible to change without chemical admixtures are as below:

- Change the setting time of concrete: For example, in hot weather conditions, we need to retard the setting time and in cold weather conditions, we need to accelerate the setting time of concrete. So, we should use retarder and accelerator admixtures.
- Accelerate the Hardening process of concrete: For example, in precast concrete plants, we need to accelerate the hardening process of concrete to release the molds and accelerate the production of elements. So, we should use hardening accelerator admixtures.
- Increase the compressive strength of concrete: To increase the compressive strength, we need to decrease the water-to-cement ratio. Increasing the cement content is limited. So, we should decrease the amount of water. It can cause drastically decreasing of the workability. So, we should use plasticizers and super-plasticizers to decrease the water in concrete without any change in workability.

DOI: 10.1201/9781003384243-5

- Improve the workability of concrete: For example, in congested concrete elements, we need to use a high slump concrete to prevent any defect in the structure. So, we should use plasticizers and super-plasticizers to produce a workable high slump concrete without using more water.
- Improve the resistance of concrete elements against the freeze-thaw cycle: If we can entrain special air bubbles with the same size and shape and with uniform distribution inside the concrete element, the resistance of this element will increase against the freeze thaw. So, we should use air-entraining admixtures.
- Waterproof concrete: In many cases, such as swimming pools, we need a waterproof concrete to avoid water penetration. For this concrete, we need to make a dense structure concrete with low water-to-cement ratio. On the other hand, we need to block the pores of the concrete to control water penetration. So, we should use super-plasticizers and waterproofing admixtures to make a waterproof concrete structure.

In the next part of this chapter, we discuss the most important chemical admixtures.

5.2 ACCELERATOR ADMIXTURES

One of the most important factors that is very important in the projects is the control of setting and hardening time of concrete. Cement manufacturers announce an initial and final setting time for their cement. As mentioned before, this is the time that we can achieve that by Vicat needle test according to ASTM C191. We should do this test in the laboratory standard temperature which is about 20°C. But in real project and different ambient conditions, the situation could be much different. In winter and in cold weather conditions the real ambient temperature could be less than 0°C and in summer hot weather conditions the real ambient temperature could be more than 30°C. So, the setting time of concrete will be much different by the time announced by the cement factory.

According to the abovementioned, we should make changes to the setting time of cement in different ambient conditions to achieve the targeted results from concrete. We can do it by using accelerators and retarders. In this part, we discuss the accelerators.

We have two types of accelerators in the market:

- Set accelerators: They can accelerate the seeing time of concrete. So, they can prevent the concrete from freezing
- Hardening accelerators: They can accelerate the hardening process of concrete. But they don't have any effect on the setting time of cement.

We have some accelerator admixtures which are set and hardened with accelerators together. For better information about the type of accelerators, you should contact the manufacturers and see the material technical data sheet[1].

5.2.1 SET ACCELERATORS

As mentioned above, set accelerators can accelerate the setting time of cement (Figure 5.1). They can cause slump loose in concrete after production. The most important uses of these admixtures are:

- Anti-freezing admixtures: The anti-freezing admixtures are the set accelerators that can protect concrete against freezing in cold weather conditions (Figure 5.2). In some countries with very cold weather, the anti-freezing admixtures contain set accelerators and anti-freezing materials which can

FIGURE 5.1 Production of concrete in cold weather. ("Volvo cement mixer truck in Jyvaskyla, Finland" by Antti Leppanen.)

FIGURE 5.2 Tunnel shotcrete device. ("Putzmeister Wetkret 5.")

decrease the melting point of water, because in these harsh conditions, there is a need to control the freezing of water inside concrete and also the freezing of water inside the anti-freezing admixture. But in most parts of the world, anti-freezing admixtures only contain set accelerators.

As these types of admixtures can cause slump loose in concrete, you should use them in the project at the time of final pouring of concrete. If you use them in the batching of the ready mixed plants, the slump of concrete in the project and at the time of pouring will be very low and it can cause many problems for you.

- Shotcrete admixtures: Sometimes we need to perform the concrete as shotcrete (Figure 5.3). For example, for the stabilization of excavated walls or some tunnel walls, we need to use concrete as the shotcrete. When we shot the concrete into the walls. It should remain and get enough strength after a short period of time to stabilize the soil of the wall. To do that, we need to use strong set accelerators. We should add the admixture at the start of the nozzle because shotcrete admixtures can cause the setting of concrete within about 1 minute. You can find shotcrete admixtures in liquid and powder form on the market.

In Figure 5.4 you can see the shotcrete application. For overhead shotcrete, using a high-quality accelerator is very important because it can reduce the amount of concrete rebound.

The effect of these admixtures is on the C_3A of cement. They will accelerate the reaction of C_3A and water. So, it will cause rapid setting. The most important chemical bases of set accelerator admixtures are calcium chloride, calcium nitrate, sodium nitrate, calcium nitrite, sodium nitrite, and many other mineral salts.

Although the calcium chloride is a very good accelerator, using it is forbidden in the reinforced concrete because it contains chloride ions that can cause corrosion for the steel rebars inside the concrete.

FIGURE 5.3   Applying shotcrete. ("Koln shotcrete application" by Raimond Spekking.)

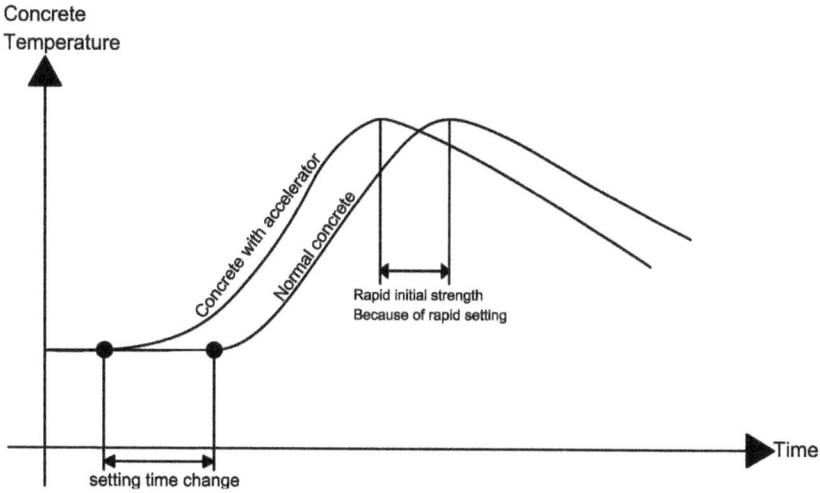

FIGURE 5.4   Performance of a set accelerator in concrete. (Photograph created by the author.)

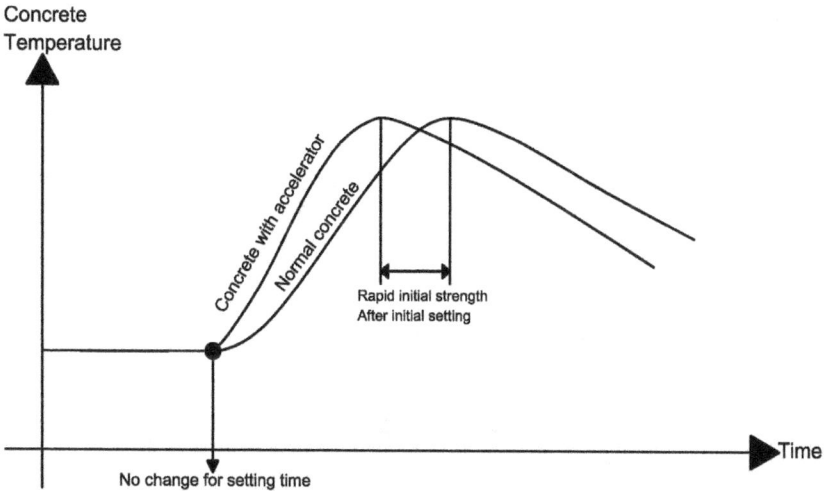

FIGURE 5.5   Performance of a hardening accelerator in concrete. (Photograph created by the author.)

5.2.2   HARDENING ACCELERATORS

Hardening accelerators are the second type of the accelerator admixtures. These types of admixtures will improve the compressive strength of concrete at early ages without any difference in the setting time (Figure 5.5). We can use hardening accelerators for precast concrete plants that we need to release the molds as soon as it is possible. So, using the hardening accelerators is helpful for the rapid hardening of concrete. For some other projects, we may need high initial strength for some special loading purposes. For these cases also, it is recommended to use hardening accelerators. There is no difference for the final compressive strength of concrete if we use a good formulated hardening admixture.

Most important chemical bases of these admixtures are some of the soluble inorganic salts like bromides, fluorides, silicates, and some soluble organic compounds like triethanolamine. Sometimes for anti-freezing admixtures, we can mix set and hardening accelerators to get the best result.

5.3 RETARDER ADMIXTURES

We should use retarders to retard the setting time of concrete. When the setting time of the concrete will be longer, we can work with the concrete for a longer period of time. This property is very good for hot weather conditions, because at these weather conditions, the high temperature will accelerate the hydration reaction. So, we need to retard and control it to work with concrete. We can name the most important usages of retarder admixtures as below:

- Control the slump loose of concrete in hot weather conditions: As mentioned above, you can use these admixtures to control the rapid hydration reaction in summer. This rapid reaction can cause a drastic slump loss of concrete. Most of the time, we should use these retarders mixed with the good slump retention super-plasticizers to get a good result.
- Ready mixed plants (Figure 5.6): When the transportation time of concrete will increase, the need for slump retention will increase much. So, in ready mixed concrete industry, we can use retarder admixtures with a good super-plasticizer to get the defined slump and transportation time.
- Control the heat of hydration in mass structures (Figure 5.7): In a huge structure with mass elements, the temperature of concrete will raise too much because of the hydration heat. This high temperature can cause decrease in the mechanical properties and also it can cause cracking inside concrete elements. So, it is very important to control the temperature inside these huge

FIGURE 5.6   Ready mixed plant. (Photograph by the author.)

FIGURE 5.7   A mass structure. (Photograph by the author.)

elements. When we use retarders, they can reduce the speed of the hydration reaction. So, we can control the temperature inside the concrete elements.

- Pouring concrete in different layers: When we would like to pour a concrete element with different layers (e.g., a huge foundation that we should pour in two or more days). We should protect the elements from the cold joints. When we pour the upper layer of concrete, if the lower layer had set, then we will have a cold joint in the border between these two layers. To avoid this, we should use retarders with different dosages in the lower layer. When we are going to pour the upper layer, the concrete of the lower layer did-not set. So, by using a good vibrator we can mix the upper and lower layer and prevent the cold joint.

This kind of work is a very hard and complicated to perform. So, you should make trials before doing this work to check the effect of different dosages of admixture with different ambient temperatures on the retarding behavior of concrete.

We can use many different chemicals as the retarders like: sodium gluconate, sugar, dextrin, and tartaric acid. The admixture manufacturers can formulate retarders with different amounts of the above chemicals mixed with some other materials.

5.3.1   DOSAGE OF RETARDER ADMIXTURES

One of the most important facts about the retarder admixtures is the optimum dosage for each purpose (Figure 5.8). Choosing the best dosage is more complicated than the other types of admixtures because the retarding effect on concrete depends on below factors:

- Type of cement: As you know, cements like ASTM type I or III are more accelerated than the cements such as type II, IV and V. So, if you are using for example, cement type V you may need less retarder admixture than cement type I.
- Ambient temperature: As the activity of cement in hot weather will accelerate, you may need more retarder admixture in hot weather conditions. When

FIGURE 5.8 Making trials in a laboratory is very important to estimate the optimum dosage of retarders. (Photograph by the author.)

the temperature will raise, you will need more retarders. So, it is possible for you to need different dosages of a retarder admixture during 1 day in summer because of different temperatures during the day.

- Type of retarder: As mentioned before, there are too many chemicals with the retarding effect. So, admixture manufacturers can produce many different types of retarders with different retarding power, and most of the time, they don't let you know the exact chemical composition of their admixture. So, you should see their technical data sheet and decide about the starting dosage for trials.
- Period of time for retarding effect: The purpose of using retarder admixture is very important for making a good decision about the dosage of retarder. Sometimes, you may use retarders to improve the slump retention effect of a super-plasticizer. In this case, you may need less dosage. Sometimes, you may need a retarder to pour a foundation with different layers without any cold joints. In this case, you may need more retarder.
- Types of other admixtures: When you would like to use two or more types of admixtures, it is very important to contact the admixture manufacturer and make trials before using them in the project. For retarder admixtures

also like any other types of admixtures, you should do the above. For example, if you use a type of plasticizer with the chemical base of lignosulfonate that we will talk about it in the future, as this type of plasticizer has the retarding effect itself, you may need less retarder.

- Use of supplementary cementitious materials: When you use these materials besides the Portland cement in concrete, you may need less dosage of retarder than the time you use only Portland cement because the activity of supplementary cementitious materials is less than the pure Portland cement. They will start their reaction after the production of enough $Ca(OH)_2$.

As you can see from the abovementioned, decision about the best dosage of retarder admixture for each purpose is more difficult than the other types of admixtures. So, the most important suggestion is to make trials in the condition as same as the real project and then choose the dosage.

If you used a retarder with higher dosage than the need for your purpose, it is possible that your concrete will not be set at the time you defined. In this case, you should cure the concrete with the best techniques according to the weather conditions and continue the curing till the concrete is set. For example, if you expect that your concrete will set after 24 hours and it didn't happen. You should cure the concrete for another 24 hours and check it again. Most of the time, the setting will happen at the time two or three times of expected setting time. In this case, you should check the mechanical properties of the concrete at the age of 28 days. It is possible that you see lower mechanical properties at the lower ages. But most of the time, the 28 days mechanical properties are the same as the testimonial concrete.

5.4 PLASTICIZERS AND SUPER-PLASTICIZERS

The most important and useful chemical admixtures are plasticizers and super-plasticizers. They are the inseparable constituent materials of modern concrete. We can use plasticizers and super-plasticizers for below reasons:

- Decrease the amount of water in concrete production
- Increase the workability of concrete without increasing the water
- Decrease the amount of water and at the same time, increase the workability of concrete

If we use super-plasticizers with the optimum dosage, we can decrease the amount of water and increase the workability of concrete. For example, we can produce a concrete with water-to-cement ratio of 0.3 and slump of more than 200 mm. So, you can see that the production of high-strength concrete (HSC) is only possible by using super-plasticizers.

When you use a plasticizer, you can reduce the amount of water up to 12% and when you use a super-plasticizer you can reduce the amount of water more than 12%. We have very high-quality super-plasticizers with the ability to reduce the amount of water more than 40%. So, it depends on the type of concrete and defined properties to choose a plasticizer or super-plasticizer. But it is better to use a high-quality super-plasticizer to produce a high-quality concrete.

One of the most important properties of plasticizers and super-plasticizers is the slump retention ability. Especially, when you would like to transport the concrete from a ready mixed plant (Figure 5.9). This property depends on the chemical base and polymer type of the super-plasticizers. But you can improve this property by using retarders across the super-plasticizers. Most of the time, the manufacturers give you a super-plasticizer with some amount of retarder inside it. You should check the technical data sheet for the exact properties of all types of admixtures.

As mentioned above, manufacturers dilute some retarders inside the super-plasticizers to get better slump retention. But sometimes for some of the admixture, they may use accelerators inside the super-plasticizers. So, if you would like to choose a suitable super-plasticizer for your concrete, first you should contact the technician of the manufacturer and second, you should make trials before using the admixture in the real project.

FIGURE 5.9 You may transport concrete for a long distance with truck mixers. (Photograph by the author.)

5.4.1 Admixtures in ASTM Standard

You can check the specifications of concrete admixtures according to ASTM C494. In this standard, you can see seven types of chemical admixtures:

- Type A: Water reducer admixtures
- Type B: Retarder admixtures
- Type C: Accelerator admixtures
- Type D: Water reducer and retarder admixtures
- Type E: Water reducer and accelerator admixtures
- Type F: Water reducer, high-range admixtures
- Type G: Water reducer, high range and retarder admixtures

You can see defined specifications for admixtures according to ASTM C494 in Table 5.1.

The specifications mentioned in Table 5.1 are the minimum rates for each type. For example, for high-range water reducers the minimum water reduction rate should be 12%. You can find high-range water reducers with the water reduction rate of more than 40%.

5.4.2 Chemical Bases of Plasticizers and Super-Plasticizers

As the using of plasticizers and super-plasticizers is very important in concrete, you should know the chemical bases of these admixtures. In the 21st century, we have

TABLE 5.1
Standard Specifications for Chemical Admixtures According to ASTM C494

	Type A	Type B	Type C	Type D	Type E	Type F	Type G
Max amount of water (%)	95	-	-	95	95	88	88
Min initial setting time (h:min)	-	After 1:00	Before 1:00	After 1:00	Before 1:00	-	After 1:00
Max initial setting time (h:min)	Before 1:00	After 3:30	Before 3:30	After 3:30	Before 3:30	Before 1:00	After 3:30
Min final setting time (h:min)	-	-	Before 1:00	-	Before 1:00	-	-
Max final setting time (h:min)	Before 1:00	After 3:30	-	After 3:30	-	Before 1:00	After 3:30
1 day comp strength (%)	-	-	-	-	-	140	125
3 days comp strength (%)	110	90	125	110	125	125	125
7 days comp strength (%)	110	90	100	110	110	115	115
28 days comp strength (%)	110	90	100	110	110	110	110
3 days flextural strength (%)	100	90	110	100	110	110	110
7 days flextural strength (%)	100	90	100	100	100	100	100
28 das flextural strength (%)	100	90	90	100	100	100	100

some new chemical bases of super-plasticizers. But the usage of them is not wide in the world. Now, we discuss the four most important types:

- Plasticizers based on lignosulfonate: This type of admixture is the first generation developed around the year 1930. The technology of these admixtures is based on the sulfonation of lignin derived from the paper and wood industry. They are dark brown liquid admixtures with a special odor (Figure 5.10). You can also find the powder of lignosulfonate on the market. Powdered lignosulfonate can solve in water by about 40%. These admixtures can formulate as the sodium salt or calcium salt. You can see different specifications for the sodium and calcium salt, but it refers to the concrete admixture formulators. There is no need for a concrete technologist to know the differences.

  The molecules of lignosulfonate will adsorb by the cement particles and give the cement particle negative charge. So, the electrostatic repulsion between the cement particles causes better moving of them during concrete and it will cause better workability.

FIGURE 5.10 Lignosulfonate plasticizer. (Photograph by the author.)

One of the most important properties of lignosulfonates is the retarding effect. It is because of the sugar that will remain inside their molecules and we cannot refine it. The higher quality of lignosulfonate means the higher refining degree of sugar. So, you should check the retarding effect. Especially, when you would like to use this type of admixture in winter. Overdosing of this admixture can cause high retardation effect in cold weather conditions.

You can find lignosulfonate admixture with different solid content from 20% to 40% in the market. It means that the active matter inside the liquid admixture could be 20% to 40%. It is possible to find a mix of this admixture with other types of chemical admixtures like accelerators, retarders, and other types of super-plasticizers.

The maximum water reduction rate of a pure lignosulfonate admixture is 12%. For higher water reduction rate, admixture manufacturers add some other types of super-plasticizers to it and sell it with the other brand name.

Although the lignosulfonate is a weak plasticizer for concrete, as we are producing lignin from wood and paper industry, we can produce lignosulfonate continuously and we can use it in the concrete industry for several purposes like the plasticizing and retarding effect.

- Super-plasticizers based on poly naphthalene sulfonate (PNS): Using naphthalene as a super-plasticizer for concrete goes about the year of 1930. But the polynaphthalene sulfonate super-plasticizers we are using nowadays were used in about the year 1960. The technology of this admixture is based on the sulfonation of naphthalene and then condensation reaction with the formaldehyde. These admixtures can formulate as the sodium salt or calcium salt. The quality is more or less the same but the most often used one is the sodium salt. The use of formaldehyde as a harmful material in the production process of PNS super-plasticizers is one of the reasons that the usage of them is going to be limited every day.

Pure PNS super-plasticizer is a dark brown liquid with a special odor (Figure 5.11). An experienced person can differentiate between a lignosulfonate and PNS liquid from the odor.

The mechanism of action for PNS super-plasticizers is the electrostatic repulsion between the cement particles. Molecules of PNS polymers adsorb by the cement particles and give them a negative charge. So, the electrostatic repulsion between the cement particles causes better movement inside the concrete and it will increase the workability.

Pure PNS super-plasticizers don't have any retarding or accelerating effect in concrete. On the other hand, the slump retention effect of them is very low. So, most of the time, admixture manufacturers add some retarder to improve the slump retention of this admixture. Sometimes, they mixed lignosulfonate and PNS to formulate a special product with a higher water reduction rate than the lignosulfonate and better slump retention than the naphthalene sulfonate.

FIGURE 5.11   Poly naphthalene sulfonate super-plasticizer. (Photograph by the author.)

You can find PNS super-plasticizers with a solid content of 20% to 40% in the market. But as mentioned before, admixtures based on PNS contain some other additives like retarders and lignosulfonates for better properties.

The maximum water reduction rate of a pure PNS super-plasticizer with 40% solid content is about 25%. Although the water reduction rate is very good for concrete production, the using of formaldehyde, high energy consumption in the production process, and low slump retention of this admixture reduced the use of them in the last decades.

- Super-plasticizers based on polymelamine sulfonate (PMS): This type of admixture came to the market around the year 1990. They are the same products as the PNS super-plasticizers with using of melamine instead of naphthalene in the production process. The plasticizing power of this type is also the same as PNS. In the chemical admixtures market, we can call PNS and PMS polycondensates.

Pure PMS liquid is a light yellow to colorless liquid with a special odor (Figure 5.12). Mechanism of action for this type of super-plasticizer is exactly the same as the PNS.

FIGURE 5.12   Poly melamine sulfonate super-plasticizer. (Photograph by the author.)

These admixtures have a little accelerating effect on concrete. So, the slump retention is weaker than the poly naphthalene sulfonates. On the other hand, the price of PMS is higher than the PNS super-plasticizers in most parts of the world. So, using of these admixtures in concrete is less than the lignosulfonate and poly naphthalene sulfonates in the market.

You can find PMS super-plasticizers with 20% to 35% solid content in the market because you cannot solve more than 35% of them in the water, it will cause weaker formulated admixtures than the PNS ones.

The maximum water reduction rate for a PMS super-plasticizer is 22% with a very low slump retention effect. So, it can be a good choice for some projects without transportation time.

- Super-plasticizers based on polycarboxylate ether (PCE): They are the newest type of super-plasticizers with the best quality and performance. They came to the market around the year 1995 with the first generation. But from that date till now, they improved day by day to achieve the best performance according to the needs of any type of project.

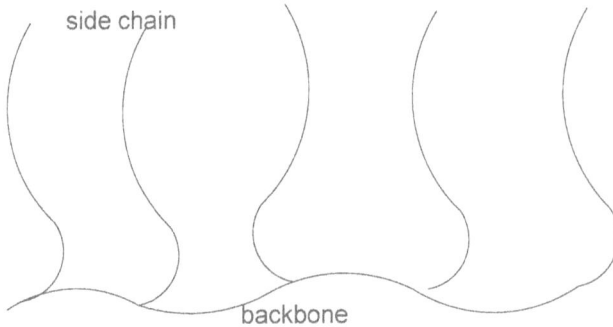

FIGURE 5.13 The shape of polycarboxylate ether polymer. (Photograph created by the author.)

They are made by different procedures. The most common is the polymerization of alcohol ethoxylates with acrylic or methacrylic acid in a special condition. They are comb-shaped polymers with a backbone and side chain (Figure 5.13).

The mechanism of action for this type of super-plasticizers is the electrostatic repulsion and steric hindrance of polymers. The polymers remain between the cement particles and it causes a very good water reduction rate for this admixture.

Pure polycarboxylate admixture is a colorless to dark yellow liquid with a very light special odor (Figure 5.14). The viscosity of PCE admixtures is higher than the viscosity of PNS or PMS.

You can find PCE super-plasticizers with 20% to 50% solid content. But as this type of admixture is very strong and overdosing will cause segregation and bleeding of concrete (Figure 5.15), it is better to use lower solid contents. Also, you can mix PCE admixtures with sodium type of lignosulfonate admixtures and different types of retarders. But you cannot mix PCE admixtures with PNS or PMS super-plasticizers.

PCE super-plasticizers can reduce the amount of water by more than 40%. It is much more than the other types of super-plasticizers. When you use it in concrete it will be unbelievable for you.

The other important specification of PCE super-plasticizer is the good slump retention ability. You can design PCE polymers with high water reduction and moderate slump retention and you can design a polymer with moderate water reduction rate and high slump retention. A good slump retention type PCE can keep concrete slump for more than 3 hours in harsh conditions.

PCE super-plasticizers are more expensive than the other types, but according to their quality, performance, and flexibility, they are the most commonly used admixtures in the world.

5.4.3 MINI SLUMP TEST

It is very important to check the performance of super-plasticizers for choosing a suitable product to use in concrete production. Most of the time, we would like to check two specifications of the super-plasticizers:

FIGURE 5.14 Polycarboxylate ether super-plasticizer. (Photograph by the author.)

- Power of plasticizing
- Slump keeping

The best test is checking them by making trial concrete with the same cement and aggregates of the final project. Sometimes, we need some simpler tests for the control of admixtures. One of the simplest tests for the performance check of super-plasticizers is the mini slump test according to the DIN EN1015. This is a very common and simple test in many texts and lectures. We can check and compare the plasticizing and slump keeping of super-plasticizers by this test.

In fact, this is a slump test with a little amount of mortar with a special mini slump cone as you can see in Figures 5.16 and 5.17. The upper circle diameter of this cone is 19 mm, the lower circle diameter is 38 mm and its height is 57 mm.

To test a super-plasticizer with this test method, first you should make a cement paste by using cement, water, and a defined dosage of super-plasticizer. After hand mixing and mixing with the electrical mixer (you can use kitchen mixers for this purpose), you will have a soft flow paste. You should use enough amount of water and super-plasticizer to achieve a soft flow cement paste.

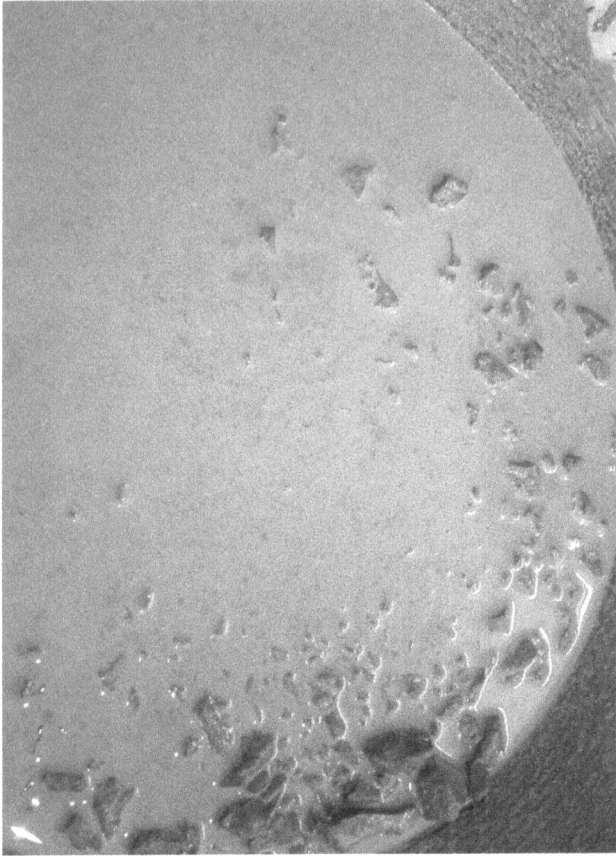

FIGURE 5.15   A segregated concrete with very low quality. (Photograph by the author.)

After that, you should fill the mini slump cone with this paste and measure the paste flow in two perpendicular lines. You can compare different super-plasticizers by using this test method. On the other hand, you can check the super-plasticizer slump keeping by repeating the test at different times. This will not be the same as the performance of the super-plasticizer in concrete. But you can give a good idea about the behavior of the super-plasticizer during time.

It is strongly recommended that you can only compare super-plasticizers with the same chemical base with each other. Nevertheless, you will not get a useful result. Especially, when you are using PCE base super-plasticizer, you should compare it with the same chemical base only.

Now you can see an example of mini slump test in Table 5.2 and Figure 5.18.

As you can see from the table and figure above, PCE-A is a super-plasticizer with a lower water reduction rate but better slump retention than the PCE-B. If you use this super-plasticizer in concrete, you will see the same results as the mini slump test.

FIGURE 5.16   Mini slump cone. (Photograph by the author.)

FIGURE 5.17   Mini slump test instruments. (Photograph by the author.)

TABLE 5.2

Comparison of Two PCE Base Super-Plasticizers

Cement paste mix design:

Cement: 200 g

Water: 60 g

Super-plasticizer: $0.7\% = 1.4$ g

Super-Plasticizer Type	Initial Flow (mm)	Flow After 30 Minutes (mm)	Flow After 60 Minutes (mm)	Flow After 90 Minutes (mm)
PCE-A	140	140	125	120
PCE-B	160	160	140	110

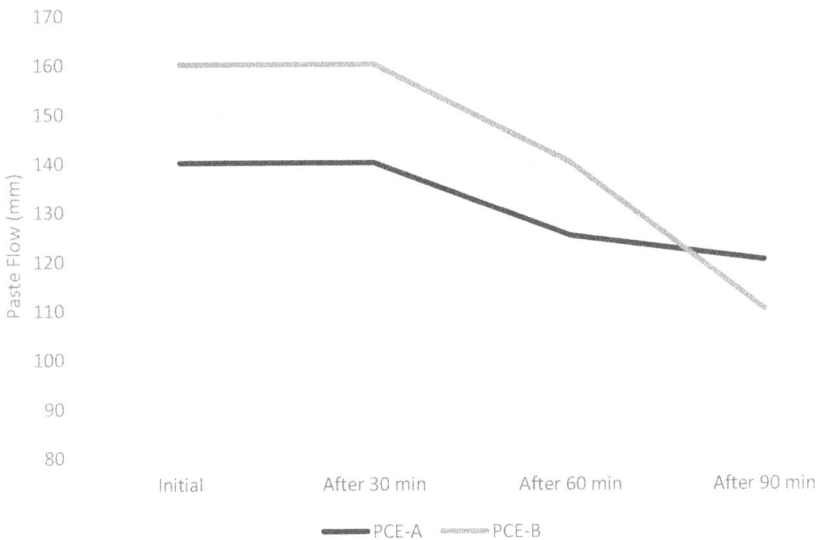

FIGURE 5.18 Mini slump test for two PCE-type super-plasticizers. (Graph created by the author.)

5.4.4 MARSH FUNNEL TEST

This test is a simplified test for the evaluation of the power and slump retention for super-plasticizers. The accuracy of this test is more than the mini slump test. But as mentioned before, the best test for the evaluation of plasticizers and super-plasticizers is making trial concrete.

This test is according to the ASTM D6910 the standard test method for marsh funnel viscosity of construction slurries. You can see the dimensions and picture of the marsh funnel that you should use for this test in Figure 5.19.

For this test, you should make a cement paste like the mini slump test. But the paste should be more flowable. So, you should adjust the amount of super-plasticizer and water. It is better to make about 1.2 L of cement paste for this test.

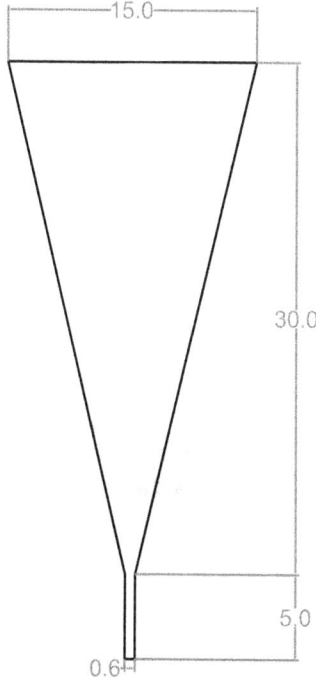

FIGURE 5.19 Dimensions in (cm) of the marsh funnel. (Photograph created by the author.)

TABLE 5.3

Comparison of Two PCE Super-Plasticizer With Marsh Funnel Test

Cement paste mix design:
Cement: 1550 g
Water: 470 g
Super-plasticizer: 0.8% = 12.5 g

Super-Plasticizer Type	Initial Flow Time (s)	Flow Time After 30 Minutes (s)	Flow Time After 60 Minutes (s)	Flow Time After 90 Minutes
PCE-A	58	69	78	90
PCE-B	50	65	83	107

After making the cement paste and mixing it enough with an electrical mixer, you should pour the marsh funnel with the slurry and let it discharge from the funnel (Figure 5.19). The time of discharging is the critical factor for the evaluation of the super-plasticizer power. On the other hand, you can compare different super-plasticizers with the same dosage and water for the water reduction rate. For the slump keeping, you can repeat the test at different times to evaluate the total behavior of the super-plasticizer. You can see an example of the marsh funnel test in Table 5.3 and Figure 5.20.

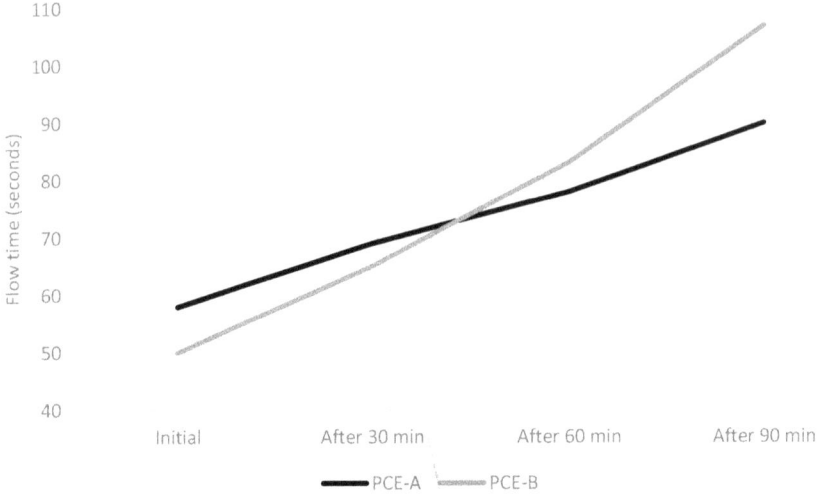

FIGURE 5.20 Marsh funnel test for two PCE-type super-plasticizers. (Graph created by the author.)

Another property of the super-plasticizers that we can evaluate by this test is the saturation point. This is the maximum amount of admixture that can increase the flowability of cement paste or concrete. In fact, more super-plasticizer than the saturation point, don't have any effect on the flowability of concrete. So, it is a very important factor for the evaluation of the optimum dosage of super-plasticizers in the concrete mix design.

To evaluate the saturation point, you should repeat the test with the same water and different dosages of super-plasticizer. When you increase the dosage, you will see that from a special dosage, increasing the amount of super-plasticizer will not cause the decrease of flow time. This is the saturation point for that super-plasticizer. You can see more or less the same behavior in concrete with the same water and increasing the super-plasticizer dosage.

5.4.5 Evaluation of Water Reduction Rate

According to the abovementioned, you can see Table 5.4 as a guide for using plasticizers and super-plasticizers with different chemical bases. This table could be useful when we don't have enough time to evaluate the water reduction rate of super-plasticizers.

It is better to evaluate the water reduction rate of a super-plasticizer before using it for concrete mix design because it is possible to have a brand name of super-plasticizer which is a mix of two chemical bases. On the other hand, choosing the optimum dosage and evaluating the water reduction rate of super-plasticizers is very important for concrete mix design.

TABLE 5.4
Simple Guide to Use for Pure Super-Plasticizers

Type of Super-Plasticizer	Min Amount of Use (%)	Min Water Reduction Rate (%)	Max Amount of Use (%)	Max Water Reduction Rate (%)
Weak lignosulfonate	0.6	4	1.1	8
Strong lignosulfonate	0.5	6	0.9	12
Weak poly naphthalene sulfonate	0.5	10	2<	15
Moderate poly naphthalene sulfonate	0.4	12	2<	18
Strong poly naphthalene sulfonate	0.4	14	2<	22
Weak poly melamine sulfonate	0.5	8	2<	12
Moderate poly melamine sulfonate	0.4	10	2<	15
Strong poly melamine sulfonate	0.4	12	2<	20
Moderate poly carboxylate ether	0.3	15	1.5<	30
Strong poly carboxylate ether	0.3	25	1.5<	40

TABLE 5.5
Mortar Test for Water Reduction Rate of PCE-A Super-Plasticizer

Silica Sand (g)	Cement (g)	Super-Plasticizer	Water (g)	Flow (mm)	Water Reduction Rate (%)
400	300	0	210	135	0
400	300	0.5% = 1.5 g	170	140	19.0
400	300	1.0% = 3.0 g	150	150	28.6
400	300	1.5% = 4.5 g	135	150	35.7

- Water reduction rate with mortar: To evaluate the water reduction rate of a super-plasticizer with mortar, you should use the mini slump test. First, you should make a special mortar with cement, silica sand, and water. After enough mixing you should test its flowability with the mini slump cone. It is better to achieve the flowability of more than 100 mm. Then you should make a mortar with different dosages of super-plasticizer with the same flowability and different amounts of water. So, you can evaluate the amount of water reduction for each dosage of the super-plasticizer. You can see an example of this test in Table 5.5 and Figure 5.21.
- Water reduction rate with concrete: As mentioned before, making trial concrete is the best test for the evaluation of a super-plasticizer in concrete. So, you can make a testimonial concrete with a special amount of water and a defined slump. Then you should use different dosages of super-plasticizer and reduce the amount of water to achieve the same slump. So, you can evaluate the amount of water reduction for the super-plasticizer. You can see an example of this test in Table 5.6 and Figure 5.22.

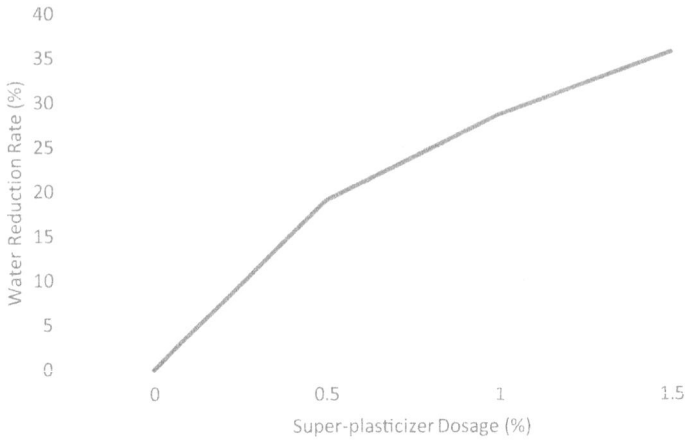

FIGURE 5.21 Water reduction rate for PCE-A super-plasticizer with mortar test. (Graph created by the author.)

TABLE 5.6
Concrete Test for Water Reduction Rate of PCE-A Super-Plasticizer

Cement (kg)	Gravel (kg)	SSD Sand (kg)	Super-Plasticizer	Water (kg)	Slump (mm)	Water Reduction Rate (%)
400	650	1150	0	225	150	0
400	650	1150	0.5% = 2.0 kg	185	160	17.8
400	650	1150	1.0% = 4.0 kg	165	160	26.7
400	650	1150	1.5% = 6.0 kg	150	150	33.3

5.5   AIR-ENTRAINING ADMIXTURES

Freeze thaw is a destructive process for concrete. If a concrete element will be in this condition, it will be destroyed after a little time. So, we should protect the structures in the cold climate conditions against the freeze-thaw cycles, especially when the amount of moisture inside the element will be high. The most important example is the precast tables or new-jerseys in the places with a temperature of less than zero at night time and more than zero at day time because these elements are exposed to the water of rain in high ways and also the water of gardens.

It was confirmed about the year of 1930 that concrete elements with little size air bubbles can resist against the freeze-thaw cycles. After that, chemical admixture manufacturers started to produce the air-entraining agents to produce uniform small-size bubbles inside concrete to improve the resistance of concrete elements against freeze-thaw cycles (Figure 5.23).

These air bubbles reduce the compressive strength of concrete. But they are necessary for the resistance against freeze-thaw cycles. We will have between 1% and 2% of the air inside any kind of concrete without using any air-entraining agents. We

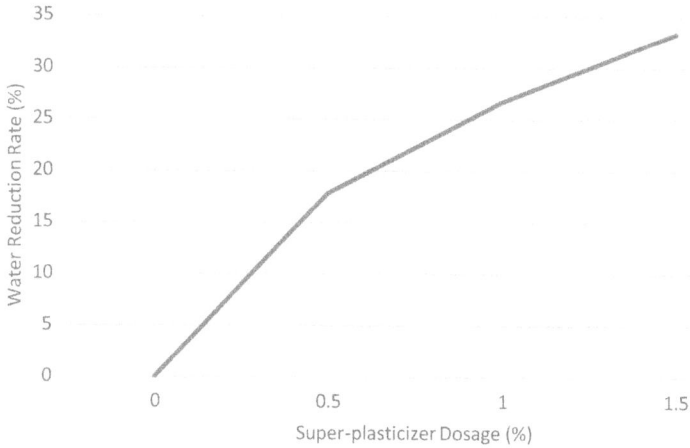

FIGURE 5.22 Water reduction rate for PCE-A super-plasticizer with the concrete test. (Graph created by the author.)

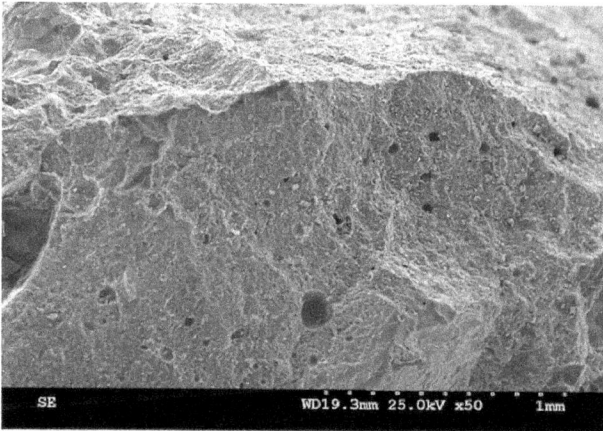

FIGURE 5.23 A section of an entrained air concrete under a microscope. ("Pore structures in fresh concrete and air entrained concrete" by Fangzhi Zhu, Zhiming Ma, Tiejun Zhao.)

call that the entrapped air. Technically, we cannot eliminate this little amount of air from concrete. On the other hand, it is very good for the workability and pumpability of concrete. But each one more percent of air will cause about 5% to 7% of strength reduction in concrete. So, if we are producing a concrete without any worry about the freeze-thaw cycles, we should control the entrapped air less than 2% (20 L in 1 m³ of concrete). We should check the cement, aggregates, sand and super-plasticizers to control the amount of entrapped air because all of the above materials have effect on the entrap air in concrete. But if we are going to produce a concrete with considerations about the freeze-thaw cycles, we should use air-entraining admixtures to entrain more air into the concrete. We can produce a concrete with more than 6% of air to improve the resistance of concrete against freeze-thaw cycles. On the other hand, you should take care that 4% of more air can reduce the compressive strength about 20% to 28%.

The moisture can remain inside the capillary pores of concrete. When the temperature falls below zero°C the water inside the pores started to freeze and it will cause the increase of volume of water. This volume increasing will apply internal stress in concrete elements. When the temperature goes above the zero°C, the water started to melt and at night the process will repeat again. This internal stress can drastically decrease the element mechanical properties and also its durability. So, we can see some of the elements in this condition with life cycle of a few years.

When we use air-entraining admixtures in concrete production, the air bubbles acted as the safety valves to control the volume increase of water during the process of freezing. So, there is no added pressure and we can control the internal stress.

You can use different standards and codes for the evaluation of air needed to increase the freeze-thaw resistance of concrete. For example, you can see Table 5.7 from the ACI 202.2R. It is better to use local standards in each country of the world for this purpose.

The most common chemical base of air-entraining admixtures is sodium En-lauryl ether sulfate. But you can find other different chemical bases in the market. They are in the liquid form. The amount of use is about 0.5% by weight of cement or less.

As the amount of air in concrete depends on many different factors, you should make trials before using the air-entraining admixtures in the production of industrial concrete. Some of the most important factors are as below:

- The quality of air-entraining admixture: Higher quality of admixtures will cause more air bubbles in concrete. So, before using any admixture you should test it by making trials.
- Cement: Using a very fine cement (higher blain) can cause less air bubbles in concrete. On the other hand, more alkalis in the cement will cause more air bubbles production in concrete element. Using more cement in concrete mix design will cause less air production in concrete.

TABLE 5.7

ACI Recommendations for the Amount of Air in Concrete for the Resistance against Freeze-Thaw Cycles

Max Size of Coarse Aggregates (mm)	Percent of Entrained Air for Intense Conditions (%)	Percent of Entrained Air for Moderate Conditions (%)
9.5	7	6
12.5	7	5
19	6	5
25	6	4

Intense conditions: When the concrete element is in the free air and with high moisture in the environment. The deicing salts also is available.

Moderate conditions: When the concrete element is in free air but with a low amount of moisture and there are no deicing salts in the environment.

- Fine aggregates: Using a sand with high amount of passing by sieve No. 100 will cause less air production. On the other hand, increasing the amount of passing by sieve No. 6 and remaining on sieve No. 30 will cause more air production. Clay in sand will cause less air production and also it can decrease the consolidation of air bubbles inside the concrete element.
- Coarse aggregates: If you have stone powder in the crushed aggregates, it can decrease the amount of air bubbles. Using crushed aggregates will cause more air production compared with the natural circular aggregates.
- Water: Using hard water can decrease the air production. On the other hand, if you have any type of washing liquid or powders in the water, it can increase the amount of air bubbles inside the concrete.
- GGBS and other Pozzolans: All types of Pozzolans can impress on the amount of air bubbles inside concrete. But for each case, you should make trials to check their positive or negative effect on the production of air bubbles.
- Other types of chemical admixtures: Using any type of chemical admixture other than the air entraining has an effect on the amount of air in the concrete. So, you should call the technicians of the manufacturers and make trials before mixing two types of chemical admixtures for any purpose.
- Concrete slump: For concrete with the slump less than 100mm, you need more admixtures for the same amount of air because forming of air in stiff concrete is more difficult. On the other hand, for concrete with the slump more than 180mm, it is possible to decrease the amount of air during time because of the unstability of the air bubbles in very soft concrete. For concrete with the slump between 100 and 180mm the dosage of air-entraining admixture will be optimum.
- Concrete temperature: Higher concrete temperature will cause less air forming in concrete and lower concrete temperature will cause more air forming in concrete. So, you should check the dosage of air-entraining admixture and the amount of air in the same temperature situation because the ambient temperature has an effect on the concrete temperature.
- Type of concrete mixer: The type of concrete mixer is very important for the formation of air in concrete. So, you should make trials with the same mixer type before using concrete in the real project. On the other hand, if you would like to transport the concrete, the spirals of the truck mixer are very important for the stability of the air bubbles in the concrete.

5.6 WATER PROOFING ADMIXTURES

Making water tight and water proof concrete is one of the most important needs of many projects. As the other methods of making waterproof structures are more difficult and expensive and sometimes it is very hard to repair a concrete structure with a permeable concrete.

Making a watertight structure (Figure 5.24) is not only dependent on the concrete quality and water tightness. The good accomplishment and performance of

FIGURE 5.24   A swimming pool as a watertight structure. ("Tuen Mun Swimming Pool in Tuen Mun, Hong Kong.")

the project are very important for this purpose. The below considerations are very important for making a waterproof concrete structure:

- Using a waterproof concrete.
- Good implementation of concrete to avoid any type of voids and defects on the surface and inside the concrete element.
- Control of cracks with high-quality curing.
- Using PVC water stops in the joints or any type of discontinuous implementation.

In this part, we discuss the production of a waterproof concrete by using different chemical admixtures. The important role is decreasing the capillary pores inside concrete. So, water cannot transit from the concrete element.

To make a water proof concrete you should apply below considerations:

- Decrease the water-to-cement ratio: By decreasing the water-to-cement ratio, you can decrease the capillary pores and make a dense structure concrete. It will help you to make a watertight concrete. As mentioned before, the best way to reduce the water-to-cement ratio is by using high-quality super-plasticizers.
- Using supplementary cementitious materials: By using these materials, you can increase the C-H-S products which can reduce the capillary pores inside the concrete. On the other hand, these materials can act as the fillers and pore blockers in concrete.

- Using micronized hydrophobia fillers: This type of powder materials can block the capillary pores inside the concrete structure. So, they help you to produce a waterproof concrete. Some of the chemical admixture manufacturers sell these powders as the waterproofing admixtures. So, this is the first type of waterproofing admixture in the market. Using these powders is not enough for the production of a waterproof concrete.
- Adjustment of the aggregates sieve analysis: To make a waterproof concrete, it is very important to adjust the total grading of aggregates in concrete. Lack of any size of aggregates can increase the capillary pores in concrete. It is very important for the passing by sieve No. 16 and less. So, using high-quality sand is very important for the production of a waterproof concrete.
- Using enough amount of cement: the amount of cement in concrete depends on the targeted compressive strength and the usage of supplementary cementitious materials. But for the production of a waterproof concrete, it is very important to use more cement than normal, because it can increase the amount of C-H-S especially in the earlier age and on the other hand, it can act as a good filler.
- Adjust the flowability of concrete: if you use a free-flow concrete, it will pour the total structure better than a stiff concrete and it can reduce the risk of the formation of any big void and pore inside concrete. So, it is recommended to use a high-quality super-plasticizer for the production of concrete to make a concrete with high flowability.
- Perfect curing of concrete: It is very important to cure the concrete to prevent any crack formation for waterproofing. So, you can use curing compound admixtures or you can cure the concrete with water. Also, you should control the concrete temperature for crack control and curing should continue for at least 48 hours and for the best result 1 week.
- Using high-technology pore blocker admixtures: There are some water proof admixtures that can block the pores inside the concrete. These are the second type of waterproofing admixtures. But you cannot make a watertight concrete only by using these admixtures. You should try to accomplish all of the above considerations to produce a waterproof concrete.

We will talk about the waterproof concrete and structures in the later chapters.

5.7   CURING COMPOUNDS

Curing is very important for any type of concrete to achieve the best result of mechanical properties and durability in concrete structures. The production of a high-quality concrete is more difficult than the good curing, but unfortunately most of the time, the contractors forget to protect the high-quality concrete with curing.

If we don't cure the concrete as well as possible, we will see below problems:

- Decreasing the growth of concrete compressive strength: Production of C-H-S needs water. So, if we don't protect the water inside concrete from the evaporation, we cannot achieve the potential compressive strength.

- Penetration of water into the concrete element: As the production of C-H-S will decrease, the permeability will increase. So, water and other hazardous chemicals can penetrate into the concrete element and they can cause severe destruction.
- Increasing the probability of cracking: As all types of shrinkage in concrete depend on the curing quality, if we don't cure concrete as well as it is possible, concrete will shrink and cracks started to grow on the surface and inside the concrete element. So, the quality, impermeability, and loading capacity of the element will decrease.

Although the cost of curing is very low compared with the other costs of the construction industry, for many years, attention to the curing is not enough by the contractors because the most usual process for curing is the curing with water. As this type of curing should continue for several days and during day and night times, checking it by the supervisors is difficult. So, chemical admixture producers made the curing compound admixtures for a simpler curing procedure. Although the curing compounds will not act as good as the water curing, but for many projects, they are enough to guarantee the quality of concrete.

The curing compounds are some liquid chemicals that you should spray them on the surface of concrete element (Figure 5.25). After spraying, their solvent will evaporate and the remained material on the surface of the concrete can protect the water inside the element. They will not let the water to evaporate and because of the lighter color of them, they can control the temperature of concrete in the condition of sunlight.

When you apply these chemicals on the surface of the concrete element, it is very important that all of the surface should cover with the curing compound and also enough layer should remain on the surface to protect it from sunlight and wind which can cause the evaporation of water.

FIGURE 5.25 Application of the curing compound. ("Applying Curing Compound and Stripping the Wall Forms at the Yard Lead Reception Pit" by MTA Construction & Development Mega Projects.)

One of the most important points for using these admixtures is the cleaning of the surface before new concrete application. For example, if you covered a concrete roof with curing compounds and after some days you would like to apply the concrete of the columns on the roof, it is very important to clean the remained curing compounds from the base of the columns. The best way for cleaning is washing with plenty of water because if the curing compounds remained on the surface, they can make problems for the cohesion between two concrete layers.

Sometimes, at very difficult climate conditions, like very hard wind or direct sunlight at hot weather conditions, there is a need for the reapplication of curing compounds several times because they will chip in contact with the sunlight and you should apply them again to protect concrete against this hard climate condition.

5.8 VISCOSITY MODIFIER ADMIXTURES

We should use the viscosity modifier admixtures (VMA) for the protection of concrete against segregation and bleeding (Figure 5.26). They can help the concrete to remain homogenous. Most of the time, the need for VMA admixtures is only when you use PCE-type super-plasticizers and at the time that we have some of the problems below:

- Aggregates without good gradation or high amount of crushed sand.
- Lack of fillers in the aggregates
- Using low amount of cement and other supplementary cementitious materials in the concrete mix design
- Using high dosage of strong water reducer PCE super-plasticizers
- Trying to produce self-compacting concrete.

Sometimes, you can protect the concrete from segregation and bleeding with other cheaper methods. For example, you can add some stone powders or high-quality

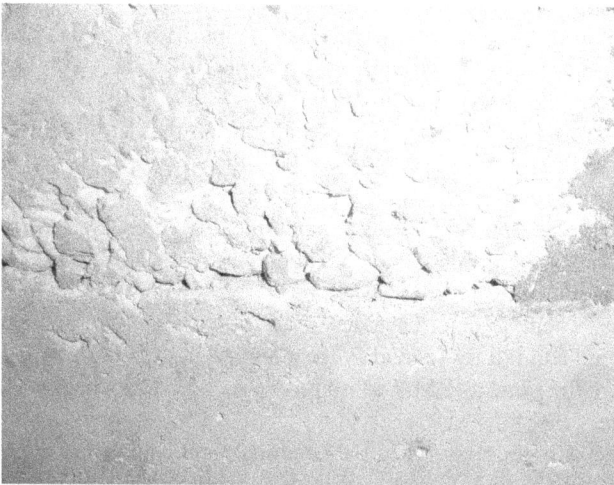

FIGURE 5.26 The result of using a segregated concrete. ("Concrete segregation.")

dune sand to the concrete. But maybe you don't have access to these additional fillers in a project. In this case, the best way is using VMA admixture.

Sometimes, the chemical admixture producers add VMA materials to their formulated polycarboxylate super-plasticizers. But they can give a negative effect on the water reduction rate of super-plasticizer. So, it is recommended to use the VMA and PCE super-plasticizer separately.

Recently, some of the producers developed new types of super-plasticizers with the chemical base of poly aryl ether and phosphate PCE which can control the viscosity of concrete with high slump. But the price of these admixtures is much more than the normal polycarboxylate super-plasticizers. So, it is recommended to use a PCE super-plasticizer with good aggregate gradation. Nevertheless, you can use VMA admixtures to modify the viscosity of concrete, especially when you would like to produce a self-compacting concrete.

We defined an index for the stability of SCC concrete which we call that visual stability index (VSI) which shows the grade of segregation and bleeding in SCC concrete. The best SCC concrete is the VSI0 and the worse is the VSI3. We can say the VSI3 concrete is a failed SCC. So, you should try to make a better VSI self-compacting concrete by modifying the aggregates gradation, using more fillers in the sand, and finally by using VMA.

5.9 PUMPING AID ADMIXTURES

One of the most important specifications of concrete is the pumpability because we should pump most of the concrete types nowadays. If we use a pumpable concrete, we will have below advantages:

- More concrete output from the pump and higher speed of concreting.
- Decrease the pump pressure and better maintenance of concrete pump instruments.
- Decreasing friction in the pump pipes.
- Prevention of blockage of concrete pump pipes is a very bad phenomenon and can cause very much retardation in concreting time.
- Decreasing the energy consumption
- Lower danger for the laborers especially when they are on the top stories of buildings.

There are too many factors affecting the pumpability of concrete. One of the most important factors is the good gradation of aggregates which can cause good viscosity and stability of concrete.

Sometimes, when we don't have access to the good quality aggregates, we can use pumping aid admixtures to modify the stability and viscosity of concrete. These admixtures are the same as VMA admixtures which they can help better pumpability of concrete.

You can use these admixtures in the batching plant and also in the hopper of the concrete pump (Figure 5.27) because there is a mixer inside the hopper of pump and it can mix the pumping aid admixtures with concrete.

FIGURE 5.27 Concrete pump hopper and its mixer. ("Aldi, Cosne, concrete pump.")

Before using the pumping aid admixtures, you should contact the manufacturer to ask about its adaptability with the different types of super-plasticizers. For example, you cannot use most of the pumping aids with naphthalene sulfonate super-plasticizers.

Sometimes, concrete admixtures manufacturers promote another kind of product with the name of pumping aid. It is in fact the starting slurry for wetting and lubricating the inside of concrete pump pipes for better pumping turnover. You can use a mixture of cement and water as the starting slurry, but if you use the pumping aid slurries for this purpose it will be much better for your concrete pumping instruments and also it can improve the pumping capacity, especially for the beginning part.

5.10 FOAMING AGENT ADMIXTURES

Foaming agent admixture is a special product for the production of cellular lightweight concrete (CLC) which we can use for the production of nonstructural concrete segments like wall blocks (Figure 5.28). Also, you can use CLC as a lightweight material for filling the roofs before performing the final surface.

CLC is a concrete with the specific gravity of between 300 and 800 kg/m^3. So, it is a very light concrete with very low mechanical properties. So, you cannot use it as a structural concrete.

The materials that you should use for the production of CLC (Figure 5.29) are the dune sand, cement, water, and foaming agent admixture. You can use only cement, water, and foaming agent for the production of super lightweight CLC with a specific gravity of less than 500 kg/m^3. The foaming agent entraps too many air bubbles inside the concrete and make it like a foam. So, it will be a very light type of concrete or mortar.

FIGURE 5.28   CLC blocks. ("Concrete blocks from rawpixel.")

FIGURE 5.29   Cellular lightweight concrete. (Photograph by the author.)

The foaming agent admixtures are a special liquid that you should mix with water inside the very high-speed mixer. Then a white foaming material like fire-fighting agents will produce that you can use in the concrete mixer to produce the CLC.

NOTE

1. TDS.w

REFERENCES

Aitcin P.C, High Performance Concrete, E&FN SPON, 2004.

American Society for Testing and Materials, Standard Practice for Making and Curing Concrete Test Specimens in the Field, ASTM C31-00.

American Society for Testing and Materials, Standard Practice for Sampling Freshly Mixed Concrete, ASTM C172-99.

American Society for Testing and Materials, Standard Practice for Making and Curing Concrete Test Specimens in the Laboratory, ASTM C192-00.

American Society for Testing and Materials, Standard Practice for Capping Cylindrical Concrete Specimens, ASTM C617-98.

American Society for Testing and Materials, Standard Specification for Ready-Mixed Concrete, ASTM C94-00.

American Society for Testing and Materials, Standard Specification for Air-Entraining Admixture for Concrete, ASTM C260-00.

American Society for Testing and Materials, Standard Specification for Light-weight Aggregates for Structural Concrete, ASTM C330-00.

American Society for Testing and Materials, Standard Specification for Chemical Admixtures for Concrete, ASTM C494-99.

American Society for Testing and Materials, Standard Specification for Coal fly Ash and Raw or Calcined Natural Pozzolan for Use as a Mineral Admixture in Concrete, ASTM C618-00.

American Society for Testing and Materials, Standard Specification for Fiber Reinforced Concrete and Shotcrete, ASTM C1116-00.

American Society for Testing and Materials, Standard Specification for Use of Silica Fume as a Mineral Admixture in Hydraulic Cement Concrete, Mortar and Grout, ASTM C1240-00.

American Society for Testing and Materials, Standard Test Method for Air Content of Freshly Mixed Concrete by the Volumetric Method, ASTM C173-01.

American Society for Testing and Materials, Standard Test Method for Air Content of Freshly Mixed Concrete by the Pressure Method, ASTM C231-97.

American Society for Testing and Materials, Standard Test Method for Flow of Hydraulic Cement Mortar, ASTM C1437-99.

Bertolini L, Elsener B, Pedeferri P, Polder R, *Corrosion of Steel in Concrete, Prevention, Diagnosis, Repair*, WILEY-VCH, 2004.

Cjp24, "Aldi, Cosne, concrete pump." Retrieved from: https://commons.wikimedia.org/wiki/File:Aldi, _Cosne, _concrete_pump_(5bis).jpg.

"Concrete blocks from rawpixel." Retrieved from: https://www.rawpixel.com/image/6034770/photo-image-public-domain-concrete-free.

Connor Jerome J, Faraji Susan, *Fundamentals of Structural Engineering*, Springer, 2016.

Ervanne Heini, Hakanen Martti, *Analysis of Cement Super-plasticizer and Grinding Aids: A Literature Survey*, Posiva Oy, 2007.

European Standard Organization, Admixtures for Concrete Mortar and Grout, EN934 Series.

European Standard Organization, Admixtures for Concrete, Mortar and Grout Test Methods, EN480 Series.

European Standard Organization, Concrete-Part 1: Specification, Performance, Production and Conformity, EN206-1, 2000.

European Standard Organization, Testing Fresh Concrete, EN12450 Series.

European Standard Organization, Testing Hardened Concrete, EN12390 Series.

Exploringlife, "Tuen Mun Swimming Pool in Tuen Mun, Hong Kong." Retrieved from: https://commons.wikimedia.org/wiki/File:Secondary_pool_of_Tuen_Mun_Swimming_Pool.JPG.

Gjorv E.Odd, Durability Design of Concrete Structures in Severe Environments, Taylor & Francis, 2009.

Gjorv Odd E, *Durability Design of Concrete Structures*, Taylor & Francis, 2009.

Hauschild Michael, Rosenbaum Ralph K, Olsen Sting Irving, *Life Cycle Assessment, Theory and Practice*, Springer, 2018.

Heinrichs Harald, Martens Pim, Michelsen Gerd, Wiek Arnim, *Sustainability Science: An Introduction*, Springer, 2016.

Iranian Institute for Research on Construction Industry, 9th Topic of National Rules for Construction, "Concrete Structures", 2009.

Iranian Institute for Research on Construction Industry, National Concrete Mix Design Method, 2015.

Iranian National Management and Programming Organization, National Handbook of Concrete Structures, 2005.

Iranian Standard Organization, Concrete Admixtures, Specification, ISIRI2930, 2011.

Iranian Standard Organization, Concrete Specification of Constituent Materials, Production and Compliance of Concrete, ISIRI2284-2, 2009.

Iranian Standard Organization, Standard Specification for Ready Mixed Concrete, ISIRI6044, 2015.

Janamian Kambiz, Aguiar Jose, A Comprehensive Method for Concrete Mix Design, Materials Research Forum LLC, 2020.

Lamond F.Joseph, Pielert H.James, *Significance of Tests and Properties of Concrete and Concrete Making Materials*, ASTM International, 2006.

Leppanen, Antti, "Volvo cement mixer truck in Jyväskylä, Finland." Retrieved from: https://commons.wikimedia.org/wiki/File:Volvo_cement_mixer_truck_in_Jyv%C3%A4skyl%C3%A4.jpg.

Mahmood Zadeh Amir, Iranpoor Jafar, Concrete Technology and Test (Farsi), Golhaye Mohammadi, 2007.

Mostofinejad Davood, Concrete Technology and Mix Design (Farsi), Arkane Danesh, 2011.

MTA Construction & Development Mega Projects photostream, "Applying Curing Compound and Stripping the Wall Forms at the Yard Lead Reception Pit." Retrieved from: https://www.flickr.com/photos/mtacc-esa/7415006130.

Newman John, Choo Ban Seng, *Advanced Concrete Technology, Concrete Properties*, Elsevier, 2003.

Pmhmarketing, "Putzmeister Wetkret 5." Retrieved from: https://commons.wikimedia.org/wiki/File:Putzmeister_Wetkret_5.jpg.

Popovics Sandor, *Concrete Materials, Properties Specification and Testing*, NOYES Publications, 1992.

Ramachandran V.S, Beaudion James, *Handbook of Analytical Techniques in Concrete Science and Technology, Principles, Techniques and Applications*, William Andrew Publishing, 2001.

Ramachandran V.S, *Concrete Admixtures Handbook, Properties, Science and Technology*, NOYES Publications, 1995.

Ramachandran, Paroli, Beaudion, Delgado, *Handbook of Thermal Analysis of Construction Materials*, NOYES Publications, 2002.

Ramezanianpoor Aliakbar, Arabi Negin, *Cement and Concrete Test Methods (Farsi)*, Negarande Danesh, 2011.

Richardson M, *Fundamentals of Durable Reinforced Concrete*, SPON Press, 2004.

Richardson M, *Fundamentals of Durable Reinforced Concrete*, Spon Press, 2002.

Safaye Nikoo Hamed, *Introduction to Concrete Technology (Farsi)*, Heram Pub, 2008.

Shekarchizade Mohammad, Liber Nicolas Ali, Dehghan Solmaz, Poorzarrabi Ali, Concrete Admixtures Technology and Usages (Farsi), Elm & Adab, 2012.

Spekking, Raimond, "Koln shotcrete application." Retrieved from: https://commons.wikimedia.org/wiki/File:Bauarbeiten_%C3%B6stliches_Domumfeld-K%C3%B6lner_Dom-Spritzbeton–8306.jpg.

Tux-Man, "Concrete segregation." Retrieved from: https://commons.wikimedia.org/wiki/File:S%C3%A9gr%C3%A9gation_b%C3%A9ton.jpg.

Zandi Yousof, Advanced Concrete Technology (Farsi), Forouzesh Pub, 2009.

Zandi Yousof, Concrete Tests and Mix Design (Farsi), Forouzesh Pub, 2007.

Zhu, Fangzhi, Zhiming Ma, Tiejun Zhao, "Pore structures in fresh concrete and air entrained concrete." Retrieved from: https://commons.wikimedia.org/wiki/File:Pore-structures-in-fresh-concrete-and-air-entrained-concrete.jpg.

.

6 Water for Concrete

Water is a critical material for the hydration reaction of cement. So, it is a very important material for the mechanical properties of concrete. On the other hand, it is important for the rheological behavior of concrete. So, we should check the quality of water before using it in concrete production.

We should use water in two phases in the concrete industry. The first phase is the production process which we use water for the hydration reaction and for the workability of concrete. The second phase is the curing process of concrete which we use water to control the hydration reaction and prevention of shrinkage and cracking. On the other hand, in most parts of the world, we are facing the problem of shortage in the sources of drinking water because of the global warming (Figure 6.1). So, we should try to consume less water in the industry.

In this chapter, we are going to talk about the suitable water for concrete and its specification. Also, we will talk about the impurities in the water sources and their effect on the properties of fresh and hardened concrete. Finally, we will talk about suitable water for concrete curing and the processes for controlling water consumption for concrete production and curing.

FIGURE 6.1 Sources of drinkable water which is limited in most parts of the world. ("Clear River water Bled Vintgar Slovenia".)

DOI: 10.1201/9781003384243-6

6.1 WATER FOR CONCRETE PRODUCTION

The best water for concrete production is drinkable water. Sometimes, we cannot reach high-quality drinkable water. On the other hand, sometimes, there are some impurities in the drinking water which is not good for the quality of fresh or hardened concrete. In these cases, we should test the water and control it with the standards. Sometimes, we may use a purifier for water before using it in concrete. Because some of the impurities can destroy the concrete during time and their effect on the properties of concrete will be very dangerous. So, we should pay the price for the purification of water to prevent the price of destroying the structure in the near future.

Now, we are going to talk about the different impurities in water and their effect on the properties of fresh and hardened concrete. Also, we will talk about the maximum amount of each impurity to use water in concrete production.

- Chloride: The most dangerous impurity in water is chloride ion. As you know, chloride will attack the structure rebars and you will see dangerous corrosion inside concrete, if you use a water with high amount of chloride ions (Figures 6.2 and 6.3). Chloride ion in water is because of the solution of different types of salts like sodium chloride or calcium chloride inside water. Some of these salts like calcium chloride can accelerate the concrete setting and it will cause drastic decrease in the workability during time. The maximum amount of chloride ion in water is 500 ppm according to the British standard.
- Sulfate ion: The other important dangerous impurity in water is the sulfate ion. As mentioned before, it can attack the C_3A of cement and will cause the corrosion of concrete structures during the time. To control this reaction, you can use type II or V cement according to the amount of sulfate ion in water or soil. But for concrete water, you should not use water with high amount of sulfate. For fresh concrete, there is no considerable effect of the sulfate ion on the properties of fresh concrete.

FIGURE 6.2 Corrosion of a concrete bridge. ("The second bridge [picture 46] crosses a canal inside the City of Amsterdam, Netherlands" by Achim Hering.)

FIGURE 6.3 Corrosion in concrete structures. ("Photograph taken in grounds of Newton Park Technical High School, Port Elizabeth, South Africa".)

- Carbonate and bicarbonate alkalis: These salts have different effects on the setting time of fresh concrete which depends on the solution density. So, you should take care of the setting time of concrete when you use water with carbonate and bicarbonate alkalis. On the other hand, they can accelerate the alkali aggregate reaction in hardened concrete. So, it is better to prevent using water with more than 1000 ppm of carbonate and bicarbonate alkalis.
- Low pH: Low pH itself don't have any considerable effect on the properties of fresh or hardened concrete. But you should check the reason for the low pH of water. For example, some of the organic acids can increase or decrease the setting time of concrete. On the other hand, as you know the pH of concrete is high because of the alkali effect of the Portland cement and it can protect the rebars inside the concrete itself. Using water with low pH (for example less than 3) can cause decreasing the pH of total concrete and it can accelerate the corrosion of rebars in severe conditions. So, it is better not to use water with a pH lower than 3.

FIGURE 6.4   Organic impurities in water. ("Aquatic plants" by Hagerty Ryan, USFWS.)

- High pH: High pH of water means there are some OH ions inside water from the dilution of NaOH or KOH. These impurities can decrease the final compressive strength of concrete. On the other hand, they can increase the probability of alkali aggregate reaction. So, it is recommended not to use water with high pH for concrete production.
- Organic impurities: Most of the time, organic impurities will retard the set-ting time of concrete. On the other hand, the pieces of organic impurities in water can decrease the compressive strength of concrete in a region where impurities will stand and they can decrease the final quality of the surface in the structures. Some of the sewage waters contain high amounts of alkalis and impurities So, you should not use them for concrete production (Figure 6.4).
- Sugar: Sometimes, you may find some amount of sugar or other materials containing saccharides diluted in water. They can drastically retard the con-crete setting time and can cause many problems for concrete structures like the need for a very long time for mold release.
- Oil in water: Sometimes, you may find different types of oils in water. Most types of oils can decrease the compressive strength of concrete. So, it is bet-ter not to use these waters for concrete production.

According to the above mentioned, you can understand that using the sources of non-drinkable water could be acceptable if you check the impurities of water. Some other sources like sea water with high amount of chloride and sulfate are not acceptable for use in concrete production.

6.2   WATER FOR CONCRETE CURING

Using water with lower quality could be acceptable for concrete curing depending on the type of water and the type of concrete structure. For example, using of water with some organic impurities for the curing of an exposed structure could not be accept-able because it can cause color changes on the surface of structure.

TABLE 6.1
Sample Analysis Test for Water

Parameter	Test Result	Parameter	Test Result (ppm)
Color	Colorless	Sulfate	1
Odor	Odorless	Sodium	88
pH	6.5	Nitrate	1
Taste	Tasteless	Organic impurities	3
Chloride	140 ppm	Dissolved solid	375

Using water with high amount of chloride or sulfate (like sea water) for the curing of concrete, especially important concrete structures is not acceptable because you will cure the concrete in the earlier ages with water. At the earlier age, the permeability of concrete is high and water with a high amount of chloride and sulfate can filtrate in to the structure and remain beside the steel bars and they can cause the corrosion in the future. So, you should not use sea water for the curing of concrete structures.

It is better not to use water with a high amount of alkalis for concrete curing because like the chloride and sulfate, they can filtrate inside the concrete structure and decrease the final compressive strength.

6.3 TEST OF WATER

You should test the water for the analysis of impurities that should be harmful for concrete. You should send the sample of water to a chemistry laboratory and they can analyze the water and give you the final result as you can see in Table 6.1.

As you can see in the table above and compare it with the descriptions of this chapter, the water mentioned in the table is acceptable to use in concrete production.

6.4 DECREASE THE AMOUNT OF WATER FOR CONCRETE PRODUCTION AND CURING

As mentioned before, decreasing the amount of water for concrete production and curing is necessary according to the lack of drinkable water resources in most parts of the world. So, we should try to use less water for concrete production which can cause better performance for hardened concrete in the structures also. On the other hand, we should use less water for concrete curing without any damage to the quality of concrete structures. According to the abovementioned, we can name below considerations for decreasing the water consumption in the concrete industry:

- Using high-quality super-plasticizers instead of the old ones to decrease the amount of water we need for the production of a high-quality concrete.
- Using concrete industry sewage water with filtration and purification. This is the sewage water that we use for washing concrete production and transportation machines and the water we may use for washing the aggregates (Figure 6.5).

- Control the temperature of concrete in summer to protect it from the evaporation of water. You can use ice instead of water for concrete production in hot weather conditions. We will talk about concrete implementation in hot weather conditions later.
- Using curing compounds instead of water for curing concrete structures when it is possible without any damage to the quality.
- Control the curing with water when it is necessary, by using a suitable water curing method. For example, the use of cotton textile (Figure 6.6) for the curing of columns and the use of fogging instead of immersing the concrete surface with water.

FIGURE 6.5   A concrete recycling system that it can recycle aggregates and water from the concrete sewage. (Photograph by the author.)

FIGURE 6.6   Special cotton textile for use in water curing for the protection of water. (Photograph by the author.)

REFERENCES

amanderson2, "Clear River water Bled Vintgar Slovenia." Retrieved from: https://commons. wikimedia.org/wiki/File:Clear_River_water_Bled_Vintgar_Slovenia_(8095482709). jpg.

American Society for Testing and Materials, Standard Practice for Sampling Freshly Mixed Concrete, ASTM C172-99.

American Society for Testing and Materials, Standard Test Method for Compressive strength of Cylindrical Concrete Specimens, ASTM C39-01.

Ervanne Heini, Hakanen Martti, Analysis of Cement Super-plasticizer and Grinding Aids: A Literature Survey, Posiva Oy, 2007.

European Standard Organization, Concrete-Part 1: Specification, Performance, Production and Conformity, EN206-1, 2000.

European Standard Organization, Testing Fresh Concrete, EN12450 Series.

European Standard Organization, Testing Hardened Concrete, EN12390 Series.

Gjorv Odd E, *Durability Design of Concrete Structures*, Taylor & Francis, 2009.

Hering, Achim, "The second bridge (picture 46) crosses a canal inside the City of Amsterdam, Netherlands." Retrieved from: https://commons.wikimedia.org/wiki/File:Qew_bruecke_ nf_beton_kaputt_33_von_46.jpg.

Iranian Institute for Research on Construction Industry, 9th topic of National Rules for Construction, "Concrete Structures", 2009.

Iranian Institute for Research on Construction Industry, National Concrete Mix Design Method, 2015.

Iranian National Management and Programming Organization, National Handbook of Concrete Structures, 2005.

Iranian Standard Organization, Concrete Admixtures, Specification, ISIRI2930, 2011.

Iranian Standard Organization, Concrete Specification of Constituent Materials, Production and Compliance of Concrete, ISIRI2284-2, 2009.

Iranian Standard Organization, Standard Specification for Ready Mixed Concrete, ISIRI6044, 2015.

JonRichfield, "Photograph taken in grounds of Newton Park Technical High School, Port Elizabeth, South Africa." Retrieved from: https://commons.wikimedia.org/wiki/File:Concrete_wall_ cracking_as_its_steel_reinforcing_cracks_and_swells_9061v.jpg.

Lamond F.Joseph, Pielert H.James, *Significance of Tests and Properties of Concrete and Concrete Making Materials*, ASTM International, 2006.

Mostofinejad Davood, Concrete Technology and Mix Design (Farsi), Arkane Danesh, 2011.

Newman John, Choo Ban Seng, *Advanced Concrete Technology, Concrete Properties*, Elsevier, 2003.

Popovics Sandor, *Concrete Materials, Properties Specification and Testing*, NOYES Publications, 1992.

Ramezanianpoor Aliakbar, Arabi Negin, Cement and Concrete Test Methods (Farsi), Negarande Danesh, 2011.

Ryan, Hagerty, USFWS, "Aquatic plants." Retrieved from: https://pixnio.com/nature-landscapes/ wetlands-and-swamps/close-view-of-aquatic-plants-on-water-surface-in-prairie-wetland.

Zandi Yousof, *Advanced Concrete Technology (Farsi)*, Forouzesh Pub, 2009.

7 Testing of Concrete

When you make a concrete with defined properties, it is very important to check and test the properties to compare them with the defined values. In this chapter, we discuss these tests.

There are too many methods according to different standards for testing concrete. We are not going to accept only one standard test method, because in different countries, you should do the tests according to the acceptable standards for that region of the world. But you should know that most of the test methods are the same from the original concept point of view. The difference is for the details and we are not going to centralize the details.

As mentioned before, we have different phases of concrete. Some of the defined properties are for the fresh concrete and some others are for the hardened. We cannot say which one is more important. You should make a concrete with high quality in both fresh and hardened phases because you should make a workable concrete for better implementation. If you make a concrete with very high quality in hardened phase, but without enough workability, it is not an acceptable concrete.

According to the abovementioned, we discuss the tests referring to the fresh concrete and then about the hardened.

7.1 TESTS FOR FRESH CONCRETE

When we start the production of concrete by contacting cement particles to the water molecules, we are starting the fresh phase. Concrete will remain in the fresh phase till the initial setting time of concrete starts. Then the concrete goes to the jelly phase which is the border between fresh and hardened concrete. We don't have any test for the concrete in the jell phase. But for fresh concrete, we have very important tests. We can extract many useful information from the fresh concrete test results. We can reject a concrete with not acceptable fresh phase test results because it will show us that probably the hardened phase test results should be not compatible with the defined properties. Although it is very hard to revise a low-quality concrete in the hardened state, it is better to control some of the properties in the fresh phase. If we reject a fresh concrete transported to the project by a truck mixer, we should pay only the price of that rejected concrete. But if there is a need to repair a structure because of the use of low-quality concrete or sometimes if there is a need to destroy the structural element, we should pay more. So, you should take care of the production quality and then we should check the fresh concrete accurately. Then we can trust the quality of hardened concrete.

In this part, we discuss the slump test, flow table test, and rheometer for concrete which are the tests for the flowability and workability of concrete. Then we discuss the temperature of fresh concrete, which is very important for quality control. Density and air content of concrete will be the next tests that we discuss. They will

DOI: 10.1201/9781003384243-7

show us the amount of entrapped air in concrete which is very important for the evaluation of the compressive strength in the hardened phase.

7.1.1 SLUMP TEST

We talked about the slump test in the first chapter of this book. As mentioned before, this is a test to evaluate the workability and flowability of fresh concrete (Figure 7.1). We should do this test according to ASTM C143.

The testing equipment for the slump test is:

- Slump cone: which is a standard cone with an upper circle diameter of 102 mm and down circle diameter of 203 mm and a height of 305 mm.
- Tamping rod: This is a rebar with a diameter of 16 mm with the top a circular shape, and we should use it for tamping the concrete inside the slump cone.
- Plate: This could be a metal or glass plate, to which we should tighten the slump cone before starting the test.
- Measurement ruler: This is a ruler or any other instrument that you can measure the diameter with.

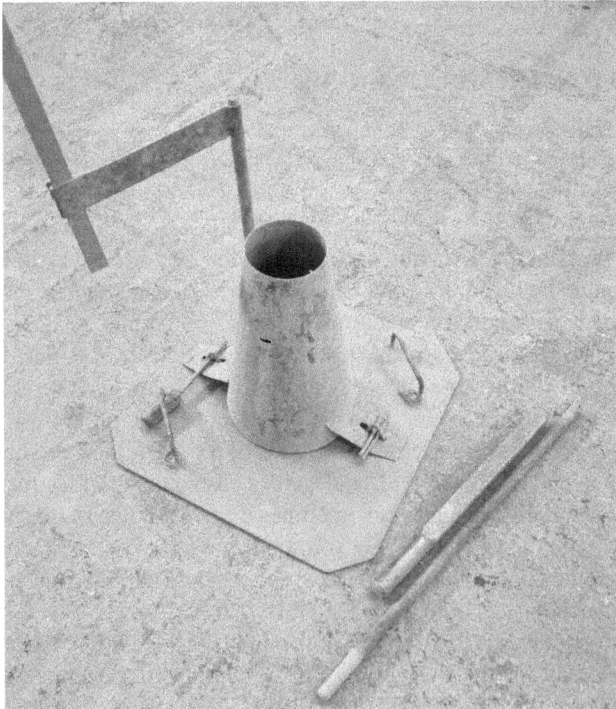

FIGURE 7.1 Slump test instruments. (Photograph by the author.)

FIGURE 7.2 Measuring concrete slump. ("Conducting a slump test on the concrete" by MTA Construction & Development Mega Projects.)

To start the test, first you should take a sample of concrete from the truck mixer or any other type of mixer. Then you should tighten the slump cone on the plate as you can see in Figure 7.1. (Some other types of instruments should have different types of tightening methods.) Then you should pour the slump cone in three volume layers. It means that you should pour first about 67 mm of the height and then to 155 mm of the height and finally to the top surface of the slump cone. For each layer, you should tamp the concrete with the rod 25 times uniformly. For upper layers, you should tamp the layer to the depth of that layer plus a little part of the below layer. For the upper layer, you should pour the concrete more than needed, because after tamping, you should make the surface flat by using the tamping rod. The surface of the concrete should be the same as the upper point of the slump cone. Finally, you should pull up the cone vertically without any horizontal or torsional moving. The concrete will fall down according to its flowability. Now, you should measure the amount of falling by measuring the distance between the upper surface of the slump cone to the upper surface of the falling concrete as you can see in Figure 7.2.

As the slump of concrete could decrease over time, because of the evaporation of water from concrete or instrument surface, you should do the test in less than 2.5 minutes from the starting time to the end. On the other hand, as you will take a little amount of concrete from a truck mixer for the slump test and this little amount of concrete could lose its slump rapidly, because of the evaporation of water, especially at high temperature, you should do the slump test as soon as it possible after taking the sample from the truck mixer.

7.1.2 Flow Table Test

Some of the people believe that the slump test is not accurate enough to show the workability of concrete. We can compare two concretes with the same slump, but with different workability performance, for example, different pumpability or malleability.

FIGURE 7.3   Flow table test of fresh concrete. ("Flow test before".)

So, we should define another test method for the evaluation of concrete workability. One of the most famous tests is the flow table test (Figure 7.3).

This test is according to the European standard EN12350-5 which described the exact procedure for this test. Here we are going to study a brief of it.

First, you should pour the cone shape mold with fresh concrete. After tamping with the special tamper, which you can see in Figure 7.3, you should pull up the cone till the concrete will collapse. After final collapsing of concrete, you should manually raise the plate to the upper place and then drop it to the bottom place for 15 times. This will cause the special vibration of concrete, so the concrete will spread through the table. Now you should measure the spread diameter in two perpendicular lines and report it as the flow table spread.

We cannot say that this test is a better or more accurate test compared with the slump test. It depends on the type of concrete. For softer concretes and more flowable ones, we can say that this is a more accurate test. But for stiffer concretes the slump test will be more accurate. The best results will be reachable if you do the slump and flow table tests together.

7.1.3   SLUMP FLOW TEST

Slump flow test is for concretes with high flowability like self-compacting concrete. You can use this test for other kinds of concrete with slump higher than 200 mm despite it is not an SCC concrete. We can call concrete with the slump of higher than 200 mm easy-compacting concrete because we can compact it easier than normal concrete with less vibration.

The slump flow test is the same as the slump test. But as the concrete is more flowable than normal, after pulling up the slump cone, the concrete will flow through the plate (Figure 7.4). So, you should measure the flow circle on the table instead of the falling distance. For SCC concrete, there is no need for tamping the concrete into three layers.

FIGURE 7.4 Slump flow test. ("Flow test after".)

You should just pour the cone and pull it up. But for easy-compacting concrete instead of SCC, it is optional to tamp or not tamp the concrete.

SCC concrete will have a slump flow of more than 600 mm. But it is not enough to name concrete as self-compacting concrete. We should do some other tests like L-box test or V-funnel test, which are exactly designed for the quality control of SCC concrete. We discuss the SCC concrete and its special test procedure in the later chapters of this book.

7.1.4 RHEOMETER FOR CONCRETE

As mentioned before, most concrete technologists believe that the evaluation of con-crete workability is very difficult and by using some tests like slump or flow table, we cannot give a reasonable result for concrete workability. So, they are working on a new concept of rheology for concrete. In fact, they are trying to make some instru-ments that can calculate the amount of shear force needed to mix a concrete. They call this instrument rheometer (Figure 7.5).

Rheometer is a device to measure the concrete flows in response to an applied force. We have a rheometer for different kinds of viscous liquids in laboratory tech-nology. But for concrete, it is a new device. So, there is no special standard for it. Many companies are working to make a special rheometer for concrete with reason-able results.

Like other types of tests for the workability of concrete, using the rheometer is not reliable for all types of concrete. Its results are better for more flowable concretes, especially for SCC. For example, we may get different results for the workability of two SCC concretes with the same slump flow, L-box, and V-funnel test results and we can see this difference in real implementation of SCC concrete. But for stiff con-crete mixes, the results of the rheometer will not be reliable.

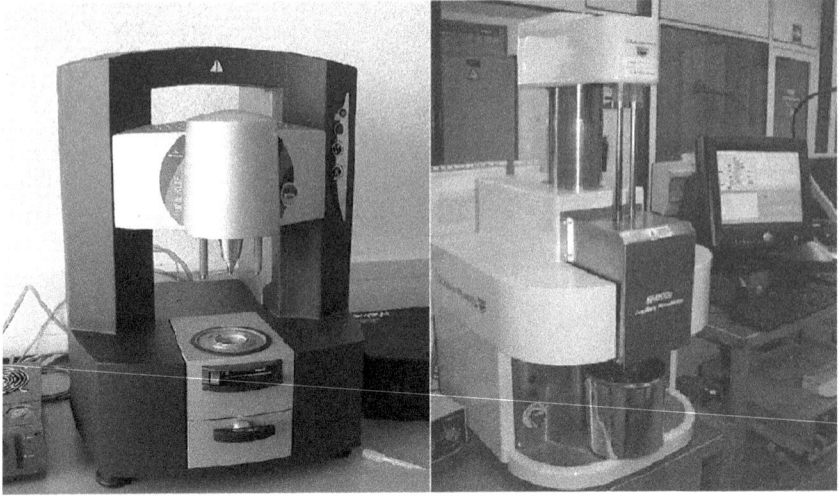

FIGURE 7.5 Two different types of rheometers. ("rheometer" by Olivier Cleynen used under a Creative Commons Attribution-Share Alike 4.0 International license), ("Capillary rheometer" by Cjp24 used under a Creative Commons Attribution-Share Alike 3.0 Unported license.)

The other important factor for the rheometer is that you should use this instrument for the comparison between two concrete with more or less the same aggregate type because the results depend on the sieve analysis of aggregates. For example, it is possible to take different results for two concrete with the same slump test but different aggregates gradation. So, we should work on this instrument to calibrate it specially, for concrete to get better and reliable results in the future. If we do it, it could be the best test for the evaluation of concrete workability in the future.

7.1.5 TEMPERATURE OF FRESH CONCRETE

One of the important parameters for the evaluation of concrete quality is its temperature. We discuss the concrete implementation in hot and cold weather conditions in the future chapters of this book. But for now, you should know that there is no ambient temperature limitation for concreting. But the limitation is for the concrete temperature.

Concrete temperature should not exceed 32°C, in hot weather conditions, and should not fall down less than 5°C, in cold weather conditions. On the other hand, the behavior of concrete at temperatures more than 25°C and less than 10°C in the case of slump keeping and growth of compressive strength should be different as described below:

- The slump keeping behavior of concrete at the temperature less than 10°C is much better than normal and it is more difficult at temperatures more than 25°C.

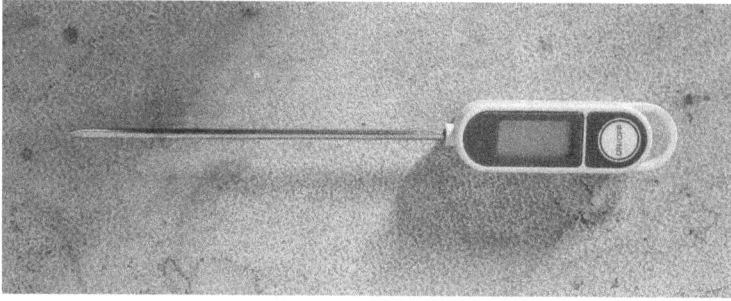

FIGURE 7.6   Special concrete thermometers. (Photograph by the author.)

- The compressive strength growth of concrete at a temperature less than 10°C is much more than concrete at a temperature more than 25°C. It means that at higher temperatures, concrete will take much of its final strength at earlier ages like 7 days. But at lower temperatures the growth of strength after the age of 7–28 days is more.

So, you can find many useful information from the temperature of concrete to make the decision about the structure.

To measure concrete temperature, you should use special thermometers for concrete as you can see in Figure 7.6. This test should be in accordance with the ASTM C1064 as the standard test method for temperature of freshly mixed Portland cement concrete.

To control the temperature of concrete between 5°C and 32°C in cold and hot weather conditions you should read the future chapter of this book about hot and cold weather concreting.

7.1.6   DENSITY OF FRESH CONCRETE

Density or weight per volume of fresh concrete is important for the control of fresh concrete quality because it can show us below specifications:

- Amount of air inside concrete which can be entrained air or entrapped air: We can understand the amount of air is related to the compressive strength of hardened concrete. For example, if the density of concrete without using of air-entraining admixture will be very low, it shows that the amount of entrapped air is very high. Sometimes, it is because of uncontrolled air entraining of the super-plasticizer. This concrete will give us a compressive strength much lower than before. So, it is better to reject this concrete.
- Density of aggregates: If we are using the same mix design as before and the density of concrete will decrease, it can show us the lower density of aggregates. Sometimes, it is because of the difference in the source of aggregates. If the density of aggregates will be very low, it can affect the strength of aggregates. So, you should take care of this problem.

- Approximately amount of cement and water used for the production of concrete: When you design a mix for a defined concrete by defined constituent materials, you will have a defined theoretical density for your concrete.
  So, if the density of concrete will be lower than the theoretical density, it
  shows that because of a problem that you may not control, you are using
  less cement or more water in the production process and it is a very bad sign
  that you should control. If the density will be very low, you should reject
  the concrete.

  For some types of concrete like light weight or heavy weight concrete,
  you may need the exact density before using them in the structure. So, deter-
  mining the density of concrete is very important.

The standard test method for the density of fresh concrete is ASTM C138. This is a
very simple test. You only need a special cylinder container with a defined volume
that you can see in Figure 7.7 and you need the rod that you used for tamping the
concrete in slump test. You should put concrete in three layers in the container and
tamp it with 25 strokes each time. Then you should use a rubber hammer for some
strokes beside the container for better compaction. Finally, you should plain the front

FIGURE 7.7 Container with defined volumes for the test of density. (Photograph by the author.)

layer and clean the side of the container from concrete. Then you should weight the container poured with concrete. Before starting the test, you should know the volume and weight of the container. Now, you can calculate the weight of concrete inside the container and divide the weight of concrete by the volume of the container to calculate the density of concrete. As mentioned before, the density of normal concrete using normal aggregates is about 2.4 kg/L.

7.1.7 AIR CONTENT OF FRESH CONCRETE

As mentioned in the previous part of this chapter, we may calculate the density of fresh concrete and see that the concrete has a lower density than we expected. At this time, one of the dangerous problems could be a high amount of air inside the concrete which should cause a drastic decrease in the compressive strength. But several other problems also could decrease the density of concrete. To be sure, the best way is to measure the air content of concrete that we discuss it in this part.

On the other hand, we may use concrete with air-entraining admixtures. At this time, it is very important to measure the exact amount of air inside the concrete, because the amount of air should not exceed the defined percent because it can cause decrease in compressive strength and should not be less than the defined percent because it is necessary for the resistance of concrete against freeze thaw cycles.

We can derive many useful information about the quality of concrete that we are going to use in the structural elements by measuring the air content.

We have two standard test methods for the air content of fresh concrete:

- ASTM C173: standard test method for the air content of freshly mixed concrete by the volumetric method. This method is based on the substitution of air bubbles inside concrete by a mixture of water and isopropyl alcohol. This method is more difficult and time consuming than the next method. So, most of the time, we recommended to use the next method. If you buy the instrument for this test method, they exactly described the procedure that you should do for this test in their catalog.
- ASTM C231: standard test method for the air content of freshly mixed concrete by the pressure method. This method is based on the substitution of water into air bubbles inside concrete by using pressure. This is a more precise and simpler/faster test method for the evaluation of air contents in concrete. So, we recommended this method for use. We have two types of instruments for this test method that you can see in Figure 7.8. If you buy each of them, the manufacturer will exactly explain the procedure that you should use for the evaluation of air content inside fresh concrete.

Concrete with a higher amount of air will be more workable in all specifications like the malleability and pumpability. But it will give you less compressive strength. So, you should adjust the amount of air inside concrete. Most of the time, you can do it by the adjustment of the super-plasticizer formulation.

According to the above mentioned, for normal concrete, we recommended to maintain the air content inside concrete at less than 2% and more than 1.5%.

FIGURE 7.8 Instrument for ASTM C231 method. (Photograph by the author.)

But when you are using air-entraining admixtures, it is possible to use between 4% and 6% of air inside the concrete. For this case, concrete will give you less compressive strength. Each 1% of air will decrease the compressive strength by about 5%.

7.2 TESTS FOR HARDENED CONCRETE

The properties of hardened concrete are very important for using it on the structures. In fact, we tested and controlled the properties of fresh concrete to make sure about the properties of hardened concrete. Because, when we use a concrete in a structure, improvement or any other change in the quality of concrete is not easy. We should test the properties of hardened concrete to check the achievement of the final properties that we designed before.

When we design a structure with any method, it is not easily possible for us to check and control the accuracy of the design because we cannot check the structure with the final loads like earthquake load to check the design. But for concrete it is different. We can design a concrete and then we can check all of the properties of fresh and hardened phases. Then we can make some corrections if there is a need and finally we will be sure about the use of concrete in the structure if we have good quality control. So, testing the hardened concrete is very important for the checking of concrete mix design and then for the quality control of the structure.

In this part, we discuss four important tests of hardened concrete. First is the most important compressive strength which is important for the checking of structure. Then we discuss the elastic modules of concrete which are related to the compressive strength and the microstructure of concrete. Density of hardened concrete is the next test that we discuss which is a little different from the density of fresh concrete and the permeability of concrete will be the final test for hardened concrete which is very important for the evaluation of the concrete and structure durability.

7.2.1 COMPRESSIVE STRENGTH TEST

The compressive strength test is the most important test for hardened concrete. When we would like to design a mix for concrete, we should focus on the desired compressive strength. On the other hand, for each concrete structure, we designed the structure with a defined concrete compressive strength. So, it is very important to achieve this target strength.

The test procedure is more or less the same in different standard test methods. But the most important difference is the type of mold that you are going to use. The ASTM C39 is the standard test method for the compressive strength of cylindrical specimens according to the ASTM standard. We should use the 15×30 cylinder molds for this test method. But in some parts of the world, we may use $15 \times 15 \times 15$ cube molds or 10×20 cylinders. On the other hand, in most parts of the world we should use the compressive strength of 15×30 cylinders at the age of 28 days for the design of concrete structures. So, if we use other types of molds, we should convert the results to the 15×30 cylinders because the cubic compressive strength is more than the cylinder one and also the 10×20 cylinder compressive strength is more than 15×30 ones (Figures 7.9 and 7.10).

If you use cubic mold for the compressive strength test, you need to convert the result to the cylindrical compressive strength, because as mentioned before, we used cylinder compressive strength in the structural design. We have many different

FIGURE 7.9 Cylindrical mold at left and cubic mold at right. (Photograph by the author.)

FIGURE 7.10 Cube specimens at left and cylinder specimens at right. (Photograph by the author.)

TABLE 7.1

Conversion Coefficients for the $15 \times 15 \times 15$ Cube to 15×30 Cylinder Compressive Strength

$15 \times 15 \times 15$ cm cube specimen	25 MPa	30 MPa	35 MPa	40 MPa	45 MPa	50 MPa	55 MPa
Coefficient	0.8	0.833	0.857	0.875	0.889	0.9	0.909
15×30 cm cylinder specimen	20 MPa	25 MPa	30 MPa	35 MPa	40 MPa	45 MPa	50 MPa

conversion coefficients in different standards and codes. You can use the recommended one in your country. Here you can see our suggestions in Table 7.1.

For compressive strength of more than 50 MPa we recommend the use of cylinder molds. You can see from the table above, when the compressive strength will increase, the coefficient will be larger because the cube and cylinder compressive strength will be nearer to each other.

For this test, you should pour the mold with concrete in three layers, each layer with 25 strokes. Then you should plain the surface and let it remain for 24 hours. After 24 hours you should get the molded concrete out of the mold and put it inside the water curing tank. You can test the specimens after 3, 7, 11, and 28 days.

For testing the specimens, you should put them under the pressure of the concrete testing machine (Figure 7.11). It will apply the load to the specimen till it will crush and fail. Now, you can calculate the compressive strength of concrete according to the type of specimen.

FIGURE 7.11   Concrete compressive strength testing machine. (Photograph by the author.)

For example, if a cube specimen failed after the loading of 80 tons, the compressive strength will be 35.5 MPa for the cube and with the coefficients of Table 7.1 it will be 30.5 MPa for the cylinder. But if you used a cylinder specimen and it failed after the loading of 80 tons the compressive strength will be 45.3 MPa for cylinder.

The accuracy of the compressive strength test is very important. You should use the calibrated testing machine. For cube specimens you should use stiff molds like steel or thick plastic because the plain surface for the specimens is very important for the distribution of load on the surface and it has an effect on the compressive strength as big as 50%. For cylinder specimens as you should use the upper surface which is a hand-made surface and it is not a plain surface for testing, you should use suitable capping for that. The capping will do with a special instrument that you can see in Figure 7.12 to make a plain surface for cylinder specimens. For lower strength less than 35 Mpa, you can use dental plaster as the capping material or you can use high-quality capping pads (Figure 7.13). For the compressive strength more than 35 to 70 MPa you can use a mixture of silica sand and melted sulfur as the capping material and for higher strength, it is recommended to make a plain surface with cutting (Figure 7.14).

FIGURE 7.12 Capping instrument at left and capped cylinder specimens at right. (Photograph by the author.)

FIGURE 7.13   Capping pad mold. (Photograph by the author.)

7.2.2   CONCRETE ELASTIC MODULES AND POISSON'S RATIO TEST

One of the important mechanical properties of concrete is the elastic modules and Poisson's ratio, which is used for structural analysis, especially for the drift of structure. So, it is important to test a hardened concrete to check its elastic modules.

FIGURE 7.14 Cutter for the cylinder specimen of high-strength concrete. ("Concrete cutting" by Bicanski the photograph is copyright free.)

The definition of elastic or Young module is the slope of stress-strain curve for any kind of material which shows us the amount of deformation under the special stress which could be tensile or compression. For concrete, we need the elastic modules in compression mode.

The definition of Poisson's ratio in compression is the ratio of transverse strain to the axial strain which shows us the ratio of deformation between two axles of a specimen.

The standard test method for elastic modules and Poisson's ratio of concrete in compression is the ASTM C469. In this test method, you should apply force with the concrete testing machine to a standard 15×30 cylinder specimen. During applying the force, you should measure the deformation of concrete specimen with the special instrument that you can see in Figure 7.15. Then you can calculate the elastic modules and Poisson's ratio by the data derived from this test. The data consist of the stress on the specimen during the time and the strain on two axles of the specimen during the time. By using this data and according to the above definitions, you can calculate the elastic modules and Poisson's ratio of concrete in compression.

If we are not going to do the ASTM C469 for the evaluation of the elastic modules of concrete, we can use some formulation for the calculation of the elastic modules by using the date from the compressive strength test. There are too many formulations in different texts and lectures. But the most useful formula is as below:

$$E = 5000 \sqrt{f'c}$$

In the above formulae, the f'c is the standard 15×30 cylinder compressive strength of concrete at the age of 28 days in MPa and E is the elastic modules of concrete in MPa. You can use the formulae, for concretes with a compressive strength less than 50 Mpa. For more compressive strength, we have different formulations. But as the form and type of concrete stress-strain curve will be different, it is recommended to use the ASTM C469 for the calculation of the elastic modules of concrete because

FIGURE 7.15 Instrument for testing of elastic modules and Poisson's ratio of concrete. (Photograph by the author.)

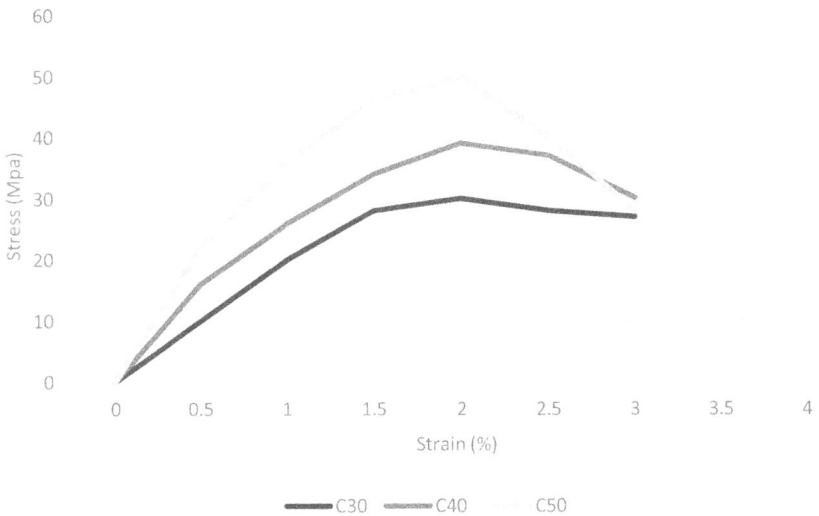

FIGURE 7.16 Stress-strain curve for different types of concrete. (Graph created by the author.)

the stress-strain curve for the high-strength concrete is different with the normal concrete. For high-strength concrete, the deformation of concrete after the final stress will be very high because the concrete will be more brittle than the normal concrete. You can see the behavior of normal and high-strength concrete in Figure 7.16.

7.2.3 Density of Hardened Concrete

Test of density is very important for fresh concrete. We can extract many useful information from this test. We can evaluate the density of hardened concrete for checking the mix design and production process of concrete.

Most of the time, there is no much difference between the fresh and hardened concrete density instead of some types of special concrete made with the aggregates with high absorption rate. You can use European standard test method for the density of hardened concrete EN12390-7 or the ASTM C642. This is a very simple test method. Just you should measure the dimensions of the specimen with a Collis (Figure 7.17). Then you should calculate the volume of specimen according to the shape. For cubic specimens, you should use the formulation of the cube volume by multiplying the length to the width to the height of the cube. For the cylindrical specimens, you should multiply the radius of the circle to itself to the π to the height of the cylinder.

After calculating the volume, you should weight the specimens accurately. Finally, you should calculate the density by dividing the weight to the volume of the specimen.

Before testing the specimens, you should take them out of the curing water and dry their surface with a napkin or let them remain in the air till the surface will be dry. But you should not let the total moisture come out of the specimen.

Most of the time, the density of hardened concrete will calculate when you would like to test the specimens for the compressive strength. So, you will have values in different ages. But there is no any difference for the density of concrete in different ages.

FIGURE 7.17 Collis for the accurate measurement of the length. (Photograph by the author.)

For example, if you have a cube specimen with the dimensions of $14.9 \times 15 \times 14.9$ cm, we will have the volume of this cube as 3330 cm^3. If this specimen weight will be 8050 g, then the density of hardened concrete for this specimen will be 2.417 kg/L.

If you have a cylinder specimen with the circle diameter of 15.1 cm and the height of 29.9 cm we will have the volume as 5352 cm^3. If the weight of this specimen will be 12930 g, then the density will be 2.416 kg/L.

7.2.4 PERMEABILITY OF CONCRETE

As mentioned before, the permeability of concrete is the key point to measure the durability of structures. Because all of the chemical ions and water should infiltrate into the concrete, then they can damage the concrete or steel bars. So, if we can make an impermeable concrete, we can protect the concrete structure from the corrosion. So, checking the permeability of concrete is very important.

To check the permeability of concrete, we have different tests in different texts and standards with different accuracy. Here we discuss two simple tests with low accuracy and then about a more complicated test with very high accuracy.

The first test is not exactly a defined test for the permeability of concrete. This is a test for the evaluation of water absorption of concrete specimens. It is the ASTM C642, standard test method for density, absorption, and voids in hardened concrete. We can call this test immersion permeability test (Figure 7.18).

For this test, first you should oven dry the concrete specimens and weigh them accurately, then you should let them cool at the room temperature and immerse them inside the water for 24 hours and again weigh them accurately. The difference between these two weights is the weight of water infiltrated into the concrete specimens. We can announce it as the percent by weight of the specimen.

For high-quality impermeable concrete, the amount of water that can infiltrate into the concrete specimen will be much less than a low-quality permeable concrete. So, the result of this test can be used as criteria for the permeability of concrete.

FIGURE 7.18 Concrete specimens immersed in water. (Photograph by the author.)

For this test, you can use any type of concrete specimen, but the best choice is the standard 15×30 cylinder specimens. If you do the test accurately, the results for different types of concrete specimens should not be much different from each other.

The other test method for the evaluation of concrete permeability is, ASTM C1585 which is a test for the measurement of increasing the weight of an oven-dried concrete specimen in contact with water from one surface. Water can penetrate the capillary pores of the concrete specimen from one surface. Then you should measure the weight increased over time till you will achieve a stable weight. This can show you the behavior of concrete in contact with water and other chemical ions. But as most of the time, the concrete is not oven dried in real projects, it is not an accurate test for the evaluation of concrete permeability.

The most useful standard test method for the measurement of concrete permeability is the EN12390-8 which is a very accurate test for this purpose. For this test, water will apply on the surface of concrete specimen with a defined pressure of 5 bars. The applied surface is a circle with a diameter of 75 mm. After 72 hours, we should measure the amount of water penetrated into the concrete specimen. To do it, we should cut the specimens in the direction of the pressure and measure the length of water penetration accurately.

This test is for high-quality low permeable concrete. For concrete with high permeability, we cannot do this test, because the water will penetrate and pass through the specimen length.

7.3   TEST OF CONCRETE IN THE STRUCTURES

Instead of testing concrete in fresh and hardened phases, sometimes there is a need to check the concrete in the structure. There are too many reasons for testing hardened concrete in the structures. Some of them are as below:

- Questionable results for hardened concrete especially in the compressive strength test.
- A very important structure on which we need more confidence.
- Not giving enough samples from concrete when we were pouring the structure.
- Objection about the hardened concrete test results from the contractor.
- Defects in the performance of the structures especially in the concrete production process.
- Change in concrete structure out of the structure sheets and there is a need for checking the structure.
- Need more loads than the designed loads of structure.
   As you can see in the abovementioned, checking the concrete in the structure is a very common request from the contractors or supervisors. So, we discuss the most important methods for checking concrete in the structure. We have two types of methods overall for the concrete checking in the structure:

- Non-destructive test methods: In these test methods, we are not going to destruct any part of the structure. These test methods are very important for the structures with high sensitivity because we cannot destroy any part of a sensitive structure. So, most of the time, the supervisors suggest us non-destructive test methods. On the other hand, the accuracy of non-destructive test methods is not highly trustable. So, we can only use them for checking the structure. You cannot trust their result for the design check or design change of structures.

  In this book, we discuss two non-destructive test methods which are the most common ones in the world. The first one is the ultrasonic test of concrete structures and the second one is the Schmidt hammer test.

- Destructive test methods: The other test method for checking concrete structures is the destructive test methods. As you can find from its name, we should destroy a little part of the structure for testing the concrete. Most of the time, this little deficiency is not very important for the performance of concrete structure. But making decision about these tests is only acceptable for the supervising system of the project. They should check the structure analysis to inform which part of the structure could be acceptable for destruction. Most of the time, for very important and sensitive structures there is no permission to use destructive test methods. For these structures, the quality control of concrete during the implementation is very important. So, in the future, there are no needs for the destructive checking of the structure.

  In this book, we discuss the most important technique for destructive testing of concrete structures which is the core test of concrete. This is a very accurate test method. You can trust the results for any purpose consisting of changing the structure.

One of the most important researches in concrete technology is on the newer non-destructive test methods with high accuracy. But till now, there is no method with high accuracy. So, if you need high accuracy for checking the structure, the only way is the destructive test methods, especially core testing.

7.3.1 ULTRASONIC TEST

One of the most common non-destructive test methods for checking concrete structures is the ultrasonic test. This method is based on the difference in the speed of ultrasonic waves in different structures according to the compaction of that. The speed of ultrasonic waves will be higher in a denser concrete than a weaker one. So, we can make a link between the concrete compressive strength and the ultrasonic wave speed because the more compressive strength will result in a denser microstructure in concrete.

Unfortunately, the only parameter for the difference in the speed of ultrasonic waves is not the compressive strength of concrete, Types of aggregates, especially the coarse ones, aggregates gradation in concrete, maximum size of coarse aggregates, and amount of air inside the concrete, are the other important factors for the ultrasonic wave speed in concrete. So, this is not an accurate test for the evaluation of compressive strength.

The other important parameter that you can check by using the ultrasonic test method is the crack depth and diameter. So, if you see a crack on the surface of the structure, you can check the depth and diameter of the cracks and evaluate the problems of the cracks.

There are too many formulations for the correlation between the ultrasonic wave speed and the compressive strength. But most of them are not useful enough instead of giving you an idea about the compressive strength of concrete. The best way is to calibrate the ultrasonic test instrument with your concrete with the same mix design, aggregate, and other constituent materials. In this way, you can check the structures poured with your concrete and compare them with the results that you have in the laboratory. This is the most accurate way to use the ultrasonic test method. If you changed any part of your concrete constituent materials, you should calibrate the instrument again.

You can see a picture of the ultrasonic test instrument in Figure 7.19. You can use the instrument two poles on two sides of the structural element for example two sides of a column. The ultrasonic wave will start from each pole and go through the concrete to receive to the other pole of the instrument. Then the instrument will calculate and show you the wave speed. Some of the instruments can contact a computer to correlate the wave speed and compressive strength. But for other instruments, you should make the correlation by your manual calculations.

The other problem with using the ultrasonic test method is the effect of steel bars on the speed of the ultrasonic wave. As the microstructure of the steel is denser than any type of concrete, the ultrasonic wave will pass it with more speed than concrete. So, the steel bars between two poles of the ultrasonic test instrument will cause an error in the evaluation of concrete quality. For example, if you are checking a concrete column, you will see this problem by checking of different heights of the

FIGURE 7.19   Multipurpose ultrasonic test instrument. ("Ultrasonic Testing Machine showing readings" by Abhijit Nandi used under a Creative Commons Attribution-Share Alike 3.0 Unported license.)

column because of different amount of the rebars. So, you should try to check the pure concrete in the structure, if it is possible. If not, you should check the structural sheets for the evaluation of the exact amount of rebars in the section and find a suitable correction coefficient for the rebars. Ultrasonic instrument manufacturers give you some tables for the rebar correction coefficient that you can use and you can make a calibration to your instrument by using different amount of steel in the checking specimens and control the effect of the steel bars.

7.3.2 Schmidt Hammer Test

Another non-destructive test method that you can use for checking concrete structures is the Schmidt rebound hammer test (Figures 7.20 and 7.21). This test is based

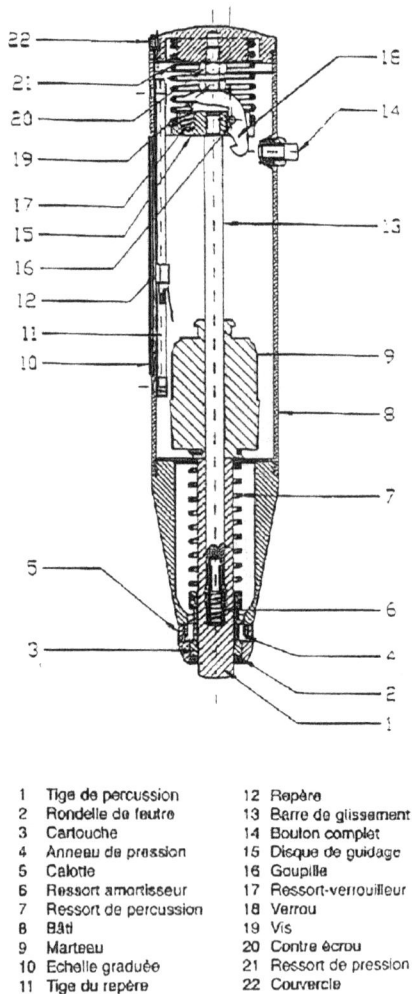

1	Tige de percussion	12	Repère
2	Rondelle de feutre	13	Barre de glissement
3	Cartouche	14	Bouton complet
4	Anneau de pression	15	Disque de guidage
5	Calotte	16	Goupille
6	Ressort amortisseur	17	Ressort-verrouilleur
7	Ressort de percussion	18	Verrou
8	Bâti	19	Vis
9	Marteeu	20	Contre écrou
10	Echelle graduée	21	Ressort de pression
11	Tige du repère	22	Couvercle

FIGURE 7.20   Schmidt test hammer mechanism. ("Concrete test hammer" by Boughattas Omar used under a Creative Commons Attribution-Share Alike 3.0 Unported license.)

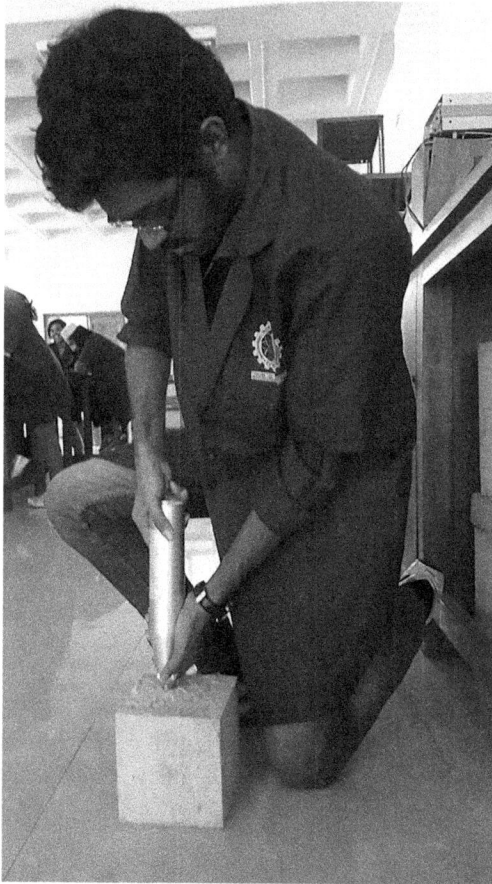

FIGURE 7.21 Testing concrete surface with Schmidt test hammer. ("Schmidt hammer testing" by Arjuncm3 used under a Creative Commons Attribution-Share Alike 3.0 Unported license.)

on the fact that a harder surface will have a higher rebound. For example, if you shoot a ball on a surface of a mattress, you will have a very low rebound. But if you shoot the ball on a concrete wall, you will see a very high rebound. Schmidt hammer was made according to this fact that a harder concrete will have a higher rebound. So, you can calibrate the hammer to give an idea about the compressive strength of concrete.

Like the ultrasonic test method, for Schmidt hammer test also the accuracy is very low. You can calibrate the test hammer with your concrete with the same constituent materials to give more accurate results from testing the structures poured with your concrete. But generally, you cannot get a trustable result.

One of the problems for this test is the limitation of the rebound value for higher strengths. So, the accuracy for high-strength concrete is much lower than normal. For example, it is possible to get the same rebound value for a 55 and 70 MPa concrete with the Schmidt test hammer.

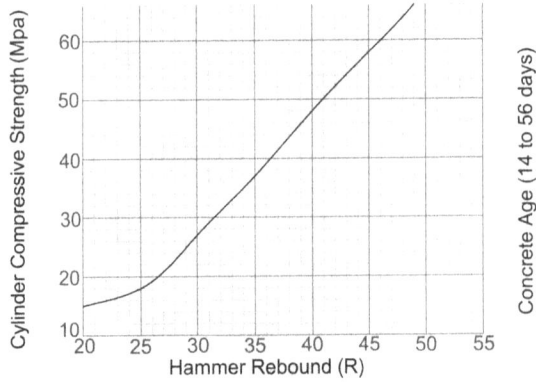

FIGURE 7.22 Calibration curve for a Schmidt test hammer. (Photograph created by the author.)

The other problem of this test method is the effect of coarse size aggregates inside concrete on the rebound value. For example, if you put the Schmidt hammer on the surface of concrete which you have 25–32 mm aggregates exactly under the surface, you will get a rebound value more than real. So, for concrete with the maximum size of coarse aggregates more than 25 mm the Schmidt rebound test will not give you good results. This problem is also important for the steel bars. If the steel bars remained exactly under the surface or with very low distance with the surface, you will see this problem.

The manufacturers of the test hammers will give you a calibration figure that you can see a sample of it in Figure 7.22. But it is strongly recommended that the calibration should check with your concrete constituent materials.

7.3.3 CONCRETE CORE TEST

As mentioned before, non-destructive tests for concrete are not accurate test methods. So, you cannot trust their results. Sometimes, you need a trustable test to check the structure. In this case, you should go for the destructive test methods.

There are different destructive test methods for concrete structures. The most important one is the core test which is a specimen of concrete you poured into the structure.

Core test will be use when you have a problem with the test results of the specimens got at the time of concreting. For example, you get the compressive strength result less than defined. So, you should check the structural elements for the exact amount of compressive strength and then check the structure design with that. Then you will find if there is a need for any repair or empowerment. On the other hand, you should make a design for repairing the defects that happened because of the cores taken from the structure.

For the core test, first you should take a core specimen with a special core test machine that can cut the concrete and bring out the cores from the concrete element (Figures 7.23 and 7.24). Then you should make a good capping for the two surfaces of the specimens as mentioned for the capping of cylinder specimens. You can use

FIGURE 7.23   Huge core taking machine for asphalt and concrete. ("Truck mounted core drill with 14" barrel" by User: Toiyabe used under a Creative Commons Attribution-Share Alike 3.0 Unported license.)

FIGURE 7.24   Core taken from the concrete structure. ("Concrete core and slices" by Yingwu Zhou, Hao Tian, Lili Sui, Feng Xing, Ningxu Han used under a Creative Commons Attribution-Share Alike 4.0 International license.)

the same procedure and materials mentioned before for this purpose. The only difference is the size of specimens. You can take cores with a diameter of 50 to 120 mm from the structure. If you used lower sizes, you should use special equipment for capping or cutting the surfaces and also use the mortar test machine for the compressive strength test. So, it is better to use cores with higher diameter like 80 mm and more if it is possible according to the structural element type and size.

After capping, you can use the cores for the compressive strength test by using a suitable testing machine according to the size of the core specimen.

If you got the core from pure concrete, you can use the compressive strength result after converting it to the standard cylinder compressive strength. But if there are any rebars on each side of the core specimen, you should use correction factors.

Most of the time, you should use core test for the compressive strength of concrete but you can use the core specimens for the evaluation of other concrete properties like density or permeability.

REFERENCES

Aitcin P.C, High Performance Concrete, E&FN SPON, 2004.

American Society for Testing and Materials, Standard Practice for Making and Curing Concrete Test Specimens in the Field, ASTM C31-00.

American Society for Testing and Materials, Standard Practice for Sampling Freshly Mixed Concrete, ASTM C172-99.

American Society for Testing and Materials, Standard Practice for Making and Curing Concrete Test Specimens in the Laboratory, ASTM C192-00.

American Society for Testing and Materials, Standard Practice for Capping Cylindrical Concrete Specimens, ASTM C617-98.

American Society for Testing and Materials, Standard Specification for Concrete Aggregates, ASTM C33-01.

American Society for Testing and Materials, Standard Specification for Ready-Mixed Concrete, ASTM C94-00.

American Society for Testing and Materials, Standard Specification for Portland Cement, ASTM C150-00.

American Society for Testing and Materials, Standard Specification for Flow Table for Use in Test of Hydraulic Cement, ASTM C230-98.

American Society for Testing and Materials, Standard Specification for Air-Entraining Admixture for Concrete, ASTM C260-00.

American Society for Testing and Materials, Standard Specification for Light-weight Aggregates for Structural Concrete, ASTM C330-00.

American Society for Testing and Materials, Standard Specification for Chemical Admixtures for Concrete, ASTM C494-99.

American Society for Testing and Materials, Standard Specification for Coal Fly Ash and Raw or Calcined Natural Pozzolan for Use as a Mineral Admixture in Concrete, ASTM C618-00.

American Society for Testing and Materials, Standard Specification for Fiber Reinforced Concrete and Shotcrete, ASTM C1116-00.

American Society for Testing and Materials, Standard Specification for Use of Silica Fume as a Mineral Admixture in Hydraulic Cement Concrete, Mortar and Grout, ASTM C1240-00.

American Society for Testing and Materials, Standard Test Method for Compressive strength of Cylindrical Concrete Specimens, ASTM C39-01.

American Society for Testing and Materials, Standard Test Method for Organic Impurities in Fine Aggregates for Concrete, ASTM C40-99.

American Society for Testing and Materials, Standard Test Method for Surface Moisture in Fine Aggregates, ASTM C70-94.

American Society for Testing and Materials, Standard Test Method for Flexural strength of Concrete, ASTM C78-00.

American Society for Testing and Materials, Standard Test Method for Compressive Strength of Hydraulic Cement Mortars, ASTM C109-99.

American Society for Testing and Materials, Standard Test Method for Chemical Analysis of Hydraulic Cement, ASTM C114-00.

American Society for Testing and Materials, Standard Test Method for Materials Finer than 75 μm in Aggregates by Washing, ASTM C117-95.

American Society for Testing and Materials, Standard Test Method for Specific Gravity and Absorption of Coarse Aggregates, ASTM C127-88.

American Society for Testing and Materials, Standard Test Method for Specific Gravity and Absorption of Fine Aggregates, ASTM C128-97.

American Society for Testing and Materials, Standard Test Method for Sieve Analysis of Fine and Coarse Aggregates, ASTM C136-01.

American Society for Testing and Materials, Standard Test Method for Slump of Hydraulic Cement Concrete, ASTM C143-00.

American Society for Testing and Materials, Standard Test Method for Autoclave Expansion of Portland Cement, ASTM C151-00.

American Society for Testing and Materials, Standard Test Method for Air Content of Freshly Mixed Concrete by the Volumetric Method, ASTM C173-01.

American Society for Testing and Materials, Standard Test Method for Heat of Hydration of Hydraulic Cement, ASTM C186-98.

American Society for Testing and Materials, Standard Test Method for Density of Hydraulic Cement, ASTM C188-95.

American Society for Testing and Materials, Standard Test Method for Time of Setting of Hydraulic Cement by Vicat Needle, ASTM C191-99.

American Society for Testing and Materials, Standard Test Method for Potential Alkali Reactivity of Cement-Aggregate Combination, ASTM C227-97.

American Society for Testing and Materials, Standard Test Method for Air Content of Freshly Mixed Concrete by the Pressure Method, ASTM C231-97.

American Society for Testing and Materials, Standard Test Method for Potential Alkali Silica Reactivity of Aggregates (Chemical Method), ASTM C289-94.

American Society for Testing and Materials, Standard Test Method for Static Modules of Elasticity and Poisson's Ratio of Concrete in Compression, ASTM C469-94.

American Society for Testing and Materials, Standard Test Method for Density, Absorption and Voids in Hardened Concrete, ASTM C642-97.

American Society for Testing and Materials, Standard Test Method for Length Change of Concrete Due to Alkali-Carbonate Rock Reaction, ASTM C1105-95.

American Society for Testing and Materials, Standard Test Method for Potential Alkali Reactivity of Aggregates (Mortar Bar Method), ASTM C1260-94.

American Society for Testing and Materials, Standard Test Method for Flow of Hydraulic Cement Mortar, ASTM C1437-99.

American Society for Testing and Materials, Standard Test Method for Compressive Strength of Hydraulic Cement Mortars, ASTM C109-99.

Arjuncm3, "Schmidt hammer testing" Retrieved from: https://commons.wikimedia.org/wiki/File:Schmidt_hammer_testing.jpg.

Bicanski, "Concrete cutting" Retrieved from: https://pixnio.com/media/concrete-cutting-saw-cutter-industrial#.

Cjp24, "Capillary rheometer" Retrieved from: https://commons.wikimedia.org/wiki/File:Capillary_rheometer.jpg.

Cleynen, Olivier, "rheometer" Retrieved from: https://commons.wikimedia.org/wiki/File:Rheometer.jpg,

European Standard Organization, Admixtures for Concrete Mortar and Grout, EN934 Series.

European Standard Organization, Admixtures for Concrete, Mortar and Grout Test Methods, EN480 Series.

European Standard Organization, Cement Composition, Specifications and Conformity Criteria for Common Cements, EN197-1: 2000.

European Standard Organization, Concrete-Part 1: Specification, Performance, Production and Conformity, EN206-1, 2000.

European Standard Organization, Methods of Testing Cement, EN196 Series.

European Standard Organization, Testing Fresh Concrete, EN12450 Series.

European Standard Organization, Testing Hardened Concrete, EN12390 Series.

European Standard Organization, Tests for General Properties of Aggregates, EN932 Series.

European Standard Organization, Tests for Geometrical Properties of Aggregates, EN933 Series.

Iranian Institute for Research on Construction Industry, 9th topic of National Rules for Construction, "Concrete Structures", 2009.

Iranian Institute for Research on Construction Industry, National Concrete Mix Design Method, 2015.

Iranian National Management and Programming Organization, National Handbook of Concrete Structures, 2005.

Iranian Standard Organization, Concrete Admixtures, Specification, ISIRI2930, 2011.

Iranian Standard Organization, Concrete Specification of Constituent Materials, Production and Compliance of Concrete, ISIRI2284-2, 2009.

Iranian Standard Organization, Standard Specification for Ready Mixed Concrete, ISIRI6044, 2015.

Janamian Kambiz, Aguiar Jose, A Comprehensive Method for Concrete Mix Design, Materials Research Forum LLC, 2020.

Knipptang, "Flow test after." Retrieved from: https://commons.wikimedia.org/wiki/File:Flow_Test_after.jpg.

Knipptang, "Flow test before." Retrieved from: https://commons.wikimedia.org/wiki/File:Flow_Test_before.jpg.

Lamond F.Joseph, Pielert H.James, Significance of Tests and Properties of Concrete and Concrete Making Materials, ASTM International, 2006.

Mahmood Zadeh Amir, Iranpoor Jafar, Concrete Technology and Test (Farsi), Golhaye Mohammadi, 2007.

Mostofinejad Davood, Concrete Technology and Mix Design (Farsi), Arkane Danesh, 2011.

MTA Construction & Development Mega Projects, "Conducting a slump test on the concrete to ensure the workability of the freshly poured concrete before using it in the future LIRR passenger concourse." Retrieved from: https://www.flickr.com/photos/mtacc-esa/50887549448.

Nandi, Abhijit, "Ultrasonic Testing Machine showing readings" Retrieved from: https://commons.wikimedia.org/wiki/File:Ultrasonic_Testing_Machine_showing_readings.jpg.

Nawy G.Edward, Concrete Construction Engineering Handbook, CRC Press, 2008.

Newman John, Choo Ban Seng, Advanced Concrete Technology, Concrete Properties, Elsevier, 2003.

Omar, Boughattas, "Concrete test hammer" Retrieved from: https://commons.wikimedia.org/wiki/File:Dessin_d%27ensemble_d%27un_scl%C3%A9rom%C3%A8tre.jpg.

Popovics Sandor, Concrete Materials, Properties Specification and Testing, NOYES Publications, 1992.

Ramachandran, Paroli, Beaudion, Delgado, Handbook of Thermal Analysis of Construction Materials, NOYES Publications, 2002.

Ramezanianpoor Aliakbar, Arabi Negin, Cement and Concrete Test Methods (Farsi), Negarande Danesh, 2011.

Richardson M, Fundamentals of Durable Reinforced Concrete, Spon Press, 2002.

User: Toiyabe, "Truck mounted core drill with 14" barrel" Retrieved from: https://commons.wikimedia.org/wiki/File:Core_drill_06_2005.jpg.

Zandi Yousof, Advanced Concrete Technology (Farsi), Forouzesh Pub, 2009.

Zhou, Yingwu, Hao Tian, Lili Sui, Feng Xing, Ningxu Han, "Concrete core and slices" Retrieved from: https://commons.wikimedia.org/wiki/File:Slice-machine-and-the-samples.jpg.

8 Durability of Concrete Structures

Durability is defined as the ability of concrete structures to resist the corrosive environment. The important factor for this ability is the quality of concrete because a high-quality concrete is an impermeable one that can control the infiltration of corrosive materials and water into concrete. So, the most important way to make a durable structure is by making and using an impermeable concrete.

In this chapter, first we discuss the different corrosive environments which destroy the concrete or steel bars inside that. Then we will talk about the minimum specifications that we should mention for concrete to control corrosion against different environmental conditions. So, we will have a minimum concrete quality for each environment that we should consider to ensure the durability of the structure.

For this purpose, we should use a defined standard. We have many different standards for the durability design of concrete structures in different parts of the world. Each of these standards is developed according to the local environmental conditions. For example, the environmental conditions in the seaside of the Persian Gulf and Oman sea are much harsher than the environmental conditions in the center of Europe. So, you should consider higher specifications for the concrete that you are going to use on the seaside of the Persian Gulf. So, some of the minimum specifications in the Persian Gulf countries should be much higher than in the European countries (Figure 8.1).

In this book, we are going to use EN206-1 standard for the minimum specifications and divisions of structures against the corrosive environment. But you can use any different standard or code for this purpose. The important point is that the specifications defined here are minimum amounts for each kind of structure against different environmental conditions. If you would like to make a durable structure with a high safety factor, it is recommended to make your concrete with higher specifications than the minimum ones.

8.1 PARAMETERS AFFECTING CONCRETE DURABILITY

To make a durable concrete structure, you should consider below specifications and parameters:

- Make an impermeable concrete: As mentioned before, corrosive ions and water cannot leach inside an impermeable concrete. So, the corrosion will control.
- Making a concrete with low water-to-binder ratio: To make an impermeable concrete, you should try to make a concrete with low water-to-binder ratio.

DOI: 10.1201/9781003384243-8

FIGURE 8.1 Persian Gulf and sea of Oman, one of the harshest parts of the world for concrete structures. ("The Gulf" by European Space Agency.)

- Using supplementary cementitious materials: To decrease the water-to-cement ratio and according to the barriers for the maximum amount of pure Portland cement, you should use supplementary cementitious materials. Using these materials will help to make more C-H-S in concrete and it can help you to make an impermeable concrete. On the other hand, supplementary cementitious materials can help us to block the pores inside concrete because they are fine powders.
- Using super-plasticizers: To decrease the water-to-binder ratio, you should use a strong super-plasticizer to decrease the amount of water you are going to use for concrete production. On the other hand, using the super-plasticizers will help you to make a workable concrete. So, you can pour the structural elements with maximum compaction.
- Using suitable type of cement: For different conditions, you should choose suitable type of cement to control the corrosion. For example, for the environment with high amount of sulfate ion, you should use type V Portland cement and for the environment with chloride and sulfate ions together, you should use type II Portland cement.

- Use more concrete cover for the steel bars: When you use more cover for the steel bars, the corrosive materials, especially the chloride ion and carbonation will arrive at the steel bars with delay. So, you can delay the start of corrosion.
- Make a high-performance concrete: If you would like to consider all of the above, you should make a high-performance concrete with high compressive strength, minimum permeability, and high workability. The strength of concrete itself don't have any effect on the durability of structures, but if you use a higher compressive strength you will have a lower water-to-binder ratio and you will have lower permeability. So, the compressive strength class of concrete will be important for the durability of the structure.

To design a concrete structure according to the durability considerations, first, we should make a category for the structures according to their importance. For this purpose, you can use Table 8.1 derived from the European standard.

8.2 DURABILITY AGAINST CARBONATION

One of the corrosive materials for the steel bars inside concrete is carbon dioxide. In big cities with high amounts of air pollution, carbon dioxide which is the most important pollutant of the air can react with moisture and make carbonic acid. This carbonic acid can leach into the concrete element, if the permeability of concrete lets it go through. The concrete environment itself is alkaline. This alkaline environment can help to protect the steel bars against corrosion. When the carbonic acid goes inside the concrete, it will decrease the pH value of the concrete environment. So, if it can reach the steel bars, the corrosion will start. After spending time, the corroded rebars will swell and will cause some signs of the rebars on the surface of the concrete. After more time, it can cause scalding on the surface of the rebars. You can see also Chapter 1 for more information about carbonation.

TABLE 8.1
Categories of Service Life for Structures in European Standard

Category	Years of Service Life	Examples
1	10	Temporary structures
2	10–25	Replaceable parts of structures like supports
3	15–30	Structures for agriculture and similar uses
4	50	Buildings and other common structures like houses, schools and hospitals
5	100	Memorial buildings, bridges, and other civil engineering structures

---

As mentioned above, carbonation could be very dangerous for concrete structures in big cities, especially the ones that are in the fresh air like bridges. So, you should make the concrete with defined properties to control the effects of carbonation.

For this purpose, first, you should use Table 8.2 for the evaluation of the environment and the harshness of it for the structures that we would like to build. Then you should use Tables 8.3 and 8.4 as the suggestions for concrete minimum specification to protect the structure against carbonation.

It is very important to mention that the specifications that you can see in the tables above are the minimum quality defined for a concrete to resist against carbonation. So, it is strongly recommended to use a higher quality concrete to get the best results. On the other hand, using supplementary cementitious materials can help you to produce a higher quality concrete.

TABLE 8.2
Exposure Classes for Corrosion Induced by Carbonation

Exposure Class	Definitions	Examples
XC_1	Dry or permanently humid	Concrete inside the building with low air humidity, concrete permanently under water
XC_2	Humid, rarely dry	Concrete surface with long time of contacting with water
XC_3	Moderately humid	Concrete inside the buildings with moderate or high humidity, exterior concrete protected from rain
XC_4	Cyclically humid and dry	Concrete surfaces contacting with water

TABLE 8.3
Minimum Specification Limits for Concrete to Protect From Corrosion Against Carbonation for the Service Life of 50 Years

Exposure Class	XC_1	XC_2	XC_3	XC_4
Minimum rebar cover (mm)	25	35	35	40
Maximum w/c ratio	0.65	0.65	0.55	0.55
Minimum cement content (kg/m^3)	260	260	300	300
Minimum compressive strength class	C25	C25	C30	C30

TABLE 8.4
Minimum Specification Limits for Concrete to Protect From Corrosion Against Carbonation for the Service Life of 100 Years

Exposure Class	XC_1	XC_2	XC_3	XC_4
Minimum rebar cover (mm)	35	45	45	50
Maximum w/c ratio	0.6	0.6	0.5	0.5
Minimum cement content (kg/m^3)	280	280	320	320
Minimum compressive strength class	C35	C35	C40	C40

8.3 DURABILITY AGAINST CHLORIDE ION

Chloride ion is very dangerous for reinforced concrete structures because it can leach to the structure and when it arrived at the steel bars, the corrosion will start rapidly. The corrosion is because of the electrochemical reaction between the chloride ion and the steel bars and it will cause a very bad and rapid corrosion of the rebars inside the concrete structure. For more information about the chloride attack, you can see Chapter 1.

The chloride ion comes from two important sources. One is from the sea water. The chloride ion from the sea water could be very dangerous because of the high concentration, especially for the structures exposed to the sea waves or tide areas. But this is not the only way that sea water could destroy the structures. Chloride ion from the sea water can transport by air to places much far from the sea side. The water contains chloride ion that will evaporate from the sea and transfer by wind and air turbulence to places far from the sea side. Then it can leach to the concrete structures and start the corrosion of the steel bars. You can see corrosion of the steel bars at places more than 100 km far from the sea side.

The other important source of chloride ions is the deicing salts that we may use in winter to protect the street surface from icing. These salts contain chloride ions. It can transfer to concrete structures like parking columns or bridge slabs by the tires of vehicles. So, this source of chloride ion could be very dangerous especially in places with very hard and cold winters or high amounts of snowing.

To protect the concrete structures from the chloride ion, first you should use Tables 8.5 and 8.6 for the category of structures exposed to the chloride ion. Table 8.5 is for the structures exposed to the sea water and Table 8.6 is for the structures exposed to the other sources of the chloride ion specially deicing salts. Then you should use Tables 8.7 and 8.8 for the minimum concrete properties for the service life of 50 and 100 years.

As you can see in the tables above, the chloride ion is a very destructive ion that can be much corrosive than the other chemicals. So, the specifications and quality of concrete exposed to the chloride ion should be much higher than the other environments. On the other hand, if you have higher concentrations of the chloride ion and/or high temperature during most of the year, you should make a higher quality and it is recommended to use a concrete with the specifications more than the tables above. The most famous example of a place with a higher concentration of chloride and sulfate and also high temperature most of the year is the countries on the sea side of the Persian Gulf and the sea of Oman as mentioned before.

TABLE 8.5
Exposure Classes for Corrosion Induced by the Chloride Ion of Sea Water

Exposure Class	Definition	Examples
XS_1	Air transported salts but not direct contact with sea water	Structures near the coastal areas
XS_2	Permanently under water	Parts of marine structures
XS_3	Tide and splash areas	Parts of marine structures

TABLE 8.6

Exposure Classes for Corrosion Induced by the Chloride Ion Instead of the Sea Water

Exposure Class	Definition	Examples
XD_1	Moderately humid	Concrete surfaces exposed to the chloride transferred by air
XD_2	Humid, rarely dry	Swimming pools and concrete exposed to industrial water contain chloride
XD_3	Cyclically humid and dry	Parts of bridges or slabs of car parking exposed to splash of water contain chloride

TABLE 8.7

Minimum Specification Limits for Concrete to Protect From Corrosion Against Chloride for the Service Life of 50 Years

Exposure Class	XS_1/XD_1	XS_2/XD_2	XS_3/XD_3
Minimum rebar cover (mm)	45	50	55
Maximum w/c ratio	0.45	0.45	0.4
Minimum cement content (kg/m^3)	360	360	380
Minimum compressive strength class	C40	C40	C50

TABLE 8.8

Minimum Specification Limits for Concrete to Protect From Corrosion Against Chloride for the Service Life of 100 Years

Exposure Class	XS_1/XD_1	XS_2/XD_2	XS_3/XD_3
Minimum rebar cover (mm)	55	60	65
Maximum w/c ratio	0.4	0.4	0.35
Minimum cement content (kg/m^3)	380	380	400
Minimum compressive strength class	C50	C50	C60

8.4 DURABILITY AGAINST SULFATE ION

Sulfate ion is one of the other corrosive chemicals which can destroy the concrete itself. It will not attack the steel bars inside the concrete. It will attack the C_3A of cement and cause a very bad corrosion from the surface of concrete exposed to the sulfate ion.

The concrete elements could be exposed to sulfate ion from the soil or water. Sometimes, the sulfate ion could be in water which is in contact with the concrete elements, like sea water. The sea water contains a high amount of sulfate. So, good protection is needed in the case of sulfate ion for the structures near the sea.

Some of the other water resources instead of sea water could contain sulfate ions with different concentrations. On the other hand, sulfate ion could be one of the chemicals in the soil. So, it can attack some structural elements like foundations that should be in contact with the soil. For more information about the sulfate attack, you can see Chapter 1.

Sulfate ion , the aggressive CO_2 dissolved in water, NH_{4+} and Mg_{2+} which can cause high acidity in the environment categorized as the chemical attack. You can see Table 8.9 for the different categories of the structures exposed to the chemical attack and then you can see Tables 8.10 and 8.11 for the minimum properties for concrete to protect the structure against the chemical attack, especially the sulfate ion.

The important parameter that you can see in Tables 8.10 and 8.11 is the type of cement according to the ASTM standard. If you have only sulfate ion in the environment you should use type V cement. But if you have sulfate and chloride ion together in the environment, the best choice is type II cement. On the other hand, the best cement for the environment with only chloride ion is type I cement.

TABLE 8.9

Exposure Classes for the Corrosion Induced by the Chemicals Specially Sulfate Ion

Exposure Class	XA_1	XA_2	XA_3
Exposed to water			
SO4 (mg/L)	>200e<600	>600e<3000	>3000e<6000
pH	>5.5e<6.5	>4.5e<5.5	>4e<4.5
Aggressive CO_2 (mg/L)	>15e<40	>40e<100	>100 and more
NH4 (mg/L)	>15e<30	>30e<60	>60e<100
Mg (mg/L)	>300e<1000	>1000e<3000	>3000 and more
Exposed to soil			
SO_4 (mg/kg)	>2000e<3000	>3000e<12000	>12000e<24000
Acidity (mL/kg)	>200	–	–

TABLE 8.10

Minimum Specification Limits for Concrete to Protect From Corrosion Against Chemical Attack, especially the Sulfate Ion for Service Life of 50 Years

Exposure Class	XA_1	XA_2	XA_3
Type of cement	Type II	Type II	Type V
Maximum w/c ratio	0.45	0.45	0.45
Minimum cement content (kg/m^3)	340	360	380
Minimum compressive strength class	C35	C40	C40

TABLE 8.11

Minimum Specification Limits for Concrete to Protect From Corrosion Against Chemical Attack, especially the Sulfate Ion for Service Life of 100 Years

Exposure Class	XA_1	XA_2	XA_3
Type of cement	Type II	Type II	Type V
Maximum w/c ratio	0.4	0.4	0.4
Minimum cement content (kg/m^3)	360	380	400
Minimum compressive strength class	C45	C50	C50

8.5 DURABILITY AGAINST FREEZE THAW

The condition that you will have the problem of freeze thaw is the temperature more than 0°C at day time and temperature below 0°C at night time. At this condition, if you have enough moisture in the environment, expansion of water during icing process will cause stress in the concrete element and as this hard process will cyclically happen during days and nights, it can destroy the concrete element. So, controlling this cycle is very important.

If you have cold winter in a region, you will have these cycles of freezing and thawing. So, you will have this condition at a minimum 50% of the places in the world. You can see more information about the freeze-thaw cycles in Chapter 1. You learned that it is very important to use a concrete with entrapped air to control the freeze-thaw cycles effect on concrete elements.

To protect the concrete structures from the effects of freeze-thaw cycles, first you should use Table 8.12 for the category of structure exposed to the freeze-thaw cycles. Then you should use Tables 8.13 and 8.14 for the minimum concrete properties for the service life of 50 and 100 years.

TABLE 8.12

Exposure Classes for Concrete Structures Exposed to Freeze Thaw Cycles

Exposure Class	Definition	Example
XF_1	Moderately saturated with water without deicing salts	Vertical concrete surfaces exposed to rain and ice
XF_2	Moderately saturated with water with deicing salts	Vertical concrete surfaces in road structures exposed to ice and deicing salts transported with air
XF_3	Highly saturated with water without deicing salts	Horizontal concrete surfaces exposed to rain and ice
XF_4	Highly saturated with deicing salts	Roads and bridges exposed to rain and ice with deicing salts or splash of deicing salts

TABLE 8.13

Minimum Specification Limits for Concrete to Protect From Freeze Thaw Cycles for Service Life of 50 Years

Exposure Class	XF$_1$	XF$_2$	XF$_3$	XF$_4$
Maximum w/c ratio	0.55	0.55	0.5	0.45
Minimum cement content (kg/m³)	300	300	320	340
Minimum compressive strength class	C30	C30	C30	C30
Minimum air content (%)	–	4%	4%	4%

TABLE 8.14

Minimum Specification Limits for Concrete to Protect From Freeze Thaw Cycles for Service Life of 100 Years

Exposure Class	XF$_1$	XF$_2$	XF$_3$	XF$_4$
Maximum w/c ratio	0.5	0.5	0.45	0.4
Minimum cement content (kg/m³)	320	320	340	360
Minimum compressive strength class	C40	C40	C40	C40
Minimum air content (%)	–	4%	4%	4%

As you can see in the tables above, deicing salts in the environment will cause more destruction of the structure because the synergy between the scalding of concrete against the deicing salts and the cracks in the structure of concrete element will cause faster corrosion of the concrete element.

We have other codes and standards with different recommendations for concrete against freeze-thaw cycles. But most of the time, they focused on the amount of air inside the concrete element. For example, the ACI recommended using air-entraining admixtures to get at least 6% of air inside a concrete element exposed to the condition of high saturation with deicing salts. So, you should use the codes and standards in your area to get the best results for each purpose because the local standards are designed according to the local environment and constituent materials.

8.6 EXAMPLE FOR THE DURABILITY OF A CONCRETE STRUCTURE

In this part, we are going to check the durability characteristics of two projects in two different parts of the world and use the recommendations of this chapter to design a suitable concrete.

8.7 FIRST PROJECT: BRIDGE DECK IN NORTH EUROPE

The first project is a concrete bridge in north of the Europe on a highway between two cities. The project is in one of the Scandinavian countries (Figure 8.2).

FIGURE 8.2 North European countries. ("Map of Northern Europe" by Nathan Hughes Hamilton.)

For this project, we should check the different possibilities for the corrosion environment. On the other hand, as the project is a bridge, we can find from Table 8.1 that the durability of this structure should be 100 years.

- Carbonation: As this bridge is in northern Europe and outside the big cities, there is no high amount of carbon dioxide in the air. So, we don't have corrosion with carbonation.
- Chloride: We don't have the chloride of sea water here. But we have the chloride of deicing salts in cold winters of Scandinavian countries with a high amount of snowing during the year. We should check Table 8.6 and see that this project is a bridge that will be cyclically humid and dry. So, it is in the category of XD_3.

Now, we should check Table 8.8 for the durability of 100 years against the chloride ion and we can see below specifications for the XD_3 category:

- Minimum rebar cover: 65 mm
- Maximum w/c ratio: 0.35
- Minimum cement content: 400 kg/m³
- Minimum compressive strength class: C60
- Sulfate: As we are talking about the deck of bridge, we know that it will not be in contact with the soil. On the other hand, we don't have airborne sulfate of sea water or other chemicals. So, the structure will not be in the danger of corrosion by chemicals, especially sulfate ion.
- Freeze thaw: This is another important environmental condition for this bridge because the weather in northern Europe is very cold at most months of a year and we have the condition of freeze thaw in this part of the world. We should check Table 8.12 for the category of structure. As this bridge deck will be highly saturated with water according to the high amount of rain and snow in this part of the world and we have the deicing salts in winter, this will be in the category of XF_4.

Then we should check Table 8.14 for the minimum specifications of concrete for the durability of 100 years as below:

- Minimum w/c ratio: 0.4
- Minimum cement content: 360 kg/m³
- Minimum compressive strength class: C40
- Minimum air content: 4%

Now for the final specifications of concrete for this structure, you can see Table 8.15.

8.8   SECOND PROJECT: COMMERCIAL BUILDING IN A BIG CITY NEAR THE SOUTH CHINA SEA

The second project is a commercial building in an offshore big city near the south China sea (Figure 8.3).

TABLE 8.15
Concrete Specifications for a Bridge Deck in Northern Europe

Concrete Specification	Durability Against Chloride Ion	Durability Against Freeze Thaw	Final Suggestions
Minimum rebar cover (mm)	65	–	65
Maximum w/c ratio	0.35	0.4	0.35
Minimum cement content (kg/m³)	400	360	400
Minimum compressive strength class	C60	C40	C60
Minimum air content (%)	–	4	4

FIGURE 8.3   Countries near the South China sea. ("China Manchukuo Map".)

In this part of the world, we have a hot and humid weather condition in most months of the year. From Table 8.1, we understand that for this project as a commercial building we should consider minimum durability of 50 years.

- Carbonation: As the project is in a big city, we will have the air pollution and carbon dioxide in the air. So, the danger of carbonation could be high. From Table 8.2 you can find that the project is in the category of XC_3 because it is a building in humid weather condition. Then from Table 8.3, you can see the concrete specifications for the durability of 50 years as below:
- Minimum rebar cover: 35 mm
- Maximum w/c ratio: 0.55
- Minimum cement content: 300 kg/m^3
- Minimum compressive strength class: C30
- Chloride: In the cities near the coast of the south china sea we will have a high amount of airborne sea water chloride ions. So. From Table 8.5 you can see that the category for this structure is XS_1 and from Table 8.7 the concrete specifications for the durability of 50 years is:
- Minimum rebar cover: 45 mm
- Maximum w/c ratio: 0.45
- Minimum cement content: 360 kg/m^3
- Minimum compressive strength class: C40
- Sulfate: Like the chloride ion, we will have a high amount of airborne sulfate ion from the sea water. But we don't have enough information about the

TABLE 8.16

Concrete Specifications for a Commercial Building Near the South China Sea

Concrete Specification	Durability Against Carbonation	Durability Against Chloride Ion	Durability Against Sulfate Ion	Final Suggestions
Minimum rebar cover (mm)	35	45	–	45
Maximum w/c ratio	0.55	0.45	0.45	0.45
Minimum cement content (kg/m³)	300	360	380	380
Minimum compressive strength class	C30	C40	C40	C40
Type of cement	–	–	Type II	Type II

amount of ion contacted with the structure. So, we should assume that the category is XA_3 as the hardest condition. From Table 8.10 you can see the concrete specifications as below:

- Portland cement type: According to Table 8.10, we should use type V. But as we have chloride and sulfate ions together, we should use type II Portland cement.
- Maximum w/c ratio: 0.45
- Minimum cement content: $380 \, kg/m^3$
- Minimum compressive strength class: C40
- Freeze thaw: As the project is in hot and humid weather conditions, we don't have the effects of freeze thaw on concrete.

Now for the final specifications of concrete, you can see Table 8.16.

REFERENCES

American Society for Testing and Materials, Standard Practice for Making and Curing Concrete Test Specimens in the Field, ASTM C31-00.

American Society for Testing and Materials, Standard Practice for Sampling Freshly Mixed Concrete, ASTM C172-99.

American Society for Testing and Materials, Standard Practice for Making and Curing Concrete Test Specimens in the Laboratory, ASTM C192-00.

American Society for Testing and Materials, Standard Specification for Ready-Mixed Concrete, ASTM C94-00.

American Society for Testing and Materials, Standard Specification for Air-Entraining.

American Society for Testing and Materials, Standard Specification for Chemical Admixtures for Concrete, ASTM C494-99.

American Society for Testing and Materials, Standard Specification for Coal fly Ash and Raw or Calcined Natural Pozzolan for Use as a Mineral Admixture in Concrete, ASTM C618-00.

American Society for Testing and Materials, Standard Specification for Use of Silica Fume as a Mineral Admixture in Hydraulic Cement Concrete, Mortar and Grout, ASTM C1240-00.

American Society for Testing and Materials, Standard Test Method for Sieve Analysis of Fine and Coarse Aggregates, ASTM C136-01.

American Society for Testing and Materials, Standard Test Method for Air Content of Freshly Mixed Concrete by the Volumetric Method, ASTM C173-01.

American Society for Testing and Materials, Standard Test Method for Density, Absorption and Voids in Hardened Concrete, ASTM C642-97.

Bertolini L, Elsener B, Pedeferri P, Polder R, *Corrosion of Steel in Concrete, Prevention, Diagnosis, Repair*, WILEY-VCH, 2004.

Derivative work: Emok, China-Manchukuo-map.png: ErnstA, User:Kingruedi, East Asia area blank CJK.svg: Eurodollers, "China Manchukuo Map." Retrieved from: https://commons.wikimedia.org/wiki/File:China-Manchukuo-map.svg.

European Space Agency, "The Gulf" Retrieved from: https://commons.wikimedia.org/wiki/File:The_Gulf_ESA412129.jpg.

European Standard Organization, Admixtures for Concrete Mortar and Grout, EN934 Series.

European Standard Organization, Admixtures for Concrete, Mortar and Grout Test Methods, EN480 Series.

European Standard Organization, Concrete-Part 1: Specification, Performance, Production and Conformity, EN206-1, 2000.

European Standard Organization, Testing Fresh Concrete, EN12450 Series.

European Standard Organization, Testing Hardened Concrete, EN12390 Series.

Gjorv E.Odd, *Durability Design of Concrete Structures in Severe Environments*, Taylor & Francis, 2009.

Gjorv Odd E, *Durability Design of Concrete Structures*, Taylor & Francis, 2009.

Hamilton, Nathan Hughes, "Map of Northern Europe." Retrieved from: https://www.flickr.com/photos/nat507/19135266166.

Hauschild Michael, Rosenbaum Ralph K, Olsen Sting Irving, *Life Cycle Assessment, Theory and Practice*, Springer, 2018.

Heinrichs Harald, Martens Pim, Michelsen Gerd, Wiek Arnim, *Sustainability Science, An Introduction*, Springer, 2016.

Iranian Institute for Research on Construction Industry, 9th topic of National Rules for Construction, "Concrete Structures", 2009.

Iranian Institute for Research on Construction Industry, National Concrete Mix Design Method, 2015.

Iranian National Management and Programming Organization, National Handbook of Concrete Structures, 2005.

Iranian Standard Organization, Concrete Admixtures, Specification, ISIRI2930, 2011.

Iranian Standard Organization, Concrete Specification of Constituent Materials, Production and Compliance of Concrete, ISIRI2284-2, 2009.

Iranian Standard Organization, Standard Specification for Ready Mixed Concrete, ISIRI6044, 2015.

Janamian Kambiz, Aguiar Jose, A Comprehensive Method for Concrete Mix Design, Materials Research Forum LLC, 2020.

Lamond F. Joseph, Pielert H. James, *Significance of Tests and Properties of Concrete and Concrete Making Materials*, ASTM International, 2006.

Mahmood Zadeh Amir, Iranpoor Jafar, Concrete Technology and Test (Farsi), Golhaye Mohammadi, 2007.

Mostofinejad Davood, Concrete Technology and Mix Design (Farsi), Arkane Danesh, 2011.

Newman John, Choo Ban Seng, *Advanced Concrete Technology, Concrete Properties*, Elsevier, 2003.

Ramachandran V.S, Beaudion James, *Handbook of Analytical Techniques in Concrete Science and Technology, Principles, Techniques and Applications*, William Andrew Publishing, 2001.

Ramachandran, Paroli, Beaudion, Delgado, *Handbook of Thermal Analysis of Construction Materials*, NOYES Publications, 2002.

Ramezanianpoor Aliakbar, Arabi Negin, Cement and Concrete Test Methods (Farsi), Negarande Danesh, 2011.

Richardson M, *Fundamentals of Durable Reinforced Concrete*, SPON Press, 2004.

Richardson M, *Fundamentals of Durable Reinforced Concrete*, Spon Press, 2002.

Safaye Nikoo Hamed, Introduction to Concrete Technology (Farsi), Heram Pub, 2008.

Shekarchizade Mohammad, Liber Nicolas Ali, Dehghan Solmaz, Poorzarrabi Ali, Concrete Admixtures Technology and Usages (Farsi), Elm & Adab, 2012.

Zandi Yousof, Advanced Concrete Technology (Farsi), Forouzesh Pub, 2009.

9 Concrete Mix Design

The first step for designing a good mix for concrete is packing complete information about the constituent materials that were mentioned in the chapters before. Sometimes, we can choose any kind of constituent materials that should be suitable for our mix, but other times there is no choice. For example, if the project will be in a region in which you cannot find any good aggregates nearby, you should use anything that is accessible. Because the transportation cost is very high for the aggregates of concrete. It could be the same for cement and other binders. But for chemical admixtures, as we should use a little amount of them in concrete mix design, the transportation cost is not much important.

After packing enough information about the constituent materials, we should define the exact properties of concrete that we would like to design. The properties should define for both fresh and hardened concrete as mentioned in the before chapters of this book.

Finally, you should design a concrete with defined properties and defined constituent materials that we are going to talk about the process in this chapter of the book. After designing the concrete, you should make trials in the laboratory to check the properties of fresh and hardened concrete. After that you should implement the concrete mix design in the batching plant to make industrial concrete and check the mix design again.

The important point about the method that you are going to use for concrete mix design is that it is not important to mention which method you are using for designing the concrete. When you design a structure, it is very important to mention the method that you are using because you cannot test the structure with the worse conditions that is possible for the structure to collapse. For example, you cannot test the structure with the earthquake load by mixing different types of live and dead loads. So, you should mention the method and other engineers should check your structure design to control the codes and methods that you mentioned for the design of the structure. But for concrete mix design, it is possible for you to test the mix design accurately. You can test the properties of concrete in both fresh and hardened phase and see if it is compatible with the defined properties. So, your method is not too much important. But you should know that we are going to use the German method for concrete mix design with some modifications.

We will explain the mix design method step by step in this chapter. After that, we will design three different types of concrete with this method. But, good concrete mix design needs high experience for the engineer. It is necessary for the designer to have enough experience in working with concrete in the projects or ready mixed concrete plants. So, the theory of concrete mix design method is not enough for you to design a good and trustable concrete mix. Although you can check the mix design for the properties of fresh and hardened concrete, it will take time if you would like to check too many mix designs to achieve the best results. So, with enough experience, you can design the concrete with maximum trust. Most of the time, you can use the design in the projects with some little adjustments.

DOI: 10.1201/9781003384243-9

9.1   THE GOALS OF CONCRETE MIX DESIGN

The goals of concrete mix design are the achievement of three properties for concrete:

- Workability of concrete: Your final mix should pass the tests for the work-ability of concrete like the slump test or flow table test and be compatible with the defined workability that is mentioned in the implementation sheets. For example, if the target slump is defined as 180 mm, the concrete mix should be compatible with the needs of the supervisor which could be the slump more than 180 mm or a defined limitation like 160–200 mm.
- Compressive strength: Your final mix should pass the compressive strength test of concrete as defined in the design of the structure. For example, if the designed compressive strength of project will be 35 MPa, the concrete mix should be compatible with the need of the supervisor which could be 38 MPa or more.
- Durability considerations: Your final mix should pass the tests for the dura-bility of concrete structure which should be the permeability test or some other tests that the supervising system confirmed.

So, you should design a concrete by using the defined constituent materials to achieve the above considerations.

9.2   THE STEP-BY-STEP METHOD FOR CONCRETE MIX DESIGN

Now, we are going to talk about the step-by-step method for concrete mix design. In the next parts, we are going to explain the steps one by one and then after finalizing the explanation, we will design three examples for different types of concrete with different constituent materials. So, if the explanations were not clear for you, after studying the examples, it will be more clear.

9.2.1   STEP (1): SPECIFY STANDARD DEVIATION

Standard deviation shows the evaluation of your plant for concrete production. If your plant is a high controlled plant that you can check the quality of constituent materials and produced concrete accurately, the plant standard deviation will be much lower than a plant without any quality control. On the other hand, you cannot produce high-strength concrete in a plant without any quality control or with low control on the constituent materials and the process of concrete production.

You should add the standard deviation with a special formulation to the proj-ect compressive strength to achieve the mix design compressive strength. Then you should design the concrete with the mix design compressive strength. In fact, it has the role of the safety factor for concrete mix design.

We have two methods for the evaluation of the standard deviation. It depends on the availability of data from the plant. If we have full data on compressive strength from the concrete produced in the plant before, we should use the first method.

Nevertheless, we should use the second method which is according to the planning of the production plant.

- First method for the evaluation of standard deviation:

  If you have at least 30 compressive strength data from before, you can use equation 9.1 for the calculation of the standard deviation.

$$S = \sqrt{\frac{\sum (X - m)^2}{n - 1}} \tag{9.1}$$

  In the above equation, S is the standard deviation, X is the compressive strength of a specimen, m is the mean value of all compressive strengths, and n is the number of specimens.

  If the data from before will be less than 30, you should multiply a coefficient by equation 9.1 that you can find in equation 9.2.

$$R = 0.75 + \sqrt{\frac{2}{n}} \tag{9.2}$$

  In equation 9.2, R is the coefficient and n is the number of specimens.

  If you have two series of data that each one is less than 30 specimens, you can use equation 9.3.

$$S = \sqrt{\frac{(n_1 - 1)s_1^2 + (n_2 - 1)s_2^2}{n_1 + n_2 - 2}} \tag{9.3}$$

  In the equation above, S_1 and S_2 are the standard deviations of each series that are calculated from equation 9.1, n_1 and n_2 are the numbers of each series and S is the final standard deviation.

  If the result of the standard deviation from the above equations will be less than 2.5 MPa, you should use 2.5 MPa as the standard deviation. It means that you cannot use the standard deviation less than 2.5 MPa in the concrete mix design. A project or plant with the standard deviation of 2.5 MPa is a high controlled concrete plant.

- Second method for the evaluation of standard deviation:

  As mentioned before, you should use this method when there is no information about the before concrete production in the plant.

  First, you should find the status of the production plant from Table 9.1. This table shows you the quality control level of the plant. After that, you should find the standard deviation from Table 9.2.

TABLE 9.1
Quality Control Level of Production Plant

Quality Control Factor	Status A	Status B	Status C
Measuring of cement in production	By weight	By weight	By volume
Measuring of aggregates in production	By weight	By weight	By volume
Sieve analysis of aggregates	Controlled	Controlled	Not controlled
Moisture of aggregates	Controlled	Controlled	Not controlled
Surveillance of production	Very good	Good	Weak
Lab instruments	Full available	Available	Not available
Experiments	Continuous	Sometimes	Sometimes
Expert product manager	Available	Available	Not available

TABLE 9.2
Standard Deviation According to the Status of Plant and Designed Compressive Strength

Production Plant Status	f'c Between 20–25 MPa	f'c between 25–30 MPa	f'c between 30–35 MPa	f'c more than 35 MPa
A	3	3.5	4	4.5
B	4	4.5	5	–
C	5	–	–	–

9.2.2 STEP (2): SPECIFY MIX DESIGN COMPRESSIVE STRENGTH

As mentioned before, we evaluated the standard deviation in the last step to find the mix design compressive strength which is the strength that we should design the concrete to ensure the performance of the final concrete with a defined safety factor.

To calculate the mix design compressive strength, you should use equation 9.4.

$$f_{cm} = f_c' + 1.34s + 1.5 \qquad (9.4)$$

In this equation, S is the standard deviation from step 1, f'c is the structural design compressive strength and f_{cm} is the mix design compressive strength.

9.2.3 STEP (3): SPECIFY PERCENTAGE OF EACH AGGREGATE IN CONCRETE

One of the most important steps of concrete mix design is the evaluation of aggregates in the mix design. You should use the best mix of aggregates to ensure the quality of concrete in the case of compressive strength and workability. The most important factor which has an effect on the workability of concrete is the mix of aggregates.

TABLE 9.3

Recommendations for Choosing of Maximum Size of Coarse Aggregates

Structural Compressive Strength (MPa)	Structural Element	Recommended Maximum Size of Coarse Aggregates (mm)
Less than 30	Foundations	25
Less than 30	Floors, columns, walls	19
30–45	Foundations, floors	19
30–45	Walls, columns	12
45–70	All kinds of elements	12
More than 70	All kinds of elements	9

The first thing that you should do for this step is specifying the maximum size of coarse aggregate. There are too many recommendations in different standards and codes for this purpose. For example, ACI 211.1 recommends as below:

The maximum size of coarse aggregate should not exceed one-fifth of free distance between molds, one-third of slab diameter, and three-fourths of free distance between rebars in the structural element.

Here we recommend Table 9.3 for this purpose:

After evaluating of the maximum size of coarse aggregate, you should use equation 9.5 for the percentage of each aggregate in concrete.

$$P = \frac{100\%}{1 - \left(\dfrac{0.075}{D}\right)^{n}} \times \left[\left(\frac{d}{D}\right)^{n} - \left(\frac{0.075}{D}\right)^{n}\right] \tag{9.5}$$

In the above equation, P is the percentage passed from sieve d, D is the maximum size of coarse aggregate, and n is a digit between 0.1 and 0.7 which shows the coarseness or fineness of total aggregates in concrete. If you use smaller values for n, the concrete will be softer and if you use bigger value, the concrete will be coarser. You should know, coarser concrete will need less water or less super-plasticizer for the same flowability and also less cement for the same compressive strength. But on the other hand, the coarse concrete has low workability considerations like pumpability and malleability. So, we should use the appropriate concrete for each case. For example, for a mass foundation it is better to use coarser concrete. But for a column, with congested rebars it is better to use finer concrete. For self-compacting concrete you should use a very fine concrete to avoid segregation and bleeding. To find the best values for n you can use Table 9.4 as the recommendation.

9.2.4   STEP (4): SPECIFY FINENESS MODULE OF TOTAL AGGREGATES

The fineness module of total aggregates defined as the sum of cumulative percentage remained on sieves 37.5, 19, 9.5, 4.75, 2.36, 1.18, 0.6, 0.3, 0.15 mm divided by 100. We will need this value in the future steps for the evaluation of free water in concrete.

TABLE 9.4
Recommended Values for n Depends on the Type of Concrete

Type of Concrete	Value of n Related to Minimum Curve	Value of n Related to Maximum Curve
Fine SCC concrete	0.2	0.1
Coarse SCC concrete	0.25	0.15
High slump concrete (more than 180 mm)	0.3	0.2
Pumpable concrete with slump between 140 and 180 mm	0.35	0.25
Pumpable concrete with slump less than 140 mm	0.4	0.3
Not pumping concrete with slump between 140 and 170 mm	0.45	0.35
Not pumping concrete for huge elements	0.5	0.4

9.2.5   STEP (5): SPECIFY WATER-TO-BINDER RATIO

This step is very critical in concrete mix design from the compressive strength point of view. Because as mentioned before, the most important factor that can specify the compressive strength of concrete is the water-to-binder ratio. So, choosing an accurate w/b means the best result for compressive strength.

Before starting the explanation about the methods that you should use for the evaluation of water-to-binder ratio, you should specify the standard compressive strength of cement according to the ASTM C109. For type I cement, we have three different types that you can find in the market:

- Type I-525: It means that the compressive strength of cement mortar according to the ASTM C109 at the age of 28 days should be more than 525 kg/cm^2.
- Type I-425: It means that the compressive strength of cement mortar according to the ASTM C109 at the age of 28 days should be more than 425 kg/cm^2.
- Type I-325: It means that the compressive strength of cement mortar according to the ASTM C109 at the age of 28 days should be more than 325 kg/cm^2.

We are going to explain three methods for the evaluation of water-to-binder ratio. So, you can decide the best value by using these three methods. In two methods we are using the above types of cement for the evaluation of water-to-binder ratio. Type I-325 is not suitable for the production of concrete because it is a low-strength cement. You can use both of the others for the production of concrete. But sometimes, you may use other types of cement like Type II or V cement. In this case, you should know the minimum value for the compressive strength of cement mortar according to the ASTM C109 at the age of 28 days for a long period of time (e.g., 1 or 2 months). Then you should use linear correlation to evaluate the water-to-binder ratio for the special cement. For example, if you use a Type II cement with a minimum compressive strength of 450 kg/cm². You should use a linear correlation between the values of cement type I-525 and I-425 to evaluate the water-to-binder ratio for this cement.

As mentioned before, we have three methods for the evaluation of water-to-binder ratio:

- Method 1 for the mix design compressive strength between 25 and 55 MPa.
- Method 2 for the mix design compressive strength between 25 and 40 MPa.
- Method 3 for the mix design compressive strength between 45 and 85 MPa.

So, if the concrete mix design compressive strength will be between 25 and 45 MPa, you should use methods 1 and 2 for the best decision, and if the concrete mix design compressive strength will be more than 45 MPa, you should use methods 2 and 3 together.

- Method 1 or modified German method for concrete mix design:

  In this method, for the evaluation of water-to-binder ratio with cement type I-525 you can use Table 9.5 and with cement type I-425 you can use Table 9.6. The assumption for this table is using pure Portland cement in concrete production. We will describe using supplementary cementitious materials in the future.

TABLE 9.5
w/b Ratio for Concrete Made With Cement Type I-525 (Method 1)

f_{cm} (MPa)	w/b for Crushed Aggregates	w/b for Natural Aggregates
25	0.69	0.66
30	0.65	0.61
35	0.6	0.57
40	0.55	0.53
45	0.52	0.48
50	0.48	0.44
55	0.44	0.39

TABLE 9.6
w/b Ratio for Concrete Made With Cement Type I-425 (Method 1)

f_{cm} (MPa)	w/b For Crushed Aggregates	w/b For Natural Aggregates
25	0.65	0.62
30	0.59	0.56
35	0.54	0.51
40	0.5	0.47
45	0.46	0.42
50	0.41	0.36
55	0.36	0.34

The columns of the above tables for crushed and natural aggregates define using only crushed or natural for both coarse and fine aggregates. If you are using mix of natural and crushed aggregates, you should use a digit between two digits of one row.

• Method 2 or modified ACI method for concrete mix design:

This is a very simple method for concrete mix design. You don't have any special assumption for the type and compressive strength of cement or natural and crushed aggregates. So, this is only for concrete with mix design compressive strength of less than 40 MPa.
   In this method, you can use Table 9.7 for the evaluation of the water-to-binder ratio.

• Method 3 or high-performance concrete method:

In this method, you should use Table 9.8 for concrete made with cement type I-525 and Table 9.9 for concrete made with cement type I-425.
   In this method, we assumed that you are using crushed coarse aggregates and natural sand as fine aggregate. So, if you use other mixed for

TABLE 9.7
w/b Ratio From Method 2

f_{cm} (MPa)	Water-to-Binder Ratio
25	0.69
30	0.61
35	0.53
40	0.47

TABLE 9.8
w/b Ratio for Concrete Made With Cement Type I-525 (Method 3)

f_{cm} (MPa)	Water-to-Binder Ratio
45	0.42
50	0.4
55	0.39
60	0.37
65	0.36
70	0.34
75	0.33
80	0.32
85	0.3

TABLE 9.9
w/b Ratio for Concrete Made With Cement Type I-425
(Method 3)

f_{cm} (MPa)	Water-to-Binder Ratio
45	0.38
50	0.36
55	0.35
60	0.34
65	0.32
70	0.31
75	0.29
80	0.28
85	0.26

aggregates, you can modify the amount of water-to-binder ratio according to the mix. You should note that concrete with crushed aggregates will let you get more compressive strength with the same water-to-binder ratio. But this is only correct when you are talking about compressive strength. In the case of workability, using only crushed aggregates means that you will need more water for the same flowability.

9.2.6   STEP (6): SPECIFY FREE WATER FOR CONCRETE

Free water in concrete depends on many properties as below:

- Fineness or coarseness of total aggregates
- Fillers in the aggregates (passing by sieve No. 100)
- Passing by sieve No. 200
- Using crushed or natural aggregates
- Type and fineness of cement
- Amount of cement and other binders in concrete
- Target slump of concrete

The most important factors for the amount of free water in concrete depend on the aggregates and cement specifications and the target slump.

Free water is the total water in concrete, if the situation of aggregates will be saturated surface dry. In the future steps, we should add or minus the lack or excess water from concrete to control the final mix. You can derive the amount of free water for concrete from Tables 9.10 and 9.11.

You can see from tables above that if the fineness module of total aggregates will increase, the amount of water will be more, and if the amount of cement in the concrete increase, then the amount of free water will also be more.

TABLE 9.10

Amount of Free Water for the Slump of 90 mm (F: Fineness Module of Total Aggregates)

Amount of Cement (kg)	F=4.0	F=4.1	F=4.2	F=4.3	F=4.4	F=4.5	F=4.6	F=4.7	F=4.8	F=4.9	F=5.0	F=5.1	F=5.2	F=5.3	F=5.4	F=5.5
300	214	210	206	203	199	196	193	189	186	183	180	177	175	172	169	167
325	218	214	210	207	203	200	197	193	190	187	184	181	179	176	173	171
350	222	218	214	211	207	204	201	197	194	191	188	185	183	180	177	175
375	226	222	218	215	211	208	205	201	198	195	192	189	187	184	181	179
400	230	226	222	219	215	212	209	205	202	199	196	193	191	188	185	183
425	234	230	226	223	219	216	213	209	206	203	200	197	195	192	189	187
450	238	234	230	227	223	220	217	213	210	207	204	201	199	196	193	191
475	242	238	234	231	227	224	221	217	214	211	208	205	203	200	197	195
500	246	242	238	235	231	228	225	221	218	215	212	209	207	204	201	199

TABLE 9.11

Amount of Free Water for the Slump of 150 mm (F: Fineness Module of Total Aggregates)

Amount of cement (kg)	F = 4.0	F = 4.1	F = 4.2	F = 4.3	F = 4.4	F = 4.5	F = 4.6	F = 4.7	F = 4.8	F = 4.9	F = 5.0	F = 5.1	F = 5.2	F = 5.3	F = 5.4	F = 5.5
300	232	228	224	220	216	213	209	206	202	199	196	193	190	187	184	181
325	236	232	228	224	220	217	213	210	206	203	200	197	194	191	188	185
350	240	236	232	228	224	221	217	214	210	207	204	201	198	195	192	189
375	244	240	236	232	228	225	221	218	214	211	208	205	202	199	196	193
400	248	244	240	236	232	229	225	222	218	215	212	209	206	203	200	197
425	252	248	244	240	236	233	229	226	222	219	216	213	210	207	204	201
450	256	252	248	244	240	237	233	230	226	223	220	217	214	211	208	205
475	260	256	252	248	244	241	237	234	230	227	224	221	218	215	212	209
500	264	260	256	252	248	245	241	238	234	231	228	225	222	219	216	213

If the target slump will be less than 90 mm or between 90 and 150 mm or higher than 150 mm, you can use linear correlation to find the amount of free water from the data you can find in these two tables.

- Using plasticizers and super-plasticizers

    For concrete production, you should use plasticizers and super-plasticizers. As mentioned before, by using them we can give one of the below specifications:

- Decrease the amount of water
- Increase the flowability
- Decrease the amount of water and increase the flowability together

In concrete mix design procedure, we assume that we are using the super-plasticizers only for water reduction in concrete. So, the first thing that you should do is find the water reduction curve according to the procedure described in Chapter 5.

By using the water reduction curve, you can decide about the dosage of super-plasticizer and you can find the water reduction rate for the defined dosage. So, you can find the final free water after using the super-plasticizer. For example, if you found the free water as 230 L from the tables above and use 0.7% of a super-plasticizer with the water reduction rate of 25% for the dosage of 0.7%, the final free water after using the super-plasticizer will be 173 L.

To decide about the best dosage of the super-plasticizer, you should consider the below points:

- Water reduction curve of the super-plasticizer
- Economic considerations of the project and getting a balance between the price of cement, other binders and super-plasticizer
- Saturation point for the super-plasticizer
- Other project considerations like the target slump, pumpability, and type of aggregates
- The recommendations of the manufacturer

9.2.7 STEP (7): SPECIFY THE AMOUNT OF PORTLAND CEMENT AND OTHER BINDERS

Now, we have the amount of free water in concrete and water-to-binder ratio. So, we can calculate the amount of total binder from equation 9.6.

$$b = \frac{W}{(W/b)} \qquad (9.6)$$

In the equation above, b is the total binder, w is the amount of free water after using super-plasticizer, and w/b is water-to-binder ratio.

If we would like to use only Portland cement in concrete production, b is the amount of Portland cement. But if we are going to use supplementary cementitious

TABLE 9.12

Coefficient of Increasing b According to the Type of the Supplementary Cementitious Material

Type of Binder	Percent of Increasing b
Silica fume	0
Fly ash	10
GGBS	15
Natural Pozzolan	10–15

materials, we should go through the below procedure to calculate the amount of Portland cement and other binders separately:

- Specify b from equation 9.6.
- According to the explanations of Chapter 3 and especially Table 3.6, you should decide about the type of supplementary cementitious material that you would like to use and the percentage of use.
- Although using supplementary cementitious materials will increase the durability and performance of concrete at higher ages, they could decrease the compressive strength, especially at earlier ages even at the age of 28 days which is very important for the design of structure compared with the usage of pure Portland cement. So, if you would like to use supplementary cementitious materials as a part of the total binder, you should use the coefficients of Table 9.12 to increase the amount of b.
- The amount of supplementary cementitious materials will obtain by multiplying the percentage that we decided to use by the increased b.
- The amount of Portland cement will obtain by subtracting the amount of supplementary cementitious materials by the increased b.

9.2.8 STEP (8): SPECIFY THE TOTAL VOLUME OF AGGREGATES

Now, we have the volume of all parts of concrete instead of the aggregates. As we would like to design the mix of concrete for $1\,m^3$ or $1000\,L$, we can calculate the total volume of aggregates from equation 9.7.

$$V = 1000 - \left(\frac{C}{d_c}\right) - \left(\frac{p}{d_p}\right) - W - V_a - \left(\frac{S}{d_s}\right) \tag{9.7}$$

In the equation above, c is the weight of Portland cement, d_c is the density of Portland cement, p is the weight of Pozzolan or supplementary cementitious material, d_p is the density of Pozzolan, w is the weight of final reduced free water which is equal to its volume, s is the weight of the super-plasticizer, d_s is the density of super-plasticizer, and V_a is the volume of air inside the concrete.

TABLE 9.13

Correlation Between the Entrapped Air and Maximum Size of Coarse Aggregate

Maximum size of coarse aggregate (mm)	12.5	19	25
Percentage of entrapped air (%)	0.8–1.5	0.5–1.2	0.5–1.0

TABLE 9.14

Correlation Between the Entrapped Air and Chemical Base of the Super-Plasticizer

Chemical Base of Super-Plasticizer	Ligno-ulfonate	Naphthalene Sulfonate	Melamine Sulfonate	Poly Carboxylate Ether
Percentage of entrapped air (%)	1.1–2.0	0.9–1.6	0.5–1.2	1.0–1.8

For a concrete without using air entraining admixtures, you can find the amount of air inside the concrete by using Tables 9.13 and 9.14. Table 9.13 is the correlation between the maximum size of coarse aggregate and the amount of entrapped air inside the concrete and Table 9.14 is the correlation between the chemical base of super-plasticizer and entrapped air inside the concrete.

9.2.9 STEP (9): SPECIFY THE WEIGHT OF AGGREGATES IN SATURATED SURFACE DRY CONDITIONS

In the step before, we calculated the total volume of aggregates in concrete. On the other hand, in step 3 we specified the percentage of each aggregate to get the best gradation of aggregates in concrete. Now it is time to calculate the weight of each aggregate.

You can do it first by multiplying the percentage of step 3 to the total volume of step 8 to calculate the volume of each aggregate. Second, you should multiply the volume of each aggregate to its density to calculate the weight.

Attention that the assumption here is: "All of the aggregates are in the saturated surface dry conditions which is not the real conditions of aggregates in the projects."

9.2.10 STEP (10): SPECIFY THE REAL WEIGHT OF AGGREGATES AND WATER IN CONCRETE

The real weight of aggregates means that you should take into account the moisture (excess or shortage) of aggregates in the calculations. If an aggregate has excess moisture inside, you should add the weight of the water inside that aggregate to the weight of it and subtract it from the weight of the free water of concrete. On the other hand, if an aggregate has a shortage of water, you should subtract the weight of the water shortage from the weight of the aggregate and add it to the weight of the free water of concrete.

The excess water means moisture is more than the water absorption rate of the aggregate and shortage of water means the moisture is less than the water absorption rate of the aggregate.

Most of the time, you can find the crushed coarse aggregates with the shortage of water and the crushed and natural sands with excess water because aggregate producers wash the sand to separate the soil from it. But they are not washing the coarse aggregates. So, the water used for washing the sands will remain on the surface of sand particles and it will be more than the water absorption rate of the sand. But this is not the same rule for all parts of the world. So, you should test the aggregates before use them in concrete to calculate the water absorption rate and excess or shortage of water

9.3 EXAMPLE 1 FOR CONCRETE MIX DESIGN

In the first example of this chapter, we would like to design a C30 concrete mix to use in the foundation of a residential building with below constituent materials:

- Natural sand with the density of 2.68 kg/L and the water absorption rate of 1.8% and total moisture of 4.7%. The sieve analysis test of this sand is as below (Table 9.15):
- Crushed gravel 12–25 mm size with the density of 2.71 kg/L and water absorption of 0.9% which is dry. The sieve analysis test of this gravel is as below (Table 9.16):
- Crushed gravel 5–12 mm size with the density of 2.71 kg/L and water absorption of 0.8% which is dry. The sieve analysis test of this gravel is as below (Table 9.17):

TABLE 9.15
Sieve Analysis Test for the Natural Sand of Example 1

Sieve Size (mm)	Weight of Aggregates Remained on Sieve (g)	Weight of Aggregates Passed by Sieve (g)	Percent of Aggregates Remained on Sieve (%)	Percent of Aggregates Passed by Sieve (%)
4.75	264	1447	15.4	84.6
2.36	329	1118	19.2	65.3
1.18	372	746	21.7	43.6
0.6	231	515	13.5	30.1
0.3	224	291	13.1	17.0
0.15	206	85	12.0	5.0
Total	1711	–	100	–

TABLE 9.16
Sieve Analysis Test for the 12–25 Gravel of Example 1

Sieve Size (mm)	Weight of Aggregates Remained on Sieve (g)	Weight of Aggregates Passed by Sieve (g)	Percent of Aggregates Remained on Sieve (%)	Percent of Aggregates Passed by Sieve (%)
25	5	1860	0.3	99.7
19	493	1367	26.4	73.3
12.5	976	391	52.3	21.0
9.5	339	52	18.2	2.8
4.75	52	0	2.8	0
Total	1865	–	100	–

TABLE 9.17
Sieve Analysis Test for the 5–12 Gravel of Example 1

Sieve Size (mm)	Weight of Aggregates Remained on Sieve (g)	Weight of Aggregates Passed by Sieve (g)	Percent of Aggregates Remained on Sieve (%)	Percent of Aggregates Passed by Sieve (%)
19	0	1802	0.0	100
12.5	42	1760	2.3	97.7
9.5	695	1065	38.6	59.1
4.75	1029	36	57.1	2.0
2.36	36	0	2.0	0.0
Total	1802	–	100	–

- Portland cement type II with the density of 3.15 kg/L and the minimum compressive strength of 450 kg/cm^2 for the last 2 months.
- Polynaphthalene sulfonate super-plasticizer with the density of 1.18 kg/L that you can see the water reduction rate in Figure 9.1.
- For the plant that we would like to produce concrete, we have the compressive strength data from C25 concrete that was produced in this plant before. You can see the data in Table 9.18. Note that the production and quality control conditions of the plant are the same as before.
  Now, we should start the step-by-step process for concrete mix design:
- Step 1: specify standard deviation
  For this example, as we have data from before concrete production in the plant, we will use equation 9.1 for the calculation of the standard deviation. As you can see, we have 20 data from before production in this plant. So, we should use equation 9.2 as the correction factor.

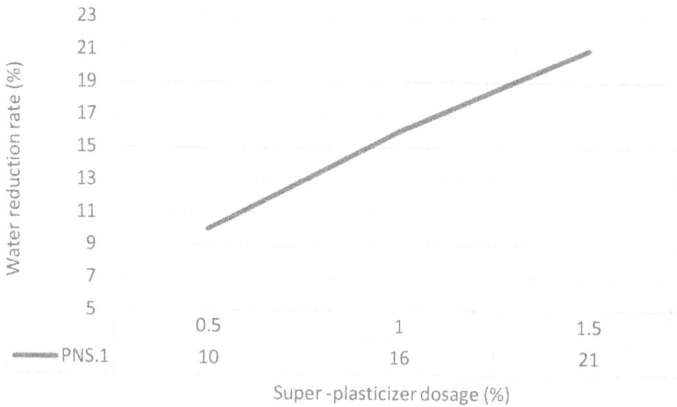

FIGURE 9.1 Water reduction curve for the PNS super-plasticizer of example 1. (Graph created by the author.)

TABLE 9.18
Compressive Strength Data for the Plant of Example 1 in kg/cm²

279	303	298	310	285	277	259	300	296	299
266	311	282	290	308	269	295	304	278	308

From equation 9.1 and the date of Table 9.18 we have:
m = 290.8 kg/cm²
S = 15.6 kg/cm² = 1.56 MPa
And from equation 9.2 we have:
R = 1.066
By multiplying the R by the S, we will have the standard deviation as 1.66 MPa. As we cannot use the standard deviation of less than 2.5 MPa for concrete mix design, we should use 2.5 MPa as the standard deviation for this plant.

- Step 2: specify mix design compressive strength

We should use equation 9.3 for the calculation of the mix design compressive strength. So, with the f'$_c$= 30 MPa and S= 2.5 MPa we will have:

$$f_{cm} = 34.85 \text{ MPa} = 35 \text{ MPa}$$

- Step 3: specify the percentage of each aggregate in concrete
  First, we define the target slump as 180 mm. Then we should use Table 9.4 for the evaluation of n in equation 9.5. You can see from the table that for a pumpable concrete with a slump between 140 and 180 mm we should use n between 0.25 and 0.35. On the other hand, we will have a 12–25 coarse

TABLE 9.19

Values of Maximum and Minimum for the Aggregates of Example 1

Sieve size (mm)	25	19	12.5	9.5	4.75	2.36	1.18	0.6	0.3	0.15
Percent passed for n = 0.25	100	91.3	79.2	71.9	55.6	41.8	30.3	20.8	12.7	5.8
Percent passed for n = 0.35	100	89.5	75.2	66.9	49.3	35.3	24.5	16.1	9.4	4.1

TABLE 9.20

Best Mix of Aggregates for Example 1

Sieve Size (mm)	27% of Coarse 12–25	13% of Coarse 5–12	60% of Natural Sand	Mix of Aggregates
25	26.9	13.0	60	99.9
19	19.8	13.0	60	92.8
12.5	5.7	12.7	60	78.4
9.5	0.8	7.7	60	68.4
4.75	0.0	0.3	50.8	51.0
2.36	0.0	0.0	39.2	39.2
1.18	0.0	0.0	26.2	26.2
0.6	0.0	0.0	18.0	18.0
0.3	0.0	0.0	10.2	10.2
0.15	0.0	0.0	3.0	3.0

aggregate in this example. So, the maximum size of coarse aggregate or D is 25 mm. Now, it is time to use d as 25, 19, 12.5, 9.5, 4.75, 2.36, 1.18, 0.6, 0.3, and 0.15 and calculate the amount of P for each value. These values will be the maximum and minimum values for total aggregates of concrete that you can see in Table 9.19.

Now, we should use different percent of each aggregate to achieve a mix that should be between the above maximum and minimum of Table 9.19. You should achieve the best mix by trial and error. For this reason, you can see Table 9.20.

You should compare the last column of Table 9.20 with the maximum and minimum amounts of Table 9.19 to find that this is the best mixture for the aggregates to achieve the defined limits. You can see Figure 9.2 for a better comparison. Note that you should find the best mixture which could be nearest to the limits. Sometimes, it is not possible to find a mix with all passed-by sieves between the maximum and minimum limits.

At the end of this step, we will have the below mixture for the concrete of example 1:

- Gravel 12–25: 27%
- Gravel 5–12: 13%

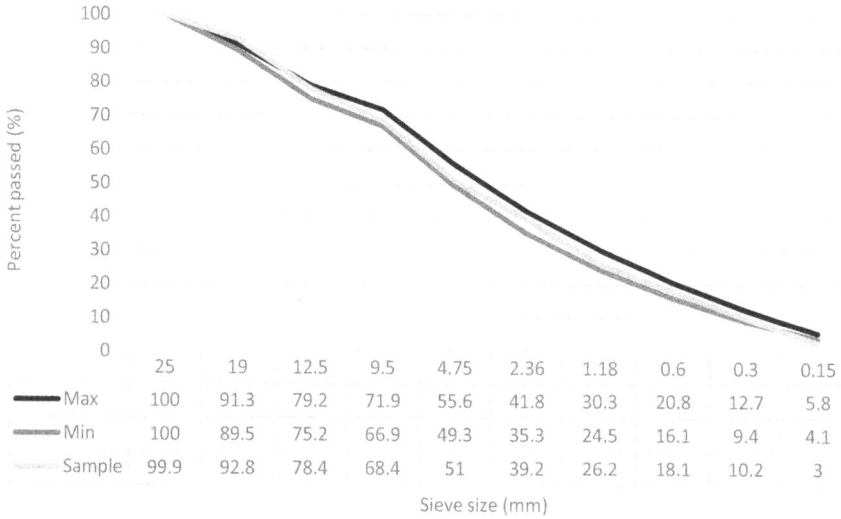

	25	19	12.5	9.5	4.75	2.36	1.18	0.6	0.3	0.15
Max	100	91.3	79.2	71.9	55.6	41.8	30.3	20.8	12.7	5.8
Min	100	89.5	75.2	66.9	49.3	35.3	24.5	16.1	9.4	4.1
Sample	99.9	92.8	78.4	68.4	51	39.2	26.2	18.1	10.2	3

Sieve size (mm)

FIGURE 9.2   The best mixture for aggregates of example 1. (Graph created by the author.)

- Natural sand: 60%
- Step 4: specify the fineness module of total aggregates

According to the definition of the fineness module for total aggregates, you can see Table 9.21 for the calculation.

So, the fineness module of total aggregates is 4.91.

- Step 5: specify water-to-binder ratio

As the mix design compressive strength for the concrete of example 1 is 35 MPa, we should use methods 1 and 2 for the evaluation of the water-to-binder ratio. On the other hand, the compressive strength of the cement in this example is 450 kg/cm². So, for method 1 we should use Table 9.5 for the compressive strength of 525 kg/cm² and Table 9.6 for the compressive strength of 425 kg/cm². Then we should find the exact amount for the compressive strength of 450 kg/cm² by the linear correlation. According to the abovementioned, we will have the below values:

- From Table 9.5 for I-525 cement and as we use a mixture of natural sand and crushed gravels, we should use the mean value for two columns correlated with the mix design compressive strength of 35 MPa which is w/b=0.585.
- From Table 9.6 for I-425 cement and as we use a mixture of natural sand and crushed gravels, we should use the mean value for two columns correlated with the mix design compressive strength of 35 MPa which is w/b=0.525.
- For method 1 as the compressive strength of the cement is 450 kg/cm², we should find the water-to-binder ratio for this cement from the linear correlation between 0.585 and 0.525 which is w/b=0.54.

TABLE 9.21

Calculation of the Fineness Module of Total Aggregates of Example 1

Sieve Size (mm)	Percent Remained (%)	Cumulative Percent Remained (%)
37.5	0.0	0.0
19	7.2	7.2
9.5	24.4	31.6
4.75	17.4	49.0
2.36	11.8	60.8
1.18	13.0	73.8
0.6	8.1	81.9
0.3	7.9	89.8
0.15	7.2	97.0
Total	–	491.2

- Now, we should use Table 9.7 for method 2. From this table, we can find that for the compressive strength of 35 MPa the water-to-binder ratio should be 0.53.

  According to the above values, we decided to use the lower value which is w/b = 0.53.

- Step 6: specify free water in concrete

  To find the amount of free water in concrete, we should use Tables 9.10 and 9.11. In these tables, we should consider two parameters. The first one is the fineness module of total aggregates we have it as 4.91 from step 4. The other parameter is the amount of cement. In this case, you should assume the amount of cement and find the water and go ahead to the next step. In the next step you will find the exact amount of cement. If your assumption of this step was not true, you should come back and try this step again with the right amount of cement. Here we assume that the amount of cement will be 375 kg/m³. So, we will have:

- From Table 9.10 for the target slump of 90 mm and fineness module of 4.9 and cement of 375 kg, we will have 195 L for the amount of free water.

- From Table 9.11 for the target slump of 150 mm and fineness module of 4.9 and cement of 375 kg, we will have 211 L for the amount of free water.

- As the target slump is 180 mm, we should find the amount of free water in concrete with a linear correlation which is 219 L.

- Now, we should decide about the dosage of the super-plasticizer. According to the economic considerations, we decide to use 0.7% of the PNS super-plasticizer that you can find in its water reduction rate in Figure 9.1. The water reduction rate for the dosage of 0.7% is 12.5%. We will use 12% for more assurance.

- We should decrease the amount of water by 12%. Twelve percent of 219 L is 26 L and the amount of final free water after using the super-plasticizer will be 193 L.

- Step 7: specify the amount of cement and other binders
  You should find the amount of cement by using equation 9.6:

$$C = 193/0.53 = 364\,kg$$

As we should use only Portland cement in this example, we will have the amount of cement as $364\,kg/m^3$. We can round it to $370\,kg$ for more assurance.

You can see that the amount of cement that we assumed in the before step as $375\,kg$ was a little more than the exact amount of cement which is better for the assurance of the mix design.

The amount of super-plasticizer is 0.7% of $370\,kg$ which is $2.6\,kg$.

- Step 8: specify the total volume of aggregates in concrete
  For this step, you should use equation 9.7. We have all of the information for use in this equation instead of the Va which is the amount of air. For this reason, we should use Tables 9.13 and 9.14. From Table 9.13 with the maximum size of coarse aggregate of 25 mm, we will have the amount of air between 0.5% and 1.0%. From Table 9.14 and for PNS super-plasticizer, we will have the amount of air between 0.9% and 1.6%. So, we can assume the amount of air for this concrete as 1.3%. Now from equation 9.7, we have:

$$V = 1000 - (370/3.15) - (193) - (13) - (2.6/1.18) = 674\,L$$

- Step 9: calculating the weight of aggregates in SSD condition
  For this step, you can use Table 9.22 as below:
- Step 10: calculating the real weight of aggregates and water
  In this step, we should calculate the real weight of aggregates and water of concrete according to the water absorption and the moisture conditions of aggregates as below:
- For coarse 12–25, we have the water absorption of 0.9% and dry conditions. So, $493\,kg$ of this gravel needs 0.9% water to be in SSD conditions which is $4.4\,kg$. It will round up to $5\,kg$. This weight should add to the weight of concrete water and subtract from the weight of aggregate. So, the real weight of coarse 12–25 is $488\,kg$ which can be rounded to $490\,kg$ and the weight of water is $198\,kg$.

TABLE 9.22

Calculating the Amounts of Aggregates in SSD Conditions for Example 1

Type of Aggregate	Percent in Total Mixture (%)	Volume (L)	Density (kg/L)	Weight of SSD (kg)
Coarse 12–25	27	182	2.71	493
Coarse 5–12	13	88	2.71	238
Natural sand	60	404	2.68	1083
Total	100	674	–	1814

TABLE 9.23
Final Concrete Mix Design for Example 1

Constituent Material	Mix Design for 1 m³ (kg)	Mix Design for 30 L (g)
Gravel 12–25	490	14700
Gravel 5–12	235	7050
Natural sand	1115	33450
Cement Type II	370	11100
PNS super-plasticizer	2.6	78
Water	168	5040
Total weight	2381	71418

- For coarse 5–12, we have water absorption of 0.8% and dry conditions. So, 238 kg of this gravel needs 0.8% water to be in SSD conditions which is 1.9 kg. It will round up to 2 kg. This weight also should add to the weight of concrete water and subtract from the weight of aggregate. So, the real weight of coarse 5–12 is 236 kg which could be rounded to 235 kg and the weight of water is 200 kg.
- For natural sand, we have the water absorption of 1.8% and 4.7% total moisture inside. So, this sand has (4.7% − 1.8% = 2.9%) of excess water which is 31.4 kg for 1083 kg of sand. It will round up to 32 kg which should subtract from the weight of the concrete water and add to the weight of the sand. So, the real weight of sand is 1115 kg and the weight of water is 168 kg.

Now, you can see the final mix design for the concrete of example 1 in Table 9.23. You can see that we calculated the mix design for 1 m³ of concrete. In the last column of the mix design, you can see the calculation of the concrete mix design for 30 L of concrete which is suitable for a laboratory concrete mixer that we are going to check the mix design with it. You can also calculate the mix design for the batches of your batching plant whose capacity could be more or less than 1 m³.

9.4 EXAMPLE 2 FOR CONCRETE MIX DESIGN

In the second example of this chapter, we would like to design a C40 concrete for the roofs of a defined structure with below constituent materials:

- Natural sand with a density of 2.66 kg/L and a water absorption of 1.6% and total moisture of 3.5%. The sieve analysis test of this sand is as below (Table 9.24):
- Crushed sand with a density of 2.69 kg/L and a water absorption rate of 1.9% and total moisture of 2.6%. The sieve analysis test of this sand is as below (Table 9.25):

TABLE 9.24

Sieve Analysis Test for the Natural Sand of Example 2

Sieve Size (mm)	Weight of Aggregates Remained on Sieve (g)	Weight of Aggregates Passed by Sieve (g)	Percent of Aggregates Remained on Sieve (%)	Percent of Aggregates Passed by Sieve (%)
4.75	168	1561	9.7	90.3
2.36	373	1188	21.6	68.7
1.18	317	871	18.3	50.4
0.6	279	592	16.1	34.2
0.3	257	335	14.9	19.4
0.15	236	99	13.6	5.7
Total	1729	–	100	–

TABLE 9.25

Sieve Analysis Test for the Crushed Sand of Example 2

Sieve Size (mm)	Weight of Aggregates Remained on Sieve (g)	Weight of Aggregates Passed by Sieve (g)	Percent of Aggregates Remained on Sieve (%)	Percent of Aggregates Passed by Sieve (%)
4.75	88	1643	5.1	94.9
2.36	395	1248	22.8	72.1
1.18	352	896	20.3	51.8
0.6	286	610	16.5	35.2
0.3	260	350	15.0	20.2
0.15	239	111	13.8	6.4
Total	1731	–	100	–

- Crushed gravel 11–19 mm size with a density of 2.73 kg/L and water absorption of 0.8% which is dry. The sieve analysis test of this gravel is as below (Table 9.26):
- Crushed gravel 5–12 mm size with a density of 2.70 kg/L and water absorption of 0.9% which is dry. The sieve analysis test of this gravel is as below:
- Portland cement type II with a density of 3.15 kg/L and a minimum compressive strength of 450 kg/cm^2 in the last 2 months.
- Ground granulated blast furnace slag (GGBS) with a density of 2.9 kg/L.
- Polycarboxylate super-plasticizer with a density of 1.07 kg/L; you can see the water reduction rate in Figure 9.3.

Now, we should start the step-by-step process for concrete mix design:

TABLE 9.26

Sieve Analysis Test for the 11–19 Gravel of Example 2

Sieve Size (mm)	Weight of Aggregates Remained on Sieve (g)	Weight of Aggregates Passed by Sieve (g)	Percent of Aggregates Remained on Sieve (%)	Percent of Aggregates Passed by Sieve (%)
25	0	1855	0.0	100
19	98	1757	5.3	94.7
12.5	1179	578	63.6	31.2
9.5	458	120	24.7	6.5
4.75	120	0	6.5	0.0
Total	1855	–	100	–

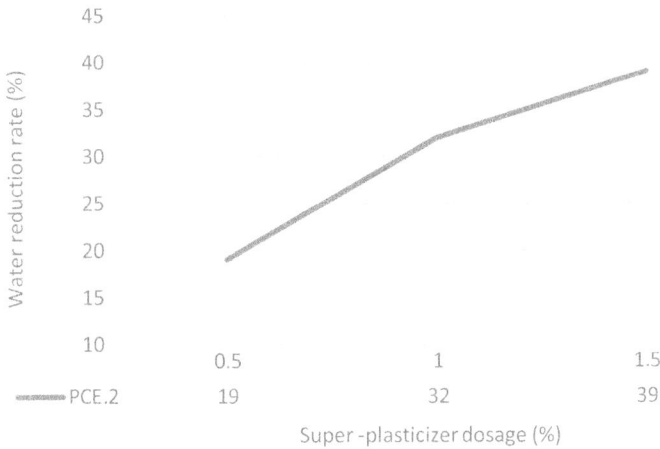

FIGURE 9.3 Water reduction curve for PCE super-plasticizer of example 2. (Graph created by the author.)

- Step 1: specify standard deviation

    For this example, we will use the same standard deviation as in example 1. Because we assumed that the plant condition is the same for the two examples. So, we will use 2.5 MPa for the standard deviation.

- Step 2: specify mix design compressive strength

    We should use equation 9.3 for the calculation of the mix design compressive strength. So, with the $f'_c = 40$ MPa and $S = 2.5$ MPa we will have:

$$f_{cm} = 44.85 \text{ MPa} = 45 \text{ MPa}$$

TABLE 9.27

Sieve Analysis Test for the 5–12 Gravel of Example 2

Sieve Size (mm)	Weight of Aggregates Remained on Sieve (g)	Weight of Aggregates Passed by Sieve (g)	Percent of Aggregates Remained on Sieve (%)	Percent of Aggregates Passed by Sieve (%)
19	0	1891	0.0	100
12.5	12	1879	0.6	99.4
9.5	339	1540	17.9	81.4
4.75	1345	195	71.1	10.3
2.36	195	0	10.3	0.0
Total	1891	–	100	–

TABLE 9.28

Values of Maximum and Minimum for the Aggregates of Example 2

Sieve size (mm)	19	12.5	9.5	4.75	2.36	1.18	0.6	0.3	0.15
Percent passed for $n = 0.2$	100	88.0	80.7	63.8	49.0	36.3	25.5	15.8	7.3
Percent passed for $n = 0.3$	100	85.4	76.8	58.0	42.6	30.2	20.3	12.1	5.4

- Step 3: specify the percentage of each aggregate in concrete

  As we would like to produce concrete with PCE super-plasticizer, we define the target slump as 210 mm. Then we should use Table 9.4 for the evaluation of n in equation 9.5. You can see from the table that for high slump concrete we should use n between 0.2 and 0.3. On the other hand, we will have a 11–19 coarse aggregate in this example. So, the maximum size of coarse aggregate or D is 19 mm. Now, it is time to use d as 19, 12.5, 9.5, 4.75, 2.36, 1.18, 0.6, 0.3, and 0.15 and calculate the amount of P for each value. These values will be the maximum and minimum values for the total aggregates of concrete that you can see in Table 9.28.

  Now, we should use different percent of each aggregate to achieve a mix that should be between the above maximum and minimum of Table 9.28. For this reason, you can see Table 9.29.

  You should compare the last column of Table 9.29 with the maximum and minimum amounts of Table 9.28 to find that this is the best mixture for the aggregates to achieve the defined limits. You can see Figure 9.4 for a better comparison.

  At the end of this step, we will have the below mixture for the concrete of example 2:

- Gravel 11–19: 20%
- Gravel 5–12: 15%
- Natural sand: 40%
- Crushed sand: 25%

TABLE 9.29

Best Mix of Aggregates for Example 2

Sieve Size (mm)	20% of Coarse 11–19	15% of Coarse 5–12	40% of Natural Sand	25% of Crushed Sand	Mix of Aggregates
19	18.9	15.0	40	25	98.9
12.5	6.2	14.9	40	25	86.1
9.5	1.3	12.2	40	25	78.5
4.75	0.0	1.5	36.1	23.7	61.4
2.36	0.0	0.0	27.5	18.0	45.5
1.18	0.0	0.0	20.2	12.9	33.1
0.6	0.0	0.0	13.7	8.8	22.5
0.3	0.0	0.0	7.8	5.1	12.8
0.15	0.0	0.0	2.3	1.6	3.9

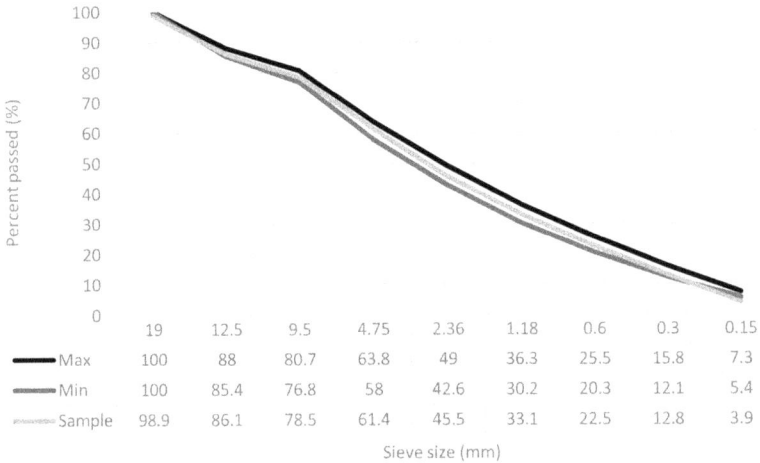

	19	12.5	9.5	4.75	2.36	1.18	0.6	0.3	0.15
Max	100	88	80.7	63.8	49	36.3	25.5	15.8	7.3
Min	100	85.4	76.8	58	42.6	30.2	20.3	12.1	5.4
Sample	98.9	86.1	78.5	61.4	45.5	33.1	22.5	12.8	3.9

Sieve size (mm)

FIGURE 9.4   The best mixture for aggregates of example 2. (Graph created by the author.)

- Step 4: specify the fineness module of total aggregates
    According to the definition of the fineness module for total aggregates, you can see Table 9.30 for the calculation.
    So, the fineness module of total aggregates is 4.43.
- Step 5: specify water-to-binder ratio
    As the mix design compressive strength for the concrete of example 2 is 45 MPa, we should use methods 1 and 3 for the evaluation of the water-to-binder ratio. On the other hand, the compressive strength of cement is 450 kg/cm². According to the abovementioned, we will have the below values:
- From Table 9.5 for I-525 cement and as we use a mixture of natural and crushed aggregates, we should use the mean value for two columns correlated with the mix design compressive strength of 45 MPa which is w/b = 0.5.

TABLE 9.30

Calculation of the Fineness Module of Total Aggregates of Example 2

Sieve Size (mm)	Percent Remained (%)	Cumulative Percent Remained (%)
37.5	0.0	0.0
19	1.1	1.1
9.5	20.4	21.5
4.75	17.1	38.6
2.36	15.9	54.5
1.18	12.4	66.9
0.6	10.6	77.5
0.3	9.7	87.2
0.15	8.9	96.1
Total	–	443.4

- From Table 9.6 for I-425 cement and as we use a mixture of natural and crushed aggregates, we should use the mean value for two columns correlated with the mix design compressive strength of 45 MPa which is w/b = 0.44.
- For method 1 as the compressive strength of cement is 450 kg/cm², we should find the water-to-binder ratio for this cement from the linear correlation between 0.5 and 0.44 which is w/b = 0.455.
- From Table 9.8 for I-525 cement, for the mix design compressive strength of 45 MPa we have w/b = 0.42.
- From Table 9.9 for I-425 cement, for the mix design compressive strength of 45 MPa we have w/b = 0.38.
- For method 3, as the compressive strength of cement is 450 kg/cm², we should find the water-to-binder ratio for this cement from the linear correlation between 0.42 and 0.38 which is w/b = 0.39.

  According to the above values, if we use the mean value between method 1 value (0.455) and method 3 value (0.39) we will have the water-to-binder ratio of 0.422 that we are going to use as 0.42.
- Step 6: specify free water in concrete

  To find the amount of free water in concrete, we should use Tables 9.10 and 9.11. In these tables, you can see that, we should check two parameters. The first one is the fineness module of total aggregates we have it as 4.43 from step 4. The other parameter is the amount of cement. Here we assume that the amount of cement will be 425 kg/m³. So, we will have:
- From Table 9.10 for the target slump of 90 mm and fineness module of 4.4 and cement of 425 kg, we will have 219 L for the amount of free water.
- From Table 9.11 for the target slump of 150 mm and fineness module of 4.4 and cement of 425 kg, we will have 236 L for the amount of free water.
- As the target slump is 210 mm, we should find the amount of free water in the concrete with a linear correlation which is 253 L.

- Now, we should decide about the dosage of the super-plasticizer. According to the economic considerations, we decide to use 0.8% of the PCE super-plasticizer; you can find its water reduction rate in Figure 9.3. The water reduction rate for the dosage of 0.8% is 26.8%. We will use 26% for more assurance.
- We should decrease the amount of water by 26%. So, 26% of 253 L is 66 L and the amount of final free water after using the super-plasticizer will be 187 L.
- Step 7: specify the amount of cement and other binders
  You should find the amount of cement by using equation 9.6:

$$C = 187/0.42 = 445 \text{ kg}$$

In this example, we would like to use GGBS beside the cement. We decided to use 20% of GGBS for the production of this concrete. So, we should increase the amount of total binder according to the Table 9.12 which is 15%. So, the amount of total binder will be 511. We should use 20% of it as the GGBS which is 102. So, after rounding the values, we will have:

- Portland cement: 415 kg
- GGBS: 100 kg
- Total binder: 515 kg

You can see that the amount of cement that we assumed in the before step as 425 is a little more than the exact amount of cement which is better for the assurance of our mix design.

The amount of super-plasticizer is 0.8% of 515 kg which is 4.1 kg.
- Step 8: specify the total volume of aggregates in concrete
  For this step, you should use equation 9.7. We have all of the information for use in this equation instead of the V_a which is the amount of air. For this reason, we should use Tables 9.13 and 9.14. From Table 9.13 and the maximum size of coarse aggregate of 19 mm, we will have the amount of air between 0.5% and 1.2%. From Table 9.14 and for PCE super-plasticizer, we will have the amount of air between 1.0% and 1.8%. So, we can assume the amount of air for this concrete as 1.6%. Now from equation 9.7, we have:

$$V = 1000 - (415/3.15) - (100/2.9) - (187) - (16) - (4.1/1.07) = 627 \text{ L}$$

- Step 9: calculating the weight of aggregates in SSD condition
  For this step, you can use Table 9.31 as below:
- Step 10: calculating the real weight of aggregates and water
  In this step, we should calculate the real weight of aggregates and water of concrete according to the water absorption and the moisture conditions of aggregates as below:
- For coarse 11–19, we have water absorption of 0.8% and dry conditions. So 341 kg of this gravel needs 0.8% water to be in SSD conditions which is 3 kg. This weight should add to the weight of concrete water and subtract from the weight of aggregate. So, the real weight of coarse 11–19 is 338 kg which can be rounded to 340 kg and the weight of water is 190 kg.

TABLE 9.31

Calculating the Amounts of Aggregates in SSD Conditions for Example 2

Type of Aggregate	Percent in Total Mixture (%)	Volume (L)	Density (kg/L)	Weight of SSD (kg)
Coarse 11–19	20	125	2.73	341
Coarse 5–12	15	94	2.70	254
Natural sand	40	251	2.66	668
Crushed sand	25	157	2.69	422
Total	100	627	-	1685

TABLE 9.32

Final Concrete Mix Design for Example 2

Constituent Material	Mix Design for 1 m³ (kg)	Mix Design for 30 L (g)
Gravel 11–19	340	10200
Gravel 5–12	250	7500
Natural sand	680	20400
Crushed sand	425	12750
Cement Type II	415	12450
GGBS	100	3000
PCE super-plasticizer	4.1	123
Water	176	5280
Total weight	2390	71703

- For coarse 5–12, we have water absorption of 0.9% and dry conditions. So, 254 kg of this gravel needs 0.9% water to be in SSD conditions which is 2 kg. This weight also should add to the weight of concrete water and subtract from the weight of aggregate. So, the real weight of coarse 5–12 is 252 kg which could be rounded to 250 kg and the weight of water is 192 kg.
- For natural sand, we have a water absorption of 1.6% and 3.5% total moisture inside. So, this sand has (3.5% − 1.6% = 1.9%) of excess water which is 13 kg for 668 kg of sand. It should subtract from the weight of concrete water and add to the weight of the sand. So, the real weight of sand is 681 kg which will round to 680 kg and the weight of water is 179 kg.
- For crushed sand, we have a water absorption of 1.9% and 2.6% total moisture inside. So, this sand has (2.6% − 1.9% = 0.7%) of excess water which is 3 kg for 422 kg of sand. It should subtract from the weight of concrete water and add to the weight of the sand. So, the real weight of crushed sand is 425 kg and the weight of water is 176 kg.

Now, you can see the final mix design for the concrete of example 2 in Table 9.32.

9.5   EXAMPLE 3 FOR CONCRETE MIX DESIGN

In the third example of this chapter, we would like to design a C60 concrete for the columns and walls of a defined structure with below constituent materials:

- Natural sand with a density of 2.71 kg/L and a water absorption rate of 1.9% and total moisture of 5.1%. The sieve analysis test of this sand is as below (Table 9.33):
- Dune sand with a density of 2.68 kg/L and a water absorption rate of 2.5% and total moisture of 1.5%. The sieve analysis test of this sand is as below (Table 9.34):
- Crushed gravel 5–12 mm size with a density of 2.76 kg/L and water absorption of 0.8% which is dry. The sieve analysis test of this gravel is as below (Table 9.35):

TABLE 9.33
Sieve Analysis Test for the Natural Sand of Example 3

Sieve Size (mm)	Weight of Aggregates Remained on Sieve (g)	Weight of Aggregates Passed by Sieve (g)	Percent of Aggregates Remained on Sieve (%)	Percent of Aggregates Passed by Sieve (%)
4.75	217	1514	12.5	87.5
2.36	352	1162	20.3	67.1
1.18	323	839	18.7	48.5
0.6	278	561	16.1	32.4
0.3	241	320	13.9	18.5
0.15	218	102	12.6	5.9
Total	1731	–	100	–

TABLE 9.34
Sieve Analysis Test for the Dune Sand of Example 3

Sieve Size (mm)	Weight of Aggregates Remained on Sieve (g)	Weight of Aggregates Passed by Sieve (g)	Percent of Aggregates Remained on Sieve (%)	Percent of Aggregates Passed by Sieve (%)
4.75	0	1447	0.0	100
2.36	0	1447	0.0	100
1.18	0	1447	0.0	100
0.6	41	1406	2.8	97.2
0.3	315	1091	21.8	75.4
0.15	854	237	59.0	16.4
Total	1447	–	100	–

TABLE 9.35

Sieve Analysis Test for the 5–12 Gravel of Example 3

Sieve Size (mm)	Weight of Aggregates Remained on Sieve (g)	Weight of Aggregates Passed by Sieve (g)	Percent of Aggregates Remained on Sieve (%)	Percent of Aggregates Passed by Sieve (%)
19	0	1798	0.0	100
12.5	59	1739	3.3	96.7
9.5	645	1094	35.9	60.8
4.75	959	135	53.3	7.5
2.36	135	0	7.5	0.0
Total	1798	–	100	–

- Portland cement type II with a density of 3.15 kg/L and a minimum compressive strength of 450 kg/cm² in the last 2 months.
- GGBS with a density of 2.9 kg/L.
- Silica fume with the density of 2.2 kg/L.
- Polycarboxylate super-plasticizer with a density of 1.07 kg/L and the water reduction curve as Figure 9.3.

Now, we should start the step-by-step process for concrete mix design:

- Step 1: specify standard deviation

  For this example, we will use the same standard deviation as examples 1 and 2. Because we assumed that the plant condition is the same for these examples. So, we will use 2.5 MPa for the standard deviation.
- Step 2: specify mix design compressive strength

  We should use equation 9.3 for the calculation of the mix design compressive strength. So, with the $f'_c = 60$ MPa and $S = 2.5$ MPa we will have:

$$f_{cm} = 64.85 \text{ MPa} = 65 \text{ MPa}$$

- Step 3: specify the percentage of each aggregate in concrete

  As we would like to produce a concrete with PCE super-plasticizer, we define the target slump as 210 mm. Then we should use Table 9.4 for the evaluation of n in equation 9.5.

  You can see in the table that, for high slump concrete we should use n between 0.2 and 0.3. On the other hand, we will have a 5–12 coarse aggregate in this example. So, the maximum size of coarse aggregate or D is 12.5 mm. Now, it is time to use d as 12.5, 9.5, 4.75, 2.36, 1.18, 0.6, 0.3, and 0.15 and calculate the amount of P for each value. These values will be the maximum and minimum values for the total aggregates of concrete that you can see in Table 9.36.

TABLE 9.36
Values of Maximum and Minimum for the Aggregates of Example 3

Sieve size (mm)	12.5	9.5	4.75	2.36	1.18	0.6	0.3	0.15
Percent passed for n = 0.2	100	91.7	72.5	55.7	41.3	28.9	17.9	8.3
Percent passed for n = 0.3	100	89.9	67.9	49.8	35.3	23.8	14.2	6.3

TABLE 9.37
Best Mix of Aggregates for Example 3

Sieve Size (mm)	25% of Coarse 5–12	70% of Natural Sand	5% of Dune Sand	Mix of Aggregates
12.5	24.2	70	5.0	99.2
9.5	15.2	70	5.0	90.2
4.75	1.9	61.2	5.0	68.1
2.36	0.0	47.0	5.0	52.0
1.18	0.0	33.9	5.0	38.9
0.6	0.0	22.7	4.9	27.5
0.3	0.0	12.9	3.8	16.7
0.15	0.0	4.1	0.8	4.9

Now, we should use different percent of each aggregate to achieve a mix that should be between the above maximum and minimum of Table 9.36. For this reason, you can see Table 9.37.

You should compare the last column of Table 9.37 with the maximum and minimum amounts of Table 9.36 to find that this is the best mixture for the aggregates to achieve the defined limits. You can see Figure 9.5 for a better comparison.

At the end of this step, we will have the below mixture for the concrete of example 3:

- Gravel 5–12: 25%
- Natural sand: 70%
- Dune sand: 5%
- Step 4: specify the fineness module of total aggregates

According to the definition of the fineness module for total aggregates, you can see Table 9.38 for the calculation.

So, the fineness module of total aggregates is 4.02.

- Step 5: specify water-to-binder ratio

As the mix design compressive strength for the concrete of example 3 is 65 MPa, we should use only method 3 for the evaluation of the

	12.5	9.5	4.75	2.36	1.18	0.6	0.3	0.15
Max	100	91.7	72.5	55.7	41.3	28.9	17.9	8.3
Min	100	89.9	67.9	49.8	35.3	23.8	14.2	6.3
Sample	99.2	90.2	68.1	52	38.9	27.5	16.7	4.9

Sieve size (mm)

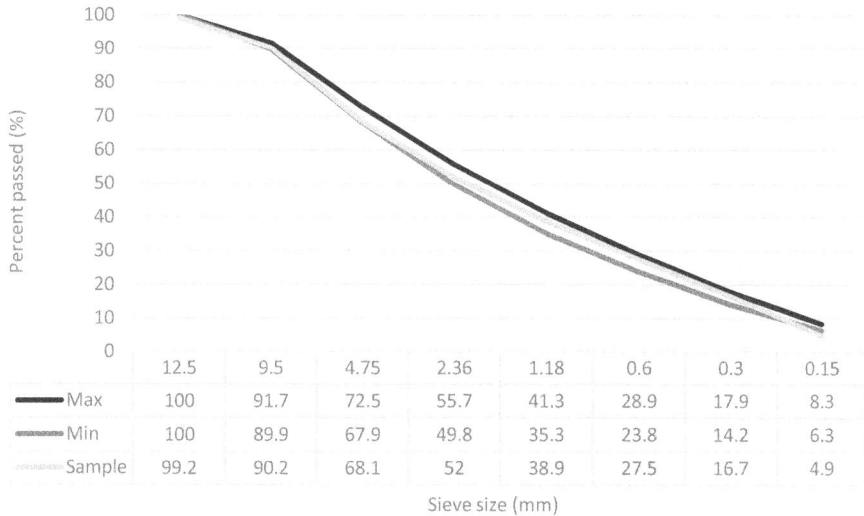

FIGURE 9.5   The best mixture for aggregates of example 3. (Graph created by the author.)

TABLE 9.38

Calculation of the Fineness Module of Total Aggregates of Example 3

Sieve Size (mm)	Percent Remained (%)	Cumulative Percent Remained (%)
37.5	0.0	0.0
19	0.0	0.0
9.5	9.8	9.8
4.75	22.1	31.9
2.36	16.1	48.0
1.18	13.1	61.1
0.6	11.4	72.5
0.3	10.8	83.3
0.15	11.8	95.1
Total	–	401.6

water-to-binder ratio. On the other hand, the compressive strength of cement is $450 \, kg/cm^2$. According to the above mentioned, we will have the below values:

- From Table 9.8 for I-525 cement and for the mix design compressive strength of 65 MPa, we will have w/b = 0.36.
- From Table 9.9 for I-425 cement and for the mix design compressive strength of 65 MPa, we will have w/b = 0.32.
- For method 3, as the compressive strength of cement is $450 \, kg/cm^2$, we should find the water-to-binder ratio for this cement from the linear correlation between 0.36 and 0.32 which is w/b = 0.33.

So, the water-to-binder ratio for this concrete is 0.33.

- Step 6: specify free water in concrete

To find the amount of free water in concrete, we should use Tables 9.10 and 9.11. In these tables, you can see that we should check two parameters. The first one is the fineness module of total aggregates that we have as 4.02 from step 4. The other parameter is the amount of cement. Here we assume that the amount of cement will be 475 kg/m^3. So, we will have:

- From Table 9.10 for the target slump of 90 mm and fineness module of 4.0 and cement of 475 kg, we will have 242 L for the amount of free water.
- From Table 9.11 for the target slump of 150 mm and fineness module of 4.0 and cement of 475 kg, we will have 260 L for the amount of free water.
- As the target slump is 210 mm, we should find the amount of free water in concrete with linear correlation which is 278 L.
- Now, we should decide about the dosage of the super-plasticizer. According to the economic considerations and the production of high-strength concrete with high slump, we decide to use 1.2% of the PCE super-plasticizer that you can find its water reduction rate in Figure 9.3. The water reduction rate for the dosage of 1.2% is 34.8%. We will use 34% for more assurance.
- We should decrease the amount of water by 34%. So, 34% of 278 L is 94 L and the amount of final free water after using the super-plasticizer will be 184 L.
- Step 7: specify the amount of cement and other binders
  You should find the amount of cement by using equation 9.6:

$$C = 184/0.33 = 557 \, kg$$

In this example, we would like to use GGBS and silica fume beside the cement. We decided to use 20% of GGBS and 8% of silica fume for the production of this concrete. So, we should increase the amount of total binder according to Table 9.12 which is 15% for the GGBS and 0% for the silica fume. So, the amount of the total binder will be 640. We should use 20% of it as the GGBS which is 130 and 8% of it as the silica fume which is 50. So, we will have:

- Portland cement: 460 kg
- GGBS: 130 kg
- Silica fume: 50 kg
- Total binder: 640 kg

You can see that the amount of cement that we assumed in the before step as 475 is a little more than the exact amount of cement which is better for the assurance of the mix design.

The amount of super-plasticizer is 1.2% of 640 kg which is 7.7 kg.
- Step 8: specify the total volume of aggregates in concrete

TABLE 9.39

Calculating the Amounts of Aggregates in SSD Conditions for Example 3

Type of Aggregate	Percent in Total Mixture (%)	Volume (L)	Density (kg/L)	Weight of SSD (kg)
Coarse 5–12	25	144	2.76	397
Natural sand	70	405	2.71	1098
Dune sand	5	29	2.68	78
Total	100	578	–	1573

For this step, you should use equation 9.7. We have all of the information for use in this equation instead of the V_a which is the amount of air. For this reason, we should use Tables 9.13 and 9.14. From Table 9.13 with the maximum size of coarse aggregate of 12.5 mm, we will have the amount of air between 0.8% and 1.5%. From Table 9.14 and for PCE super-plasticizer we will have the amount of air between 1.0% and 1.8%. So, we can assume the amount of air for this concrete is 1.7%. Now from equation 9.7, we have:

$$V = 1000 - (460/3.15) - (130/2.9) - (50/2.2) - (184) - (17) - (7.7/1.07) = 578\,L$$

- Step 9: calculating the weight of aggregates in SSD condition
  For this step, you can use Table 9.39 as below:
- Step 10: calculating the real weight of aggregates and water
  In this step, we should calculate the real weight of aggregates and water of concrete according to the water absorption and the moisture conditions of aggregates as below:
- For coarse 5–12, we have water absorption of 0.8% and dry conditions. So, 397 kg of this gravel needs 0.8% water to be in SSD conditions which is 3 kg. This weight should add to the weight of concrete water and subtract from the weight of aggregate. So, the real weight of coarse 5–12 is 394 kg which could be rounded to 395 kg and the weight of water is 187 kg.
- For natural sand, we have water absorption of 1.9% and 5.1% total moisture inside. So, this sand has (5.1-1.9=3.2%) of excess water which is 35 kg for 1098 kg of sand. It should subtract from the weight of concrete water and add to the weight of the sand. So, the real weight of sand is 1133 kg which will round to 1135 kg and the weight of water is 152 kg.
- For dune sand we have a water absorption of 2.5% and 1.5% total moisture inside. So, this sand has (1.5% − 2.5 = −1%) lack of water which is 1 kg for 78 kg of sand. It should add to the weight of concrete water and subtract from the weight of the sand. So, the real weight of dune sand is 77 kg which will round to 80 kg and the weight of water is 153 kg.

Now, you can see the final mix design for the concrete of example 3 in Table 9.40.

TABLE 9.40

Final Concrete Mix Design for Example 3

Constituent Material	Mix Design for 1 m³ (kg)	Mix Design for 30 L (g)
Gravel 5–12	395	11850
Natural sand	1135	34050
Dune sand	80	2400
Cement Type II	460	13800
GGBS	130	3900
Silica fume	50	1500
PCE super-plasticizer	7.7	231
Water	153	4590
Total weight	2411	72321

9.6 IMPLEMENTATION OF MIX DESIGN IN THE PROJECTS

The procedure for concrete mix design is the primary part of it. After that, you should check the mix design in the laboratory to evaluate it according to the fresh and hardened concrete specifications. If there is a need, you should make some modifications as below:

- If the slump of fresh concrete in the laboratory will be higher than the target slump, you can decrease the amount of the super-plasticizer and check the compressive strength. If you think that there is a need to increase the compressive strength for more assurance, you should decrease the amount of water.
- If the slump of fresh concrete in the laboratory will be lower than the target slump, you should check the water reduction rate of the super-plasticizer. You can use more super-plasticizer for the higher slump or you can change the super-plasticizer and use a stronger one. But you should also care about the slump retention effect of the super-plasticizer.
- If the amount of air in the laboratory will be higher than defined. You should check the super-plasticizer. Most of the time, if you contact with the super-plasticizer manufacturer, they can control the amount of air in concrete by modifying the formulation of the super-plasticizer without any change in the water reduction rate.
- You should check the slump keeping effect of the concrete in the laboratory. For this purpose, you should try to define the conditions as same as real project from temperature, speed of concrete mixing, time of transportation, and air and aggregates moisture conditions. If the slump keeping of the super-plasticizer will not satisfy you, contact with the super-plasticizer producer. Note that if there is a need for changing the super-plasticizer because of the slump keeping effect, you should check the water reduction rate of it and sometimes, there is a need to design your concrete again with the new

super-plasticizer. So, it is recommended to check the super-plasticizer slump keeping effect when you are checking its water reduction rate before starting the concrete mix design procedure.

- If the compressive strength of concrete in the laboratory will be less than you desired, you should check cement and other binder quality and you should design the mix again with the assumption of lower cement compressive strength. The checking of concrete compressive strength in the laboratory should be at the age of 7 days. So, if there is a need for any modification, you can do it on time. Most of the time, if you are using a cement with steady quality, the compressive strength in the laboratory will be good or even higher than you desired.

- If the compressive strength of concrete in the laboratory will be much higher than desired. Because of the economic considerations, you should check the compressive strength of the cement and if there is a need, you can check your mix design again for decreasing the amount of cement or super-plasticizer.

After finalizing the mix design in the laboratory, you can use it in the batching plant with high assurance. During the quality control of concrete in the project, it is possible to find needs for some modifications in the concrete mix design that you can do according to the above mentioned for the laboratory. But most of the time, if you do the procedure with high accuracy, there is no need for any modification instead of any change in the constituent materials.

REFERENCES

Aitcin P.C, High Performance Concrete, E&FN SPON, 2004.

Connor Jerome J, Faraji Susan, *Fundamentals of Structural Engineering*, Springer, 2016.

European Standard Organization, Admixtures for Concrete Mortar and Grout, EN934 Series.

European Standard Organization, Admixtures for Concrete, Mortar and Grout Test Methods, EN480 Series.

European Standard Organization, Cement Composition, Specifications and Conformity Criteria for Common Cements, EN197-1: 2000.

European Standard Organization, Concrete-Part 1: Specification, Performance, Production and Conformity, EN206-1, 2000.

European Standard Organization, Testing Fresh Concrete, EN12450 Series.

European Standard Organization, Testing Hardened Concrete, EN12390 Series.

European Standard Organization, Tests for General Properties of Aggregates, EN932 Series.

European Standard Organization, Tests for Geometrical Properties of Aggregates, EN933 Series.

Iranian Institute for Research on Construction Industry, National Concrete Mix Design Method, 2015.

Iranian National Management and Programming Organization, National Handbook of Concrete Structures, 2005.

Iranian Standard Organization, Concrete Admixtures, Specification, ISIRI2930, 2011.

Iranian Standard Organization, Concrete Specification of Constituent Materials, Production and Compliance of Concrete, ISIRI2284–2, 2009.

Iranian Standard Organization, Standard Specification for Ready Mixed Concrete, ISIRI6044, 2015.

Janamian Kambiz, Aguiar Jose, A Comprehensive Method for Concrete Mix Design, Materials Research Forum LLC, 2020.

Lamond F. Joseph, Pielert H. James, *Significance of Tests and Properties of Concrete and Concrete Making Materials*, ASTM International, 2006.

Mahmood Zadeh Amir, Iranpoor Jafar, Concrete Technology and Test (Farsi), Golhaye Mohammadi, 2007.

Mostofinejad Davood, Concrete Technology and Mix Design (Farsi), Arkane Danesh, 2011.

Nawy G. Edward, *Concrete Construction Engineering Handbook*, CRC Press, 2008.

Newman John, Choo Ban Seng, *Advanced Concrete Technology, Concrete Properties*, Elsevier, 2003.

Popovics Sandor, *Concrete Materials, Properties Specification and Testing*, NOYES Publications, 1992.

Ramachandran V.S, Beaudion James, *Handbook of Analytical Techniques in Concrete Science and Technology, Principles, Techniques and Applications*, William Andrew Publishing, 2001.

Ramezanianpoor Aliakbar, Arabi Negin, Cement and Concrete Test Methods (Farsi), Negarande Danesh, 2011.

Richardson M, *Fundamentals of Durable Reinforced Concrete*, Spon Press, 2002.

Safaye Nikoo Hamed, Introduction to Concrete Technology (Farsi), Heram Pub, 2008.

Zandi Yousof, Advanced Concrete Technology (Farsi), Forouzesh Pub, 2009.

Zandi Yousof, Concrete Tests and Mix Design (Farsi), Forouzesh Pub, 2007.

10 Production, Transportation, and Implementation of Concrete

In this chapter, we are going to talk about the process of concrete usage in the structures. The process consists of three parts as follows:

- Production of concrete: The production of concrete means all of the activities that we should do to produce a concrete with high quality exactly as we designed it in the before chapters. This process is very important for the quality of concrete. But it is not enough for high-quality implementation of concrete in the structure.

    We can produce the concrete with different instruments. But the most important parameter is good and accurate weighting system in the production instruments for all of the constituent materials and good mixing process.
- Transportation of concrete: When we produce a concrete, the second step is transporting it to the place that we are going to use. The transportation of concrete is much different from the other types of construction materials. So, we need special instruments for this purpose. On the other hand, you should attention to some special considerations for good transportation of concrete without any changing the quality.
- Implementation of concrete: When you transported the concrete to the place that we are going to use, we should implement the concrete in the structural elements. For this purpose, we should consider different activities. First, we should pump it to the final place. Then we should place it to the structural element and compact it as good as it possible. If there is a need for smoothing the surface, it should be done and finally, we should cure the concrete in the structure to protect quality.

In this chapter, we are going to talk about the above processes separately. We will discuss the instruments that we need for concrete production, transportation, and implementation. Also, we will talk about the important considerations to achieve the targets of the implementation with highest quality in the structure.

DOI: 10.1201/9781003384243-10

10.1 PRODUCTION OF CONCRETE IN THE LABORATORY

The first subject that we are going to talk about is the production of concrete in the laboratory. As mentioned before in the chapter on concrete mix design, after finalizing the concrete mix design, we should check it in the laboratory. So, you should produce the designed concrete in the laboratory with the maximum equality to the real project and batching plant. If the checking process of mix design in the laboratory will be good, you can implement the mix design in the batching plant with high accuracy. Nevertheless, you will give different results from the laboratory and project and this duplication will need very hard activity to finalize the mix design in the project.

The other usage of concrete production in the laboratory is checking the consistency of quality for some of the constituent materials like Portland cement and super-plasticizer. You can make an accurate concrete with the same materials and different parts of the Portland cement and check the compressive strength and either the workability of concrete to check the quality of cement. You can use the same materials with different parts of the super-plasticizer to check its performance from the water reduction rate and air entraining point of view. On the other hand, if you give a new sample of cement or super-plasticizer that you may like to use in the production process, you can check them in the laboratory to be sure about the quality and performance. So, for all of the projects, you need a good laboratory with all of the necessary instruments for the quality control of concrete that one of the most important parts of it is the concrete production instruments as below:

- Balance with minimum capacity of 30 kg and the accuracy of 1 g for weighting of aggregates and binders (Figure 10.1).
- Buckets to use for weighting aggregates and binders (Figure 10.2).
- Balance with minimum capacity of 1 kg and the accuracy of 0.1 g for weighting of water and super-plasticizer (Figure 10.1).
- Beaker for weighting of water and super-plasticizer (Figure 10.2).
- Suitable laboratory concrete mixer (Figure 10.3).

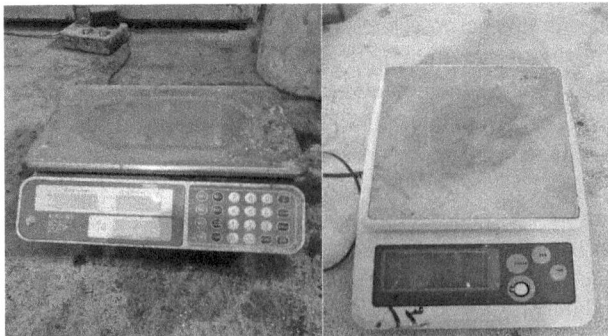

FIGURE 10.1 Two types of balance for concrete laboratory (the right one with more accuracy). (Photograph by the author.)

FIGURE 10.2 Bucket(left) and beaker(right) as the container for constituent materials in laboratory. (Photograph by the author.)

FIGURE 10.3 Laboratory concrete mixer. ("Concrete lab mixer".)

The quality of the laboratory mixer is very important for the accuracy of the test. We have different laboratory mixers in the market. The best one is pan mixers with good mixing power and different rotation speed. But if it is not possible for you to buy this type of mixer, you can use other types. On the other hand, you should use a concrete mixer with a minimum capacity of 50 L. The capacity of mixer is very important for good mixing of concrete. You should not use less than 50% of the capacity of mixer for the production of concrete. For example, if you have a 50 L capacity mixer, the minimum capacity that you can use is 25 L and it is better to use 30 L to get more accurate results.

To produce a concrete in the laboratory, it is recommended to use below procedure:

- You should weight aggregates, cement, and other binders in the separate buckets as accurate as possible.
- Weight 50% of total water.
- Put the aggregates in the mixer and add 50% of total water with the aggregates and let them mix and remain for about 5 minutes. This will help you to get the slump retention result near the real project conditions.
- Weight the remained water and add the super-plasticizer to the water and hand mix it. You should weigh water and super-plasticizer with a high accuracy balance. If you checked the aggregates moisture exactly, you may know the exact amount of water. But if you didn't check the exact amount of moisture in the aggregates at the time of test, you should continue the procedure with slump base method. So, you should weigh the minimum water that you estimate for the concrete and mix it with the super-plasticizer. Then you should weigh another amount of water in other beakers to add it to the final concrete mix till you get the target slump and weight the remained water to achieve the exact amount of water for concrete. For checking the mix design, it is recommended to use dry aggregates to calculate the exact amount of water. So, you can check all of the properties that you may need.
- Add the cement and other binders to the aggregates in the mixer and let them mix together for a few seconds.
- Add water and super-plasticizer to the mixer and let them mix for minimum one minute. Then stop the mixer and check the concrete visually. If it is good, you can continue testing. If not, you should find the problem and probably do the test again.
- Now it is time to check the concrete slump. You should stop the mixer and make a slump test on concrete.
- After checking the slump, it is time to check the slump retention. If you have a mixer with different speed, you can use the lower speed and let the concrete remain for defined time and again check the slump. Before checking the slump, first you should mix it with high speed for a few seconds and then do the slump test. If you don't have a mixer with different speeds, you should let the mixer stop and mix the concrete each 5 minutes for a few seconds and again let it remain without mixing to the defined amount of time that you would like to test the slump retention.
- Finally, you should pour the molds for the compressive strength test. You may use cubic or cylinder specimens for the test.

FIGURE 10.4 Suitable aggregate storage system with good protection. (Photograph by the author.)

10.2 PRODUCTION OF CONCRETE IN THE BATCHING PLANT

The final part of concrete production is in the batching plant. Batching plant is a special instrument made for the production of concrete in huge capacity. This instrument is made of different parts as below:

- Aggregates storage: You can store the aggregates in a suitable condition with a roof to protect them from direct sunlight and rain. But for some types of batching plant, you should store them exactly on the back of the batching system. You can see a picture of a good aggregate storage system in Figure 10.4.
- Cement and other binder storage: For the storage of powder materials, you should use silos that you can see in Figure 10.5. You can full the silos from a pipe which will go to the top of the silo by high pressure of air from the cement bunker and you can discharge it from the lowest part by using of spirals.
- Aggregates transition system: Now it is time to transit the aggregates from the storage to the weighting system. We have different types of systems in different batching plants. One using the buckets on the top of the weighting system that you can full them by using of a loader that you can see this system at the left part of Figure 10.6. The other system which is a little older is using a bucket with a towing wire to pick up the aggregates behind the valve of weighting system. You can see this system at the right part of Figure 10.6.
- Cement transition system: The transition of cement and other binders is possible by using of spirals. You can transit all kinds of powder materials with the spirals. These spirals will transit the powders by rounding in one direction and it can transit back the powder by rounding in other direction.

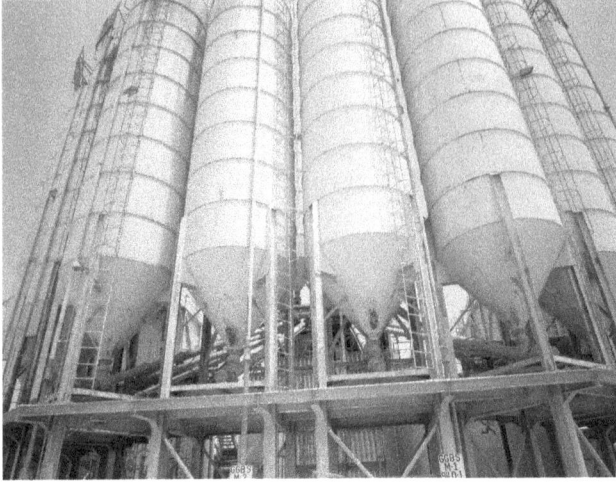

FIGURE 10.5   Cement and other binder storage. (Photograph by the author.)

FIGURE 10.6   Two types of aggregate transition systems. (Photograph by the author.)

You can see a picture of two spirals which can transit the cement from the lower part of silo to the upper part of the cement loading system in Figure 10.7.

- Aggregates weighting system: We have a bucket on a load cell. The aggregates separately will discharge from the valve to the bucket which in on the load cells till it achieves the defined weight. Then the other valve will open for the next type of aggregate. So, we can weigh all types of aggregates separately. For example, if we need 500 kg of 11–19 coarse, the valve for the 11–19 will open. When the weight is near 500 kg the valve will close a little for better controlling the discharge amount. When the weight will be 500 kg the valve will close and the other valve for 5–12 will open. If we need 300 kg of 5–12, it means that we should get to the weight of 800 kg

FIGURE 10.7 Cement and other powder materials transition system. (Photograph by the author.)

for the sum of 11–19 and 5–12. So, the valve will be open till the weight of 800 kg. Then it is time for the sand. If we need 1000 kg of sand, it means that we will have the total weight of aggregates as 1800 kg. So, when we get to the weight of 1800 kg the sand valve will close. This weighting system is accurate enough to weigh the aggregates for concrete. We can see a maximum of 5% of accuracy which is enough for the weighing of aggregates. It is possible to give more accuracy from this system but more accuracy means less speed in the weighting system. You can see a picture of aggregates weighting system in Figure 10.8.

- Sand vibration system: For moisturized soft sand, when the valve of the weighting system will be open sometimes it cannot fall down to the weighting bucket, because of the stickiness of moisturized fillers inside the sand. So, you should use a special vibration system to fall it down by the power of vibration.
- Cement and other binders weighting system: For this purpose, we need a special and bucket on loadcells that you can see a picture of it in Figure 10.9. In this system, the cement spiral will work till we get the defined weight and

FIGURE 10.8 Aggregate weighting system with different pneumatic valves. (Photograph by the author.)

FIGURE 10.9 Cement and other binders weighting system and bucket. (Photograph by the author.)

then it will stop and if we need other binders the spiral for that binder will work till we get to the final weight for the mix of binders. The transportation of cement and other binders from silo to the upper part of the bucket is with a spiral which we talked about them before.

FIGURE 10.10 Aggregates transition system to the mixer by using of conveyor belt. ("MOBILE CONCRETE BATCHING PLANT mekaglobal".)

- Aggregates transition to the mixer: For the transition of aggregates to the mixer we have two important systems. One is the mobile bucket that we will pull it up to the place of the mixer by using of towing wire. You can see a picture of this system in the right part of Figure 10.8. The other more usable system is using of conveyor belt for transition of the weighted aggregates to the mixer on the top that you can see a picture of this system in Figure 10.10.
- Cement and other binders transition to the mixer: As you can see in Figure 10.11, the cement weighting system is on the top of mixer. So, that is enough to open the valve in the lower place of the bucket to discharge the cement and other binders to the mixer.
- Cement vibration system: As mentioned before, the cement weighting system is on the top of the mixer in the batching plant. So, you can discharge it by opening the pneumatic valve. But, you know that the powder materials like cement is sticky itself according to the electrostatic loads. You can use a vibration system on the cement bucket to accelerate the cement and other binders discharging to the mixer.
- Admixture dosing system: For liquid admixtures, we have two systems. One is weighting system on the top of the mixer. The admixture will transit to the weighting system with a pump until the weight of admixture will be exactly the defined value. In the second system, we will use a volume system on the top of the mixer. The transition of admixture is with a pump or vacuum system. But the amount will adjust in volume base by signal of an electronic scale. For the liquid admixtures using of the volume system is much better than the weighting system because in the batching plant we have much dust and vibration and these two effects are very bad for the accurate loadcells. For admixtures, as we are using little amounts, we should use very accurate loadcells.

FIGURE 10.11 Cement discharging system by using the pneumatic valve. (Photograph by the author.)

These loadcells and related electronic devices will destroy in the condition of a batching plant. So, it will be out of calibration soon. You can see a picture of a volume base admixture dosing system in Figure 10.12. You can discharge the admixture to the mixer by using a valve if you put it on the top of the mixer or you can use a pump.

- Mixer: All of the constituent materials should discharge to the mixer for final mixing. First, we should discharge the aggregates to the mixer. After discharging half of the aggregates, we can start discharging of the cement and other binders. After discharging half of the aggregates, you can pump the water to the mixer and finally after pumping 80% of the water and full of cement and other binders, you can discharge the super-plasticizer. Then you should let it mix for minimum 30 seconds and if there is a need you can discharge remained water and let it mix well. Most of the batching plants will do the above procedure automatically. Some batching plants can calculate the amount of moisture in the aggregates and finalize the water automatically. For the mixer, we have two types. One is the pan mixer that

FIGURE 10.12 Volume base admixture dosing system. (Photograph by the author.)

you can see in Figure 10.13 and the other type is the twin shaft mixer that you can see in Figure 10.14. There is no much different between the production of concrete with these two types of mixers. But, some of the experts recommended using of twin shafts for better mixing. On the other hand, if you would like to use a mixer with higher capacity, you should use twin shaft system.

- Concrete discharging system: After finalizing the mixing process of concrete, you should discharge it to the truck mixer from a special cone which is below the valve of the mixer. You can see a picture of it in Figure 10.15.
- Batching control system: The loading of constituent materials and mixing and discharging of concrete should control from a control room with a control system. We have different controlling systems which are going to be fully automatic nowadays. You can see a picture of a half-automatic control system in Figure 10.16.

FIGURE 10.13   Pan mixer system. (Photograph by the author.)

FIGURE 10.14   Twin shaft mixer system. (Photograph by the author.)

FIGURE 10.15 Discharging system for concrete. (Photograph by the author.)

FIGURE 10.16 Half-automatic batching control system. (Photograph by the author.)

FIGURE 10.17   Truck mixer under the batching, the special vehicle for concrete transportation. (Photograph by the author.)

10.3   CONCRETE TRANSPORTATION WITH TRUCK MIXERS

When you produce a high-quality concrete, now it is time to transport it to the place that you would like to use. So, it is very important to protect the quality of concrete during the transportation time. Sometimes, you may use ready mixed concrete for the project. In this case, the transportation time will be more than 30 minutes, but if you use a batching plant in the project which is the case for bigger projects, you should transport concrete only from the batching plant to the final place which is less than 30 minutes and it could be even less than 15 minutes.

For concrete transportation, you should use truck mixers as you can see in Figure 10.17. It is a special vehicle for concrete transportation. It can mix concrete during the time to control the setting of concrete. The mixing process should be with a minimum speed because, if you use higher speed for mixing, it will activate the cement and it can cause severe slump loose of concrete during the transportation time.

You can mix concrete with the truck mixer engine power and gearbox with different speeds. At the time of loading truck mixer under the batching, you should use higher speed. At the time of transportation, you should use minimum speed. Before discharging you should again mix the concrete with high speed for a few minutes.

On the other hand, the truck mixer tank can rotate in two different directions. When it rotates in one direction, the spirals inside the tank will only mix the concrete. For discharging the tank should rotate in the other direction. So, the spirals can transport the concrete to the discharging point and discharge it.

In the case of slump loose, you can add a little amount of admixture as the remainder into the truck mixer and let it mix at high speed for about 5 minutes and then discharge the concrete.

FIGURE 10.18   A conveyor belt that is suitable for concrete transportation. ("Mobile conveyor belt from TRIDIC".)

10.4   OTHER INSTRUMENTS FOR CONCRETE TRANSPORTATION

For normal concrete, the only vehicle for transportation is the truck mixer. But for other cases it is possible to use other vehicles for concrete transportation as below:

- For roller compacted concrete (RCC), as the concrete is with very low slump like 30 mm, you can use normal trucks for concrete transportation. On the other hand, these concretes will use for road surface or dam structures. So, using normal trucks instead of truck mixers will be more suitable for the case.
- For the transportation of concrete in short distances, you can use conveyor belt. In this case, you cannot use high slump concrete because it is not possible to transport a concrete with slump more than 120 mm with a conveyor belt. You can see a picture of a conveyor belt system which is suitable for concrete transportation in Figure 10.18.
- Tower crane: This is a very important instrument for big projects. You need it for many purposes like any type of transportation of heavy weight materials in the project. But if you use a bucket in the tower crane system, you can use it for concrete transportation. In this case, you can transport the concrete for a short distance. For example, from the truck mixer or batching to the roof of the structure. So, you can use tower crane also instead of the concrete pump for the projects. You can see a picture of a tower crane in Figure 10.19.

FIGURE 10.19   Use of tower crane in big projects. (Photograph by the author.)

10.5   CONCRETE PUMPING

One of the most important instruments for concreting is the concrete pump (Figure 10.20) because for most of the concrete usage in the world, we have pumping process. As mentioned in the before chapters, the quality of concrete is very important for the pumpability. But the quality of the pumping machine also is very important for good and high-speed pumping of concrete.

The concrete pump consists of two parallel cylinders. As one cylinder moves forward, the other moves back. The first cylinder which is called a material cylinder moves concrete out of the hopper. The second cylinder which is called the discharge cylinder pushes the concrete out of the pump in the location where it is needed. The two pistons work in tandem, alternately pulling in and pushing out their volumes of liquid concrete. The hydraulic flow created by the continuous flow of concrete is what causes the two cylinders to alternate back and forth. Instead of the above system, there is a mixer inside the pump hopper which mixes the concrete to prevent setting.

We have two main types of the concrete pump as below:

*   Stationary pump (Figure 10.21): This is a kind of pump without any pipe extended with it. So, you should assemble it before the start of the concreting. In fact, it is only the pumping machine and you can assemble the pipes

FIGURE 10.20 Concrete pumping. (Photograph by the author.)

FIGURE 10.21 Stationary pump machine. ("Construction of the new Malolos City hall building".)

according to the needs of the project. You can find different types of stationary pumps with different powers. For high raise buildings, you should use a very powerful stationary pump with the piping system which can go to the top floor and a boom system which can rotate and move through the roof with a hydraulic system.

FIGURE 10.22   Boom concrete pump. (Photograph by the author.)

- Boom pump (Figure 10.22): This is another type of concrete pump with the pumping machine and pipes assembled on a truck. So, you can drive this type of pump to the project and raise the pipes to the place that you would like to pump the concrete. You should just attention to the pipe length that you need for the project to choose the suitable pump. On the other hand, for some projects, especially in the big cities, you cannot assemble a boom pump with long pipes and special hydraulic supports. For these cases, although the usage of boom pumps is much easier for concreting, you should use stationary pumps.

10.6   COMPACTION OF CONCRETE

After pumping the concrete, you should make sure that the concrete will compact in the element to ensure the best quality. If you are using the SCC concrete, there is no need for any compaction because SCC concrete will compact in the best form and the quality of the final concrete element will be very good. You can see a picture of concreting with an SCC concrete in which there is no need for compaction in Figure 10.23. You can see that the concrete will move through the element without any problem.

For normal concrete, you will need the compaction. The compaction of concrete in the structural elements is possible by using the vibrators (Figure 10.24). The need for vibration is different according to the below considerations:

- Concrete slump: For higher slump, you need less vibration and for lower slump, you need more vibration
- Congestion of the rebars: For congested elements, you may need more vibration.

FIGURE 10.23 SCC concrete which is moving through the element. (Photograph by the author.)

FIGURE 10.24 Concrete vibration. ("Using a vibrating rod to compact and to purge the concrete of any air bubbles".)

- Aggregates gradation and maximum size of coarse aggregates: For coarser concrete, you should control the vibration, because if you use much vibration, the concrete will segregate and it can cause decreasing the quality. For softer concrete, you can use more vibration if there is a need.
- Dimensions of the element: For bigger elements, you need to use more vibration and it is recommended to use stronger vibrators. But for smaller elements, you can use small vibrators.

10.7   SMOOTHING THE SURFACE OF CONCRETE ELEMENTS

The quality of the concrete element's final surface is very important according to the considerations of cracking and durability. For some other structures like the flooring, it is more important than normal because of the special loading form in these structures. So, one of the activities that should be done for all types of structures is smoothing the concrete surface.

For most of the structures, you can do the smoothing by using handy trowels. It is very important to protect the surface against the walking before finalizing the setting time of concrete, and curing also is very important to protect the surface from cracking.

For flooring, you should smooth the surface by using the electric trowels that you can see a picture of it in Figure 10.25. This is a special instrument for smoothing the concrete surface very well. You should use it for the parking or industrial floors. You can get the best quality for the surface, if you use silicate hardener powders on the surface of concrete before using of the electric trowel. The usage of electric trowel should be at the time after the setting time happened. You cannot use it on the surface of fresh concrete because of the heavy weight of it. Sometimes, you can use electric trowel to protect the surface against plastic shrinkage. After happening the cracks of plastic shrinkage, you can use the electric trowel for the second time to eliminate the cracks and then cure the concrete to protect it against cracking.

FIGURE 10.25   Electric trowel. ("Flattening of just poured concrete by vibration" by Wouter Hagens.)

10.8   COLD JOINT IN CONCRETE

We have different types of joints in concrete structures. Some of them are very important for the performance of the structure like the expansion joints. The implementation of these joints is not the subject of this book. You can find the implementation procedure for different types of structures in the texts and catalogs of the special products for this purpose. Here, we are going to talk about the cold joint which is not good for the performance of the structure.

Cold joint is the joint between two concrete implementations in one element which happens because of the time between two concreting which causes the setting of concrete of the first step.

You should prevent the happening of the cold joints according to the below considerations:

- For most of the elements, you can check the time of concreting and use enough concrete continuously to prevent cold joints. It is very important especially when you are using ready mixed concrete.
- For huge elements like the foundations, which continuous concreting is not possible for total element. You can divide the structure into smaller parts. You can use a special fine expanded metal mesh for this purpose (Figure 10.26).
- If dividing of the huge element is not possible, you should use special retarder admixtures with a defined dosage. So, you can retard the setting of the first step concrete. Then after pouring the second step, you can use the vibrator to mix two layers with each other.

FIGURE 10.26   Expanded metal mesh for dividing the concrete elements. (Photograph by the author.)

10.9   CURING OF CONCRETE

As mentioned before, it is very important to cure concrete to save the quality of the project. If you produce a high-quality concrete but you didn't cure it in the project after concreting, the quality will drastically decrease in the elements and you cannot be sure about the performance of the structure. So, concrete curing is very important, although in most of the projects, there is no enough attention to that.

If you don't cure the concrete in the structure, you may see one or all of the below problems:

- Decrease in the compressive strength and other mechanical properties of concrete in the project compared with the laboratory test results.
- Huge cracks in the structure which cause a decrease in the performance and loading capacity of the structure.
- Microcracking in the structure which causes the permeability of aggressive ions into the concrete elements and decreasing the durability of structure.

    You see the importance of concrete curing in structures. Unfortunately, in the projects, contractors don't cure the concrete as well as it needs because the payment in most parts of the world is for pouring the concrete into the structure. So, there is not enough inspection for the curing of concrete in the structures instead of some very sensitive projects.

    Curing of concrete in all of the ambient conditions should be as below:

- Prevent the loss of moisture inside concrete: You can use water jet, steam curing or curing compound admixtures for this purpose. In hot weather conditions, this process is harder than in cold weather conditions because of the evaporation rate of water. On the other hand, wind also could increase the rate of evaporation. So, in this case, also you should try harder to control the loss of moisture inside concrete (Figures 10.27 and 10.28).
- Control the temperature of concrete in the structure: It depends on the weather conditions. In hot weather conditions, you can control the temperature of concrete by using cold water for curing. In cold weather conditions, it is much hard to control the temperature of concrete in the structure. You should use special blankets or using some hot weather sources to control the temperature. You should also prevent the structure from ice and snow.
- Doing the actions of curing for enough time: The time of curing is very important, especially for the prevention of cracking in the structure. The best time for curing depends on the type of structure, weather conditions, and type of concrete. But it is recommended to cure concrete for one week to get the best result. In the projects, most of the times, it is not possible to cure the structure for one week because of the implementation problems for the next step. In this case, the minimum time of curing recommendation is 48 hours with the best control.

FIGURE 10.27 Using of curing compound for curing of concrete. ("Applying Curing Compound and Stripping the Wall Forms at the Yard Lead Reception Pit" by MTA Construction & Development Mega Projects.)

FIGURE 10.28 Water jet curing. ("Iraqi subcontractors use a water base curing process" by Jim Gordon.)

10.10 CONCRETE RECYCLING SYSTEM

One of the problems in the process of concrete production, transportation, and implementation is the waste concrete that remained in the instruments. We have below concrete wastes in the process:

- Concrete remaining in the batching plant mixer should be washed before finishing the work.
- Concrete remaining in the truck mixer after discharging should be washed.
- Concrete remaining in the concrete pump equipment should be washed.
- Possibly concrete in a truck mixer with some problems like rejecting or excess concrete in a project should be discharged out.

For all of these purposes, we need much water. So, it is very important if we can recycle the waste concrete and water as below:

- We should separate the aggregates from the cement slurry and grade them again for use in the production process in the future.
- We should separate the water and purify that for using in the production process in the future.
- The remained waste should be a dense cement slurry with minimum volume that we cannot recycle it. It is the final waste of concrete that is very little.

There are different special instruments that you can use for concrete recycling (Figure 10.29). In some parts of the world, the use of these instruments is mandatory in projects and ready mixed plants. But unfortunately, in some other parts of the world,

FIGURE 10.29 Concrete recycling system in a ready mixed plant. (Photograph by the author.)

the usage of these instruments is not common. So, the concrete waste is going to be very dangerous for the environment in these areas.

For the disposal of final slurry waste which is a very little amount, there are different laws in different parts of the world. One of the best options is the deactivation of cement by using some chemicals, then let the slurry dry, and finally, use the deactivated cement particles as the filler for the sand that we are going to use in concrete production.

REFERENCES

Aitcin P.C, High Performance Concrete, E&FN SPON, 2004.

Asurnipal, "Mobile conveyor belt from TRIDIC." Retrieved from: https://commons.wikimedia.org/wiki/File:TRIDIC_Stacker_-_mobile_Conveyor_belt_(mobiles_Foerderband).jpg.

"Concrete lab mixer." Retrieved from: https://pxhere.com/en/photo/653528.

Connor Jerome J, Faraji Susan, *Fundamentals of Structural Engineering*, Springer, 2016.

European Standard Organization, Concrete-Part 1: Specification, Performance, Production and Conformity, EN206-1, 2000.

European Standard Organization, Testing Fresh Concrete, EN12450 Series.

European Standard Organization, Testing Hardened Concrete, EN12390 Series.

Fsmazlum, "MOBILE CONCRETE BATCHING PLANT mekaglobal" Retrieved from: https://commons.wikimedia.org/wiki/File:MOBILE_CONCRETE_BATCHING_PLANT_mekaglobal.jpg.

Gordon, Jim, CIV, "Iraqi subcontractors use a water base curing process on the wall of a building at the construction site for the Public Order Battalion (POB) complex, located in Baghdad, Iraq." Retrieved from: https://nara.getarchive.net/media/iraqi-subcontractors-use-a-water-base-curing-process-on-the-wall-of-a-building-eecf08.

Hagens, Wouter, "Flattening of just poured concrete by vibration." Retrieved from: https://commons.wikimedia.org/wiki/File:Concrete_flattening.jpg.

Iranian Institute for Research on Construction Industry, 9th topic of National Rules for Construction, "Concrete Structures", 2009.

Iranian National Management and Programming Organization, National Handbook of Concrete Structures, 2005.

Iranian Standard Organization, Concrete Specification of Constituent Materials, Production and Compliance of Concrete, ISIRI2284–2, 2009.

Iranian Standard Organization, Standard Specification for Ready Mixed Concrete, ISIRI6044, 2015.

Judgefloro, "Construction of the new Malolos City hall building." Retrieved from: https://commons.wikimedia.org/wiki/File:1097Construction_of_the_new_Malolos_City_hall_building_34.jpg.

Lamond F. Joseph, Pielert H. James, *Significance of Tests and Properties of Concrete and Concrete Making Materials*, ASTM International, 2006.

Mostofinejad Davood, Concrete Technology and Mix Design (Farsi), Arkane Danesh, 2011.

MTA Construction & Development Mega Projects, "Applying Curing Compound and Stripping the Wall Forms at the Yard Lead Reception Pit." Retrieved from: https://www.flickr.com/photos/mtacc-esa/7415006130.

Nawy G. Edward, *Concrete Construction Engineering Handbook*, CRC Press, 2008.

Newman John, Choo Ban Seng, *Advanced Concrete Technology, Concrete Properties*, Elsevier, 2003.

Ramachandran V.S, Beaudion James, *Handbook of Analytical Techniques in Concrete Science and Technology, Principles, Techniques and Applications*, William Andrew Publishing, 2001.

Ramezanianpoor Aliakbar, Arabi Negin, Cement and Concrete Test Methods (Farsi), Negarande Danesh, 2011.

Safaye Nikoo Hamed, *Introduction to Concrete Technology (Farsi)*, Heram Pub, 2008.

Shihchuan, Taipei city, Taiwan, "Using a vibrating rod to compact and to purge the concrete of any air bubbles." Retrieved from: https://commons.wikimedia.org/wiki/File:Using_a_vibrating_rod_to_compact_and_to_purge_the_concrete_of_any_air_bubbles.jpg.

Zandi Yousof, *Advanced Concrete Technology (Farsi)*, Forouzesh Pub, 2009.

11 Usage of Fibers in Concrete

One of the newest materials for use in modern concrete is fibers. It is not necessary to use fibers in concrete. But using them can improve the performance of concrete in many cases. So, using of fibers is going to increase in the projects.

In this chapter, we discuss the different types of suitable fibers for concrete and the improvement of properties that will cause because of them (Figure 11.1). We have different types of fibers that we can use in concrete. The properties of each one are different. On the other hand, they have different effects on the properties of fresh and hardened concrete. So, we should use them according to the needs of the project.

11.1 STEEL FIBERS

The most common fibers for use in concrete is the steel fibers. Concrete technologists are trying to use steel fibers instead of using steel bars in concrete. We have two types of steel fibers in the market:

- Macro-steel fibers: They are the most common type of steel fibers. Most of the time, we should use them for the flooring structures. They can improve the mechanical properties of concrete. But the research on using them instead of steel bars is in progress. You can see a picture of the macro-steel fibers in Figure 11.2.

FIGURE 11.1 Crushed fiber-reinforced concrete. ("Ultra-High Performance Fiber Reinforced Concrete" by Bianca Paola Maffezzoli.)

DOI: 10.1201/9781003384243-11

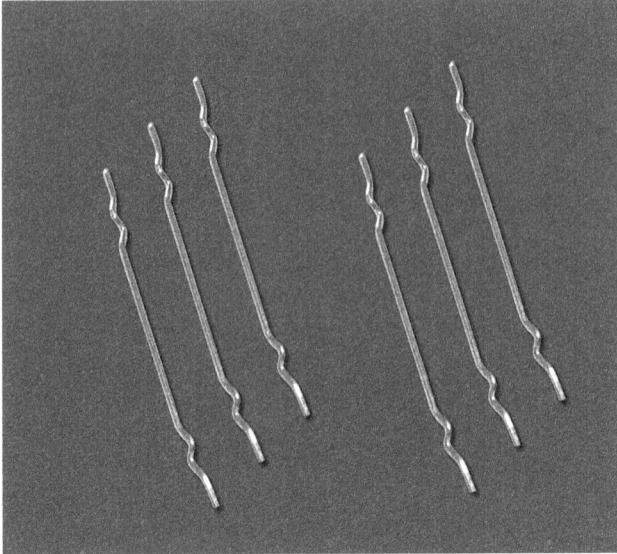

FIGURE 11.2 Macro-steel fibers. ("Stainless steel fiber.")

- Micro-steel fibers: This type of steel fiber is much different than the macro-steel ones. They are much thinner than the macro-steel fibers and the type of steel that we should use for the production of these types of fibers are different. They can improve the mechanical properties of concrete. The most important and common usage of them is the production of ultra-high strength concrete. In this case, the usage of these fibers instead of steel bars is possible. But it is not economical to use them instead of steel bars because of the high price of the micro-steel fibers in the market which is according to the high technology of the production process and the number of manufacturers in the world. You can see a picture of the micro-steel fibers in Figure 11.3 and the properties of these kind of fibers in Table 11.1.

You can see the positive and negative effects of using steel fibers in fresh and hardened concrete as below:

- Using both types of steel fibers will decrease the workability of concrete. So, most of the time, you should use more super-plasticizer for the production of concrete. The worse effect of them is on the pumpability of concrete because the fibers will increase the fraction of concrete with the pump pipes and it will cause drastically high pressure of pumping. So, we should try not to use steel fiber-reinforced concrete for the projects that we need high raise building pumping.
- When you would like to use steel fibers in concrete, you should consider the special concrete mix design according to the amount of steel fibers that you would like to use.

FIGURE 11.3 Micro-steel fibers. (Photograph by the author.)

TABLE 11.1
Normal Specifications of the Steel fibers

Type of Fibers	Density (kg/m³)	Diameter (μm)	Tensile Strength (MPa)
Macro-steel fibers	7800	500–1000	More than 500
Micro-steel fibers	7800	50–200	More than 1000

- Micro-steel fibers will increase the compressive strength of concrete. They can increase the compressive strength of concrete even more than 50%. Because of this property, we can try to use micro-steel fibers instead of steel bars in concrete structures.
- The effect of macro-steel fibers on the compressive strength of concrete is much lower than the micro-steel ones. They can increase the compressive strength by up to 10%.
- Both types of steel fibers will increase the tensile and flexural strength of concrete. This will be higher for the micro-steel fibers. They can increase the tensile strength by more than 50% and flexural strength by more than 150%. Because of this property, we can use the macro-steel fibers in the flooring structures and the micro-steel fibers instead of the steel bars in the concrete structures.
- Cracking will control in concrete by using steel fibers. It is because of the higher tensile strength of concrete.
- Reinforced concrete with steel fibers will be much resistant against the shock loads.

The most important factor you should mention for using any type of fiber to improve the properties of concrete is the homogeneous dispersion of them in the concrete element. In the case of not homogeneous dispersion, you will see even destroying the properties of concrete. The homogeneous dispersing of the fibers in concrete mainly depends on the below parameters:

- Good mix design for concrete according to the usage of the fibers. It means that using high performance super-plasticizers like polycarboxylate ether types with higher dosage and the softer concrete by using soft sand and lower size for the coarse aggregates can help the fibers to disperse homogenously.
- Enough mixing time for concrete in the batching plant. It means the mixing time is two times more than a concrete without using fibers.
- Good and suitable mixer for concrete. It realized that the pan mixers will be a little better for the production of fiber-reinforced concrete.
- It is better to decrease the transportation time for the fiber-reinforced concrete. Mixing concrete in the truck mixers too much will cause gathering of the fibers in some parts of the concrete inside the truck mixer. Using a truck mixer with good spirals can help to prevent agglomeration.

The amount of use for the steel fibers is different according to the type of concrete that you would like to produce. But you can use the below considerations:

- For both types of steel fibers to control cracking on the surface of concrete, you can use 2 to 10 kg of the steel fibers in 1 m³ of concrete.
- For the macro-steel fibers in the case of flooring with low loading capacity, you can use 10 to 20 kg of the steel fibers in 1 m³ of concrete.
- For the macro-steel fibers in the case of any type of high loading capacity flooring, you can use 20 to 30 kg of the steel fibers in 1 m³ of concrete.
- For the micro-steel fibers and for the production of ultra-high strength concrete, you can use steel fibers between 2% and 10% of the total weight of concrete which is the higher amount of use of any type of fiber in concrete.

11.2   GLASS FIBERS

Glass fibers consist of several fine fibers of glass. Glass fibers is formed when thin strands of silica-based or other formulation glass are extruded into many fibers with small diameters (Figure 11.4). There are many different types of glass fibers that you can use in different industries. You can see Table 11.2 for some of the different types of glass fibers that you can find in the market.

From the above types of fibers, the research is ongoing for using the best one to get the best properties for concrete. For example, for two types of them, you can see Table 11.3 for the specification.

The effect of glass fibers on the properties of concrete is different and it depends on the type of fiber, concrete mix design, and the element that we are going to make with the glass fiber-reinforced concrete (GFRC). But normally, we can name below properties for concrete when we would like to use glass fibers:

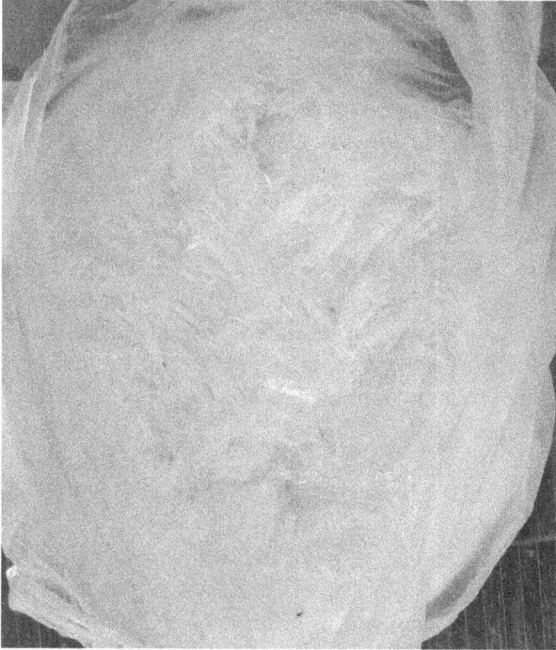

FIGURE 11.4 Glass fibers for use in the production of glass fiber reinforced concrete. (Photograph by the author.)

TABLE 11.2
Some of the Different Types of Glass Fibers

A glass	Alkali lime glass with little or no boron oxide
C glass	Alkali lime glass with high boron oxide content
D glass	Borosilicate glass with low dielectric constant
E glass	Alumino-borosilicate glass
R glass	Alumino silicate without MgO and CaO

TABLE 11.3
Specifications of Two Types of Glass Fibers

Type of Fibers	Density (kg/m³)	Diameter (μm)	Tensile strength (MPa)
E glass fiber	2500	10–15	More than 2000
R glass fiber	2600	10–20	More than 1500

- When you use glass fibers in concrete, the workability will decrease. You will have a concrete with hard moving speed. So, all of the properties referred to the workability will decrease.
- You should use a special concrete mix design when you would like to use glass fibers. It is better to use softer concrete with a soft sand and lower size of the coarse aggregates and use more super-plasticizer than normal. The best type of super-plasticizer for GFRC is the polycarboxylate base.
- Glass fibers will decrease the compressive strength of concrete. We cannot say how much you will see the decrease of compressive strength because it will be different according to the mix design, type of fibers, and other properties. So, research is ongoing to control this effect of glass fibers on concrete.
- Glass fibers will increase the tensile and flexural strength of concrete. We cannot say the exact amount of increase. But if you produce a good concrete, you will see a reasonable increase in the tensile and flexural strength.
- Because of the increase in the tensile and flexural strength of concrete, you can say that the glass fibers can help to prevent cracking on the surface of concrete elements.

  The most important usage of GFRC is the production of exterior building segments that you can use for the design of beautiful surfaces in the structures. You can see pictures of using GFRC in the exterior design of structures in Figure 11.5. On the other hand, you can make some concrete segments that can transmit the light which is very special for concrete. These segments also could be very useful as decorative materials. You can see a picture of these decorative parts in Figure 11.6.

11.3 ARTIFICIAL FIBERS

There are different types of artificial fibers in the market that you can use them in concrete production. Each type of these fibers can cause different properties in fresh and hardened concrete. The research is ongoing for the artificial fibers and the properties that you can achieve by using them in concrete. But in this book, we discuss the most important and widely used artificial fibers which is the poly propylene or PP fibers.

FIGURE 11.5　Use of GFRC in exterior design. (Photograph by the author.)

FIGURE 11.6 Light transition by GFRC concrete. ("Light transmitting concrete.")

11.3.1 POLYPROPYLENE FIBERS

Polypropylene fibers are made from the by-products of the textile industry (Figure 11.7). You can find them with different lengths in the market which is most of the time 8, 10, and 120 mm. You can see the specifications of polypropylene fibers in Table 11.4.

The effect of polypropylene fibers on the properties of concrete is different and depends on the amount of use, concrete mix design, and the element that we are going to make with the polypropylene fibers. But normally we can name below properties for concrete, when we would like to use PP fibers:

- When you use PP fibers in concrete, the workability will decrease. You will have a concrete with hard moving speed. So, all of the properties referred to the workability will decrease.
- You should use a special concrete mix design when you would like to use the PP fibers. It is better to use softer concrete with a soft sand and lower size of the coarse aggregates and use more super-plasticizer than normal. The best type of super-plasticizer for this type of concrete is the polycarboxylate base.

FIGURE 11.7   Polypropylene fibers. ("PP fiber for concrete" by Henan Botai.)

TABLE 11.4
Specifications of the Polypropylene Fibers

Type of Fibers	Density (kg/m³)	Diameter (μm)	Tensile Strength (MPa)
Polypropylene (PP)	900	20–100	More than 500

- Polypropylene fibers will decrease the compressive strength of concrete. We cannot say how much you will see the decrease in compressive strength because it will be different according to the mix design, amount of fibers, and other properties. So, research is ongoing to control this effect of PP fibers on concrete.
- Polypropylene fibers will increase the tensile and flexural strength of concrete. We cannot say the exact amount of increase. But if you produce a good concrete, you will see a reasonable increase in the tensile and flexural strength of concrete.
- Because of the increase in the tensile and flexural strength of concrete, you can say that the PP fibers can help to prevent cracking on the surface of concrete elements.
- Use of PP fibers will increase the resistance of concrete against the fire. In fact, if you would like to make a fire resistance concrete you should use PP fibers.

   The amount of use for polypropylene fibers in concrete is about 1 to 2 kg in 1 m³ of concrete. Like all other types of fibers, you should make sure that the fibers will separate from each other and spread during the total volume of concrete to get the best result.

FIGURE 11.8 Using of PP fibers in the production of CLC blocks. ("Concrete architecture.")

One of the important usages of the PP fibers is the production of cellular lightweight concrete blocks (Figure 11.8). You can use PP fibers in these blocks to prevent their shrinkage during time.

11.4 NATURAL FIBERS

There are too many types of natural fibers that you can use in concrete like straw fibers, hemp fibers, horse hair fibers, and so on (Figure 11.9). According to the recent researches and better properties of the artificial fibers, using of the natural fibers is limited these days. But in some parts of the world, you can find some types of concrete made with natural fibers.

As the properties of concrete depend on the type of natural fibers, we cannot name all of the properties you can see by using natural fibers. But the most important ones are as below:

FIGURE 11.9 Palm tree fibers. ("Palm tree.")

- The workability of concrete will decrease by using natural fibers. So, you should use a special concrete mix design and high-performance super-plasticizers like PCE types.
- The compressive strength of concrete will decrease by using natural fibers. The percentage of decrease depends on the type of fiber and amount of use.
- Most of the time, the tensile and flexural strength of concrete will increase by using natural fibers.
- Concrete cracking will control by using natural fibers according to the improvement of the tensile and flexural strength of concrete.

REFERENCES

Aitcin P.C, High Performance Concrete, E&FN SPON, 2004.
American Society for Testing and Materials, Standard Practice for Sampling Freshly Mixed Concrete, ASTM C172–99.
American Society for Testing and Materials, Standard Specification for Chemical Admixtures for Concrete, ASTM C494–99.
American Society for Testing and Materials, Standard Test Method for Specific Gravity and Absorption of Fine Aggregates, ASTM C128–97.
American Society for Testing and Materials, Standard Test Method for Sieve Analysis of Fine and Coarse Aggregates, ASTM C136–01.
American Society for Testing and Materials, Standard Test Method for Slump of Hydraulic Cement Concrete, ASTM C143–00.
American Society for Testing and Materials, Standard Test Method for Static Modules of Elasticity and Poisson's Ratio of Concrete in Compression, ASTM C469–94.
Bertolini L, Elsener B, Pedeferri P, Polder R, Corrosion of Steel in Concrete, Prevention.
"Concrete architecture." Retrieved from: https://www.rawpixel.com/image/6034770/photo-image-public-domain-concrete-free.
Ervanne Heini, Hakanen Martti, Analysis of Cement Super-plasticizer and Grinding Aids: A Literature Survey, Posiva Oy, 2007.
European Standard Organization, Concrete-Part 1: Specification, Performance, Production and Conformity, EN206-1, 2000.
European Standard Organization, Testing Fresh Concrete, EN12450 Series.
European Standard Organization, Testing Hardened Concrete, EN12390 Series.
Forgemind ArchiMedia, "Light transmitting concrete." Retrieved from: https://www.flickr.com/photos/eager/15578813800.
Henan Botai Chemical Building Materials Co., Ltd, "PP fiber for concrete." Retrieved from: https://commons.wikimedia.org/wiki/File:PP_fiber_for_concrete.jpg.
Iranian Institute for Research on Construction Industry, 9th topic of National Rules for Construction, "Concrete Structures", 2009.
Iranian Institute for Research on Construction Industry, National Concrete Mix Design Method, 2015.
Iranian National Management and Programming Organization, National Handbook of Concrete Structures, 2005.
Iranian Standard Organization, Concrete Specification of Constituent Materials, Production and Compliance of Concrete, ISIRI2284–2, 2009.
Iranian Standard Organization, Standard Specification for Ready Mixed Concrete, ISIRI6044, 2015.
Maffezzoli, Bianca Paola, "Ultra-High Performance Fiber Reinforced Concrete." Retrieved from: https://commons.wikimedia.org/wiki/File:UHPFRC_-_Ultra-High_Performance_Fiber_Reinforced_Concrete.jpg.

Mahmood Zadeh Amir, Iranpoor Jafar, Concrete Technology and Test (Farsi), Golhaye Mohammadi, 2007.

Newman John, Choo Ban Seng, *Advanced Concrete Technology, Concrete Properties*, Elsevier, 2003.

Ramachandran V.S, Beaudion James, *Handbook of Analytical Techniques in Concrete Science and Technology, Principles, Techniques and Applications*, William Andrew Publishing, 2001.

Ramachandran V.S, *Concrete Admixtures Handbook, Properties, Science and Technology*, NOYES Publications, 1995.

Ramezanianpoor Aliakbar, Arabi Negin, Cement and Concrete Test Methods (Farsi), Negarande Danesh, 2011.

Richardson M, *Fundamentals of Durable Reinforced Concrete*, Spon Press, 2002.

Safaye Nikoo Hamed, Introduction to Concrete Technology (Farsi), Heram Pub, 2008.

Shekarchizade Mohammad, Liber Nicolas Ali, Dehghan Solmaz, Poorzarrabi Ali, Concrete Admixtures Technology and Usages (Farsi), Elm & Adab, 2012.

Ulleo, "Palm tree." Retrieved from: https://pixnio.com/textures-and-patterns/tree-bark-cortex/palm-tree-palm-dry-nature-bark-fiber-brown#.

Zandi Yousof, Advanced Concrete Technology (Farsi), Forouzesh Pub, 2009.

Zs871124, "Stainless steel fiber." Retrieved from: https://commons.wikimedia.org/wiki/File:Stainless_steel_fiber.jpg.

12 Hot and Cold Weather Concreting

Concrete production, transportation and pouring highly depend on the weather conditions. Mild weather condition with an optimum temperature of around 20°C and high humidity is the best conditions for concrete. But we will have serious problems in hot and cold weather conditions that can decrease the quality of concrete and final structure too much.

In some parts of the world, we have hot weather conditions for most of the times of the year. In some other parts of the world, we may have cold weather conditions at most of the times of the year. In some other parts of the world, we may have hot weather conditions in warm seasons and cold weather conditions in cool seasons. So, you should know how to work with concrete under these conditions and how you can protect the concrete against harmful hot and cold weather problems.

Before starting the subjects about hot and cold weather concreting, you should know about the importance of concrete temperature. In concrete technology, the importance of concrete temperature is more than the weather temperature. In fact, it is very important to produce concrete with normal temperatures even in hot and cold weather conditions. If you can control the concrete temperature, you can protect the concrete against the harmful effects of hot and cold weather. You can check the concrete temperature by using the special thermometers as you can see in Figure 12.1. We will discuss this subject in the latter parts of this chapter.

12.1 CALCULATING CONCRETE TEMPERATURE ACCORDING TO THE CONSTITUENT MATERIALS TEMPERATURE

Now, it is time to calculate the concrete temperature according to the concrete mix design and the temperature of constituent materials. For this reason, you can use equation 12.1.

$$T = \frac{0.22(T_a T_a + T_c M_c) + T_W M_W + T_{Wa} M_{Wa}}{0.22(M_a + M_c) + M_W + M_{Wa}} \tag{12.1}$$

In the equation above, we will have T_a as the temperature of each aggregate, M_a is the weight of each aggregate, T_c is the temperature of the cement or any type of supplementary cementitious material, M_c is the weight of the cement or any type of supplementary cementitious materials, T_w is the temperature of water that you are using for concrete production, M_w is the weight of water, T_{wa} is the temperature of total water in the aggregates, M_{wa} is the weight of total water in aggregates and T is the temperature of concrete produces with the above constituent materials.

DOI: 10.1201/9781003384243-12

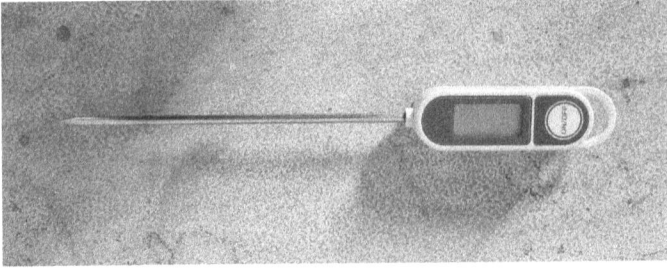

FIGURE 12.1   Special digital thermometer for concrete. (Photograph by the author.)

By using the above equation, you can calculate concrete temperature when you know the concrete mix design and the temperature of each material. We will use this equation in the latter parts of this chapter.

12.2 DEFINITIONS OF HOT WEATHER CONDITIONS FOR CONCRETE

The first thing that you should know about hot weather concreting is the maximum defined temperature for concrete or weather that you can continue concreting. In fact, we don't have any defined temperature for weather in which you are going to make concrete or pour a structural element. There is no recommendation about it in any concrete technology text or document. But we have a maximum defined temperature for concrete to protect it from any damage caused by hot weather conditions which is 32°C. So, it is very important to produce concrete with a temperature as low as possible in hot weather conditions.

We can define places with hot weather conditions as below:

- All of the places between the Tropic of Cancer or Tropic of Capricorn and the equator. These are the hottest places in the world. Most of the time, the weather conditions are like summer or spring. If you check a world map, you can see that many parts of the world are in this weather climate.
- All other parts of the world may have at least 1 month with a weather temperature of more than 30°C. By this definition, it is possible to have hot weather concreting even in a cold country in summer.

We may have two types of hot weather climates:

- Hot and dry: These are the places with high temperatures but dry weather. These are the places near deserts in the world (Figure 12.2).
- Hot and humid: These are the places with high temperatures and high humidity in the air. These are the places near the coastal areas in the world (Figure 12.3).

We cannot say which type of hot weather should be much harmful to concrete with confidence because it depends on the property of fresh or hardened concrete that we

FIGURE 12.2 A place with hot and dry weather conditions. ("Leaving traces on soft sand dunes in Tadrart Acacus" by Luca Galuzzi.)

FIGURE 12.3 A place with hot and humid weather conditions. (Photograph by the author.)

would like to talk about. For example, hot and dry weather is much harmful to the slump keeping of fresh concrete, because although the high temperature itself causes slump loose in concrete, the rate of evaporation in a hot and dry weather is very high and it can cause drastic slump loss of concrete. But in hot and humid weather, because of the high amount of humidity in the air the rate of evaporation is lower. On the other hand, the humidity in hot and humid weather conditions is the sea water with high amount of chloride and sulfate, which can attack the hardened concrete and cause serious corrosion. But in dry weather, we don't have this high amount of chloride and sulfate in the air. So, we can say that hot weather with humid or without humid could be dangerous for concrete and we should control it.

12.3 CEMENT HYDRATION REACTION AT HOT WEATHER CONDITIONS

As mentioned before, the hydration reaction is an exothermic chemical reaction that will cause the release of heat in the concrete element. High temperature will activate the chemical reactions much. For cement hydration reaction also, high temperature can activate it and cause rapid reaction between cement and water. It can cause rapid setting time and slump loose. On the other hand, it can cause high temperature inside fresh and hardened concrete at earlier ages. It will be much dangerous when we would like to pour concrete in big elements like a mass foundation.

High temperature in fresh and early age hardened concrete will cause loss of strength and concrete quality because of some unknown chemical reactions inside concrete. So, if you produce two concretes with the same constituent materials and mix design, one in high-temperature production place and another in the low temperature production place, you can see that the 28 days compressive strength of the concrete produced in the cooler place will be higher. It is possible to achieve higher initial strength for the concrete produced in higher temperatures at 3 or even 7 days. But if you compare the 28 and more day compressive strength, you will see different results. So, it is very important to control the concrete temperature for hot weather concreting. The maximum 32°C concrete temperature is defined because of the above explanations. If you produce a concrete with a temperature of 32°C, you will see higher temperature at the time of discharging in the project place, especially if you have high transportation time. So, it is strongly recommended to produce concrete with a temperature lower than 25°C in hot weather conditions.

12.4 THE EFFECTS OF HOT WEATHER ON THE PROPERTIES OF CONCRETE

As mentioned before, hot weather could be harmful to concrete. We can name the most important harmful effects as below:

- Higher water demand: The need for water in hot weather is higher than the mild or cold weather. This is because of the higher temperature of the constituent materials and higher evaporation rate. More water in concrete production means lower quality and mechanical properties for concrete.

- Rapid slump loosing during time: The slump keeping of concrete during transportation time is very hard in hot weather conditions because higher temperature will cause higher activation for the hydration reaction which will cause rapid setting time and slump loss. On the other hand, high evaporation rate in hot weather, especially in dry conditions, will cause loss of concrete water and rapid loss of slump. Unfortunately, most of the time, when the concrete slump at the project will be low, the laborers in the project will add water into the concrete before pouring to raise the slump. It will cause more water in concrete mix and lose the quality and mechanical properties of concrete. On the other hand, if the temperature of concrete will be high at the project time and you add cooler water to it, the cement molecules will shock and it will cause decreasing the quality of concrete more than expected.
- Rapid setting time: High temperature will cause more activation of the hydration reaction which will cause rapid setting time. If you continue the mixing of concrete after the initial setting time, you are destroying the strength and dense structure of the concrete which will cause lower quality in the final structure.
- Higher cracking probability: We will talk about concrete cracking in the latter parts of this chapter. But for now, you should know that higher water evaporation from the poured concrete will cause cracking on the surface of the concrete element. These cracks could be deep in the concrete element and can damage the strength of the structure.
- Higher temperature of fresh concrete: The maximum acceptable temperature for fresh concrete is 32°C. In hot weather conditions, achieving this temperature will be very hard and higher temperature will cause lower final strength of concrete.
- Different strength development patterns for concrete: In hot weather conditions, concrete will give you more early strength. But the latter strength will be much lower compared with a concrete produced at mild temperatures. So, we will have two problems. One is the lower final strength and two is the different strength development pattern which should be important for some projects. For example, you may need special compressive strength for 7 or 11 days. If you used special admixtures to achieve this strength, at higher temperature, the dosage and even type of admixtures that you are going to use should be different.

12.5 CONSIDERATIONS FOR HOT WEATHER CONCRETING

Now, it is time to talk about the considerations for the protection of concrete from the above damages. The most important activity that we should do is controlling the temperature of fresh concrete. So, we can name the most important considerations as below:

- Using cold water or powdered ice instead of concrete water or as a part of concrete water to decrease the temperature of fresh concrete: The special heat capacity for water is very high compared with the other materials.

FIGURE 12.4   Ice plant instrument besides a batching plant. (Photograph by the author.)

So, water is a very good cooler liquid for any purpose. If you use cold water for concrete production, it will cause decreasing concrete temperature very much. If you use powdered ice, the effect will be much higher because the ice will give high amount of heat to melt into the water and it will cause a higher decrease in concrete temperature. We will check the effect of water on concrete temperature by using equation 12.1.

To use powdered ice in concrete production, there are some special plants that produce powdered ice (Figure 12.4). You can use them beside the batching plant.

- Minimizing the transportation time: In hot weather climate, the truck mixer drivers should learn to minimize the transportation time. On the other hand, you should check them to control the best performance. Using GPS and traffic control systems in the truck mixers could be helpful. Adjustment of loading and sending time of the truck mixer with the discharging capacity in the project is very important to avoid wasting time in the project area. There are computer programs for this purpose.

FIGURE 12.5   Shade on the surface of aggregates to control their temperature. (Photograph by the author.)

- Using shade on the aggregates (Figure 12.5): Because of the dark color of aggregates, they absorb much heat from the sunlight. So, their temperature will be very high if we store them in an open area. This high-temperature aggregates will cause higher temperature for fresh concrete. Using shade on the surface of the aggregates can prevent high temperature of the aggregates and concrete.
- Using of retarding admixtures in concrete production: You can use retarding admixtures to retard the cement hydration reaction and control setting time and increasing of concrete temperature during time. You can use retarders separately or mixed with the super-plasticizers.
- Using strong super-plasticizers with slump-keeping effect: As mentioned before, in hot weather conditions, the water demand for concrete will raise. So, you will need stronger super-plasticizers to achieve the same slump compared with mild weather conditions. On the other hand, slump keeping of concrete is very important in hot weather conditions. So, you should use super-plasticizers with high slump-keeping effect. The best one is polycarboxylate ether-based super-plasticizers with slump retention polymers. You can use retarders mixed with the super-plasticizer to achieve better slump retention. Most of the time, you can find a special formulated retarded super-plasticizer for hot climate conditions in the market.
- Cooling the aggregates before using them in concrete production: If you cool the aggregates before using them in the concrete production process, it can decrease the temperature of concrete as you can calculate from equation 12.1. Most of the time, you can cool the aggregates by splashing water on their surface. This will cause a high amount of water consumption. So, it is not recommended to do it in most parts of the world with high worry about the water resources. On the other hand, if you use excess water for cooling

FIGURE 12.6   Liquid nitrogen tanks. ("Liquid nitrogen tanks and storage containers" by Matylda Sęk.)

the aggregates, controlling the amount of water for concrete production will be much hard.

- Using liquid nitrogen for the cooling of concrete (Figure 12.6): If you couldn't produce a concrete at low temperature, you can cool the final concrete by using liquid nitrogen. When the liquid nitrogen would like to evaporate, it will take a high amount of heat and will cool the concrete very much. The problem is the danger of using liquid nitrogen for the laborers, if they don't take care of themselves and also high price of the instruments for the usage of liquid nitrogen.
- Working at cooler times of the day: For some parts of the world with very high temperature, it is strongly recommended to work at cooler times of the day or it is better to work at night time. It can help us to protect the concrete from the high temperature of noon times and control the temperature of the concrete. On the other hand, it can help us to cure the concrete simpler.
- Using cooler cement for concrete production: Sometimes, when you transport the cement from the factory to the batching site and after discharging it, you will see cement with a high temperature in the siloes. It can highly increase the concrete temperature. So, it is strongly recommended to use the cement as cool as it is possible. You can ask the cement manufacturer to give you cooler cement. On the other hand, you can use high volume siloes and different siloes in one batching plant. So, you can use older cements with lower temperature first.
- Cure the concrete as well as possible: As mentioned before, the curing of concrete in hot weather conditions is much important than in any other case. So, you should control the temperature and moisture of concrete after pouring and you should do it as long as it is possible.

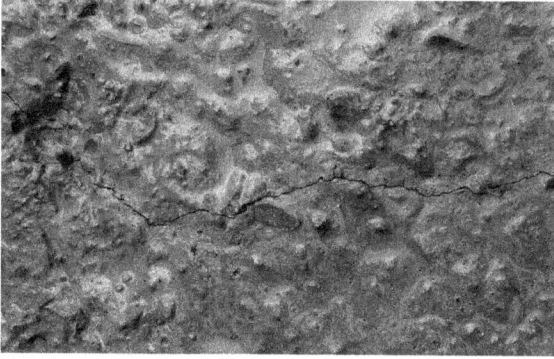

FIGURE 12.7   Plastic shrinkage and cracks. (Photograph created by the author.)

12.6   CONCRETE CRACKING AT HOT WEATHER CONDITIONS

We have different types of cracks in concrete. In this part of the book, we are going to talk about four types of cracks that should happen at any type of temperature or weather conditions. But the probability of this happening in hot weather conditions is much higher.

First, you should know that decreasing the volume of hardened concrete is called shrinkage, and if you cannot control the shrinkage, concrete will crack, especially on the surface of thinner elements. We have four main types of shrinkage that we are going to talk about:

- Plastic shrinkage (Figure 12.7): This is because of the early drying of the concrete surface. The rapid evaporating of water from concrete surface especially at high temperatures will cause rapid drying which is the reason for the plastic shrinkage. It can make cracks on the surface of concrete till the compressive strength of the concrete surface will raise to about 1 MPa.

The only way to control plastic shrinkage and prevent cracking is curing of concrete. In fact, you should prevent the evaporation of water from concrete surface. As the concrete surface is on plastic stage, you cannot use water curing. So, the best way is using the curing compounds or steam curing.

- Chemical shrinkage (Figure 12.8): It is also called autogenous shrinkage and is because of the hydration reaction of cement as the hydrated products volume is less than the volume of cement and water before starting the hydration reaction. So, it will cause a high amount of shrinkage and cracking especially in high-strength concrete. As the amount of cement in high-strength concrete is more than normal strength concrete, the danger of autogenous shrinkage will be much higher. If you do not cure the high-strength concrete, you will see a drastically cracked surface 12–24 hours after concreting. So, it is very important to cure high-strength concrete as well as it is possible.

FIGURE 12.8   Chemical shrinkage and cracks. (Photograph by the author.)

FIGURE 12.9   Concrete cracks that can cause by drying shrinkage and extension due to loading. ("Concrete Cracked" by John Harvey.)

The only way to control chemical shrinkage is water curing because you should substitute the water consumed during the early age hydration reaction to prevent lack of water inside the capillary pores to control cracking. As in the case of concretes with lower amount of cement, the autogenous shrinkage is lower, you can use curing compounds for total curing of concrete. But for concretes with a higher amount of cement, you should water cure the concrete from 4 to 10 hours after concreting for the time as long as it is possible to control autogenous shrinkage.

- Drying shrinkage (Figure 12.9): This is because of the evaporating of water during the hardening time of concrete which causes a lack of water, volume decrease, and finally shrinkage and crack in the concrete.

FIGURE 12.10 Special impermeable membrane that can use on concrete surface to prevent drying shrinkage. ("Airmen unfold a membrane tarp that will be used to cover a runway crater repaired during a quick reaction runway repair test" by TSGT Walter Perkins Jr.)

To prevent drying shrinkage, you should control the evaporating of water from the concrete element surface by water curing or using impermeable membranes specially made for this purpose (Figure 12.10).

- Thermal shrinkage: This is because of the thermal gradient between two parts of a concrete element which will cause cracking inside the concrete element. This type of shrinkage is especially for the huge concrete elements (Figure 12.11) and the only way to prevent it is controlling the temperature of concrete and using concrete with low temperature and low Portland cement content. We will discuss the considerations for mass concreting in Chapter 13.

FIGURE 12.11   A huge concrete structure with the risk of thermal shrinkage. (Photograph by the author.)

12.7   CHEMICAL ADMIXTURES FOR HOT WEATHER CONCRETING

To protect concrete from the hot weather problems, we need different types of chemical admixtures as below:

- Highly effective super-plasticizers: It means that you should use a strong super-plasticizer with high water reduction rate and good slump-keeping effect.
- Retarding admixtures: You should use retarder admixtures to control the heat release from the hydration reaction of cement.
- Curing compound: As the curing of concrete in hot weather conditions is very important, you can use curing compounds for this purpose.
- Shrinkage reducer admixtures: You can find some newly formulated admixtures for the reduction of shrinkage in the market. As the probability of shrinkage and cracks in concrete at hot weather conditions is higher than normal, you can use these admixtures for some types of elements to control cracking.

12.8   CALCULATIONS FOR CONCRETE TEMPERATURE AT HOT WEATHER CONDITIONS

In this part, we are going to check the effect of constituent materials on the temperature of concrete with a few examples. First, we should calculate the concrete temperature in a batching plant at a place with hot weather conditions without any consideration to control the concrete temperature. You can see the concrete mix design with the temperature of each constituent material in Table 12.1.

Now, we should use equation 12.1 to calculate the concrete temperature. We will use the super-plasticizer as part of water in the equation and we will have 35.5 kg water with a temperature of 29°C in the natural sand. So, the real amount of natural sand is 674.5 kg.

The numerator:

$$0.22 \ ((510 \times 45) + (260 \times 45) + (674.5 \times 29) + (330 \times 33) + (350 \times 42) + (90 \times 38))$$
$$+ (164 \times 26) + (35.5 \times 29) = 23602$$

Denominator:

$$0.22 \ (510 + 260 + 674.5 + 330 + 350 + 90) + 164 + 35.5 = 686.7$$

$$T = 23601/686.7 = 34.4°C$$

As you can see, the concrete temperature is upper than the defined 32°C. So, it is not acceptable if you make this concrete without any consideration of hot weather conditions. Also, you should know that this is the starting temperature of concrete at the time of production. You need to transport and probably pump it to the project. As the concrete age will increase, the temperature will raise much and it will be more dangerous.

TABLE 12.1

Concrete Mix Design and the Temperature of Constituent Materials for the Calculations of Concrete Temperature at Hot Weather Conditions

Constituent Material	Amount for 1 m³ (kg)	Temperature (°C)
Portland cement	350	42
GGBS	90	38
Coarse 11–19	510	45
Coarse 5–12	260	45
Natural sand with 5% moisture	710	29
Crushed sand without any moisture	330	33
Water	160	26
Super-plasticizer	4	26
Total	2414	–

In the first step, we try to cool down the aggregates by water splash on them. By using this simple method, you can decrease the temperature of aggregates by about 5–10 degrees. For the sand it is not common to use water sprinkle. So, we can use it for the coarse aggregates. Assume that we used sprinkle water for the two types of coarse aggregates. Then the temperature of them decreased to 35°C. It is the maximum amount that is possible by using this method. Now, the coarse aggregates have 4% of moisture which is 20.4 kg for the 11–19 and 10.4 kg for the 5–12 aggregates. So, the real weight of 11–19 will be 489.6 kg and the real weight of 5–12 will be 249.6 kg. Now, you can see the calculations as below:

The numerator:

$$0.22 \ ((489.6 \times 35) + (249.6 \times 35) + (674.5 \times 29) + (330 \times 33) + (350 \times 42)$$
$$+ (90 \times 38)) + (164 \times 26) + (35.5 \times 29) + (20.4 \times 35) + (10.4 \times 35) = 22748.9$$

Denominator:

$$0.22 \ (489.6 + 249.6 + 674.5 + 330 + 350 + 90) + 164 + 35.5 + 20.4 + 10.4 = 710.7$$

$$T = 22748.9/710.7 = 32°C$$

You can see that, by using this method, you can decrease concrete temperature only about 2.4°C. The concrete temperature is now on the borderline of acceptable. But, the hydration reaction will start soon and it will increase concrete temperature more than 32°C. So, it is not enough.

In the next step, instead of decreasing the temperature of coarse aggregates, we will use the binders with lower temperature for example cement and slag with a temperature of 35°C. You can see the calculations as below:

The numerator:

$$0.22 \ ((489.6 \times 35) + (249.6 \times 35) + (674.5 \times 29) + (330 \times 33) + (350 \times 35)$$
$$+ (90 \times 35)) + (164 \times 26) + (35.5 \times 29) + (20.4 \times 35) + (10.4 \times 35) = 22150.5$$

Denominator:

$$0.22 \ (489.6 + 249.6 + 674.5 + 330 + 350 + 90) + 164 + 35.5 + 20.4 + 10.4 = 710.7$$

$$T = 22150.5/710.7 = 31.2°C$$

You can see that this will decrease the concrete temperature to less than 1°C which is not too much to control the concrete temperature as low as it could be trustable.

Now, we would like to use cool water without cooling the coarse aggregates or using cooler cement. The only thing that we are going to do is using water with a temperature of 5°C.

The numerator:

$$0.22 \ ((510 \times 45) + (260 \times 45) + (674.5 \times 29) + (330 \times 33) + (350 \times 42)$$
$$+ (90 \times 38)) + (164 \times 5) + (35.5 \times 29) = 20158$$

Denominator:

0.22 (510 + 260 + 674.5 + 330 + 350 + 90) + 164 + 35.5 = 686.7

T = 20158/686.7 = 29.4°C

You can see that using cool water lonely can decrease the concrete temperature more than cooling aggregates which need wasting too much water and using cool cement which is very hard.

Now, we are going to use powdered ice with a temperature of −5°C instead of the total water of concrete.

The numerator:

0.22 ((510 × 45) + (260 × 45) + (674.5 × 29) + (330 × 33) + (350 × 42) + (90 × 38)) + (−5 × 160) + (4 × 26) + (35.5 × 29) = 18642

Denominator:

0.22 (510 + 260 + 674.5 + 330 + 350 + 90) + 164 + 35.5 = 686.7

T = 18642/686.7 = 27.1°C

You can see that the most effective way that you can use to control concrete temperature is using powdered ice instead of total water.

12.9 DEFINITIONS OF COLD WEATHER CONDITIONS FOR CONCRETE

Like hot weather conditions, the first thing that you should know for cold weather conditions is the minimum defined temperature for fresh concrete that you can be sure about the quality which is 5°C. If the temperature of fresh concrete falls down the 5°C, you should not continue pouring this concrete at the project, because at lower temperatures, the hydration reaction will stop or it will be very slow. So, the temperature of concrete will not raise and the concrete water can freeze. It can be very dangerous for the structure. So, you should take care of concrete in cold weather conditions to protect it from the damages caused by the cold weather.

You can see cold weather climate at any place in the world with a weather temperature of less than zero in winters (Figure 12.12). In some parts of the world, this kind of weather could be only for a few months and in some other parts of the world, it could be for several months. You can see this type of weather in all places with a latitude of more than 60 degrees at north and south for several months. For the latitude between 25 and 60 degrees you can see cold weather conditions for a few months. These are the places of the world that you can see hot and cold weather conditions together in 1 year. For the places with a latitude of less than 25 degrees, you don't see cold weather conditions.

FIGURE 12.12 Snowy day which is very hard conditions for concrete. ("A blacktop road in a forest during snowfall".)

Like before, you can see humid cold weather conditions or dry cold weather conditions. In this case, there are no special considerations referring to the weather humidity instead of the normal considerations for the protection of the structures against the chloride ion of sea water. When you have dry cold weather, you can continue concreting with some special considerations that we are going to talk about, but in humid weather conditions, you may have for example snowing for many days which is another problem that you should solve for concreting. So, we can say that control and protection of concrete in dry cold weather conditions could be a little simpler than the humid cold weather conditions.

12.10 CEMENT HYDRATION REACTION AT COLD WEATHER CONDITIONS

As mentioned before, the hydration reaction of cement is an exothermic chemical reaction that is good for cold weather conditions. But, as the reaction will accelerate in hot weather conditions, it can reduce in cold weather also. So, the speed of the cement hydration reaction highly depends on the ambient temperature. Although the hydration reaction can help the concrete to increase the initial temperature. But you should know that, at very low temperatures, the reaction will be very slow. So, it cannot help the concrete protect itself from the cold weather's harshly effects.

For this case, the concrete temperature is the important parameter. You should not use a concrete with a temperature of less than 5°C in any case because at the temperature of less than 5°C we can say that the hydration reaction will stop. So, concrete temperature without the action of the hydration will be lower and lower, the water inside the concrete will freeze and after passing of several hours, you will not see any effect of the setting time. In this case, even if the concrete will set after many hours, you will see drastic loss of the mechanical properties and quality. So, the effect of using concrete with a temperature of less than 5°C is very dangerous for the structure. In this case, you cannot repair the concrete element, you should completely destroy and make it again.

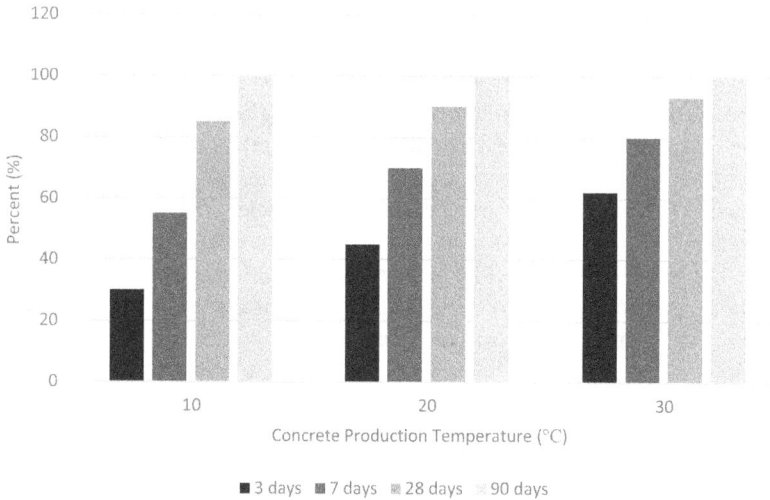

FIGURE 12.13 Compressive strength growth according to the production temperature. (Graph created by the author.)

The production temperature of concrete is very effective in the compressive strength growth over time. You can see the effect of concrete production temperature on the compressive strength growth in Figure 12.13.

In the above figure, we assumed that the compressive strength of concrete at the age of 90 days will be 100%. Then we compared the compressive strength growth at the other ages for concrete made with different temperatures. Concrete mix design and constituent materials are the same for all of the concretes. On the other hand, we produced concrete with a temperature of 10°C in cold weather conditions. So, the concrete will remain in the cold weather after pouring, but with good curing. We made concrete at a temperature of 20°C at mild weather conditions and concrete at a temperature of 30°C in hot weather conditions and good curing.

You can see from Figure 12.13 that in cold weather conditions the compressive strength growth will be much lower than the hot weather conditions. On the other hand, you can see that, if you produce the concrete at a temperature of about 20°C, you will see the normal growth of the compressive strength.

12.11 THE EFFECTS OF COLD WEATHER ON THE PROPERTIES OF CONCRETE

We can name the most important harmful effects of the cold weather on concrete as below:

- Freezing of concrete before setting: This is the most harmful effect of cold weather on concrete, which can cause completely destroying of the structure. If the concrete temperature will be near 5°C, the hydration reaction will be very slow. So, the temperature of concrete element will not raise and if you don't cure concrete in cold weather, the temperature will decrease lower than 5°C and even it can go near or lower than zero. So, the water inside the concrete will freeze and the hydration reaction will completely stop for more than several hours. After warming up the weather, hydration reaction can continue. But the compressive strength of concrete will be very low. We can call this process, freezing of concrete. In this case, it is recommended to destroy the concrete elements and made them again.

• High retarding effect: As mentioned before, in cold weather conditions the hydration reaction will be slower than normal. So, you will see a high retarding effect in concrete which could be harmful to some cases and projects. For example, it is possible to see an unset concrete after 24 hours in cold weather conditions. So, you should take care of concrete for a longer period of time.

• Lower early age compressive strength: The compressive strength of concrete at early ages like 3 or 7 days will be much lower than the concrete made in mild or hot weather conditions. This is very important for the age of mold release and some structures which needs early age strength. In this case, you may need using of some accelerating admixtures to ensure defined early age strength.

• Lower strength at the age of 28 days: Depending on the type of cement, you may achieve the lower strength at the age of 28 days in cold weather conditions. If you use retarded cement like type IV or V or blended cement, you will see lower strength at the age of 28 days in cold weather conditions. On the other hand, if you use accelerating admixture for the modification of early age compressive strength, you will have probably less strength at the age of 28 days.

• Scalding on the surface of concrete elements: Sometimes, you may see freezing of concrete on the surface layer of the concrete element but not in the internal layers. In this case, the compressive strength of the element could be good, but on the surface of concrete you will see scalding after a period of time. For example, if you use steel forms in a cold area with high wind, you may see freezing on the layers contacted with the steel forms. This phenomenon will cause dirty and ugly surface for concrete elements which could be important for some projects.

• Different strength development patterns: At cold weather conditions, you will see different strength development patterns. For early ages, you will see lower strength but for ages more than 28 days you will have more compressive strength. It could be very important for some projects that we may need higher compressive strength at early ages.

12.12   CONSIDERATIONS FOR COLD WEATHER CONCRETING

In this part, we are going to talk about the most important considerations for the protection of concrete elements against the harmful effects of cold weather. The most important activity that we should do is producing a concrete with higher temperature

FIGURE 12.14   Water warming equipment. (Photograph by the author.)

and curing concrete after pouring. So, you can see below considerations as the most important ones:

- Using warm water for concrete production: You know that, water temperature is very important to control the temperature of concrete. So, if you would like to produce concrete with higher temperature in cold weather conditions, the most important thing that you can do is warming up the water. There are special instruments that you can use to warming up the water in the batching plants (Figure 12.14). They can increase the water temperature more than 20°C. If you don't have these instruments, you should try to protect the water tank from freezing. For example, you can use under-ground tank for water storage.
- Using of accelerated cements: To protect concrete against freezing, it is recommended to use accelerated cements in cold weather conditions. For example, using type I or III cement is much better than using type V or IV. Some of the cement manufacturers can give you type II accelerated cement which is suitable for cold weather conditions.
- Using more Portland cement in concrete mix design: It is recommended to use more Portland cement in cold weather conditions because it can cause acceleration of the hydration reaction. If you use supplementary cementitious materials in the binder system, it is recommended to use more pure Portland cement to control the hydration reaction and hydration heat at early ages. On the other hand, it is recommended to use accelerated cement in the binder system.
- Using the accelerating and anti-freezing admixtures: You can use accelerating admixture to accelerate the hydration reaction and release heat at early ages. As mentioned before, the anti-freezing admixtures are also some types of accelerating admixtures. So, you can use them for winter concreting. They can increase early ages compressive strength. But unfortunately, the final compressive strength of concrete will decrease when you use these admixtures. So, using of these admixtures should be under high control.

- Deicing the aggregates before use: One of the important considerations for cold weather concreting is deicing the aggregates before use for concrete production. Sometimes, you may have snow on the surface of aggregates. You should not use these aggregates instead of deicing them before use. For the washed sand, sometimes, it is possible to see freeze sand. You should not use them instead of warming up until deicing. If you use these aggregates without deicing, they can decrease the concrete temperature very much because the ice inside these aggregates acts as a part of concrete water and you know the effect of water temperature on the temperature of concrete. You can use shades for the aggregates to protect them against the snowy weather.

- Using of super-plasticizers without retarding effects: Some of the super-plasticizer manufacturers use retarding admixtures mixed with the super-plasticizers to improve the slump retention effect. On the other hand, some types of plasticizers based on ligno-sulfonate have the retarding effect itself. You should not use these types of super-plasticizers in cold weather conditions because they will retard the concrete and the danger of freezing will be higher.

- Using cement with higher temperature: Unlike hot weather, you should use cement with higher temperature for concrete production because it will cause increasing the concrete temperature. So, it is better not to store much cement in winter in the siloes because the temperature of stored cement will be low at the time of concrete production.

- Working at the warmer times of the day: Unlike hot weather, you should work at the warmer times of the day in cold weather conditions. It means that you should complete the work before the warmest times of the day. So, the poured concrete will set soon and the danger of freezing will decrease.

- Using wooden or plastic base forms (Figure 12.15): For cold weather concreting, it is recommended to use wooden or plastic base molds and forms to control the effect of cold wind and freezing the surface of concrete elements. If you are forced to use steel forms, you should warm up them before concreting and also you should try to warm the area around the forms after concreting to ensure the setting of concrete before freezing.

- Cure concrete as well as possible: Like hot weather, curing concrete as well as possible is very important in cold weather conditions. You should take care of the moisture and temperature of concrete for a defined period of time. You should use special blankets to cover the concrete surface and make it warm. You can use a heater to warm up the environment near the concrete elements. You should protect the concrete surfaces against the snow and ice and you should control concrete elements loosing of moisture as long as it possible.

FIGURE 12.15 Wooden forms. ("Column formwork" by Bill Bradley.)

12.13 CHEMICAL ADMIXTURES FOR COLD WEATHER CONCRETING

We can use some types of chemical admixtures for the protection of concrete against the harmful effects of cold weather. Some of the most important ones are as below:

- Set accelerating admixtures: If you use set accelerating admixtures, the danger of concrete freezing decreases, because these admixtures increase the heat of hydration at the earlier hours and can increase the temperature of concrete. On the other hand, they can speed up the setting of concrete before freezing.
- Hardening accelerator admixtures: as mentioned before, cold weather will cause decreasing in the early age compressive strength, which should be very important for some projects. By using hardening accelerator admixtures, we can increase the early strength of concrete.
- Anti-freezing admixtures: These admixtures are the same as accelerators. They accelerate the setting and hardening of concrete. So, you can protect concrete against freezing. Sometimes, in very cold areas of the world, there are some chemicals in the formulation of anti-freezing admixtures that can control the freezing of concrete water.

- Curing compounds: To avoid water evaporation from concrete surface in cold weather, the best way is to use curing compounds because the usage of water for concrete curing in cold weather conditions could be dangerous according to the freezing. It can decrease concrete temperature which could be harmful to concrete.

12.14   CALCULATIONS FOR CONCRETE TEMPERATURE AT COLD WEATHER CONDITIONS

Like the hot weather, in this part, we are going to check the effect of constituent materials on the temperature of concrete in cold weather conditions by a few examples. First, we should calculate the concrete temperature in a batching plant at a place with cold weather conditions without any considerations to control the concrete temperature. The concrete mix design is the same as before. The only difference is constituent materials' temperature which is in winter conditions. You can see the concrete mix design with the temperature of each constituent material in Table 12.2.

Now, we should use equation 12.1 to calculate the concrete temperature. We will use the super-plasticizer as part of water in the equation and we will have 35.5 kg water with a temperature of 5°C in the natural sand. So, the real amount of natural sand is 674.5 kg.

The numerator:

$$0.22 \, ((510 \times 15) + (260 \times 15) + (674.5 \times 5) + (330 \times 12) + (350 \times 30) \\ + (90 \times 27)) + (164 \times 1) + (35.5 \times 5) = 7340.2$$

Denominator:

$$0.22 \, (510 + 260 + 674.5 + 330 + 350 + 90) + 164 + 35.5 = 686.7$$

$$T = 7340.2/686.7 = 10.7°C$$

TABLE 12.2

Concrete Mix Design and the Temperature of Constituent Materials for the Calculations of Concrete Temperature at Cold Weather Conditions

Constituent Material	Amount for 1 m³ (kg)	Temperature (°C)
Portland cement	350	30
GGBS	90	27
Coarse 11–19	510	15
Coarse 5–12	260	15
Natural sand with 5% moisture	710	5
Crushed sand without any moisture	330	12
Water	160	1
Super-plasticizer	4	1
Total	2414	–

As you can see, the concrete temperature is about 11°C. This temperature is acceptable but as we should pour the concrete in cold weather conditions, it is possible to decrease the temperature of the concrete during transportation and curing time according to the low-speed hydration reaction if the weather will be too cold. So, it is better to produce concrete with higher temperature for example, between 20°C and 25°C.

In the first step, we will use the binders with higher temperature, for example, cement and slag with a temperature of 42°C. You can see the calculations as below:

The numerator:

$$0.22 \,((510 \times 15) + (260 \times 15) + (674.5 \times 5) + (330 \times 12) + (350 \times 42) + (90 \times 42))$$
$$+ (164 \times 1) + (35.5 \times 5) = 8561.2$$

Denominator:

$$0.22 \,(510 + 260 + 674.5 + 330 + 350 + 90) + 164 + 35.5 = 686.7$$

$$T = 8561.2/686.7 = 12.5°C$$

You can see that this will increase the temperature by about 2°C which is not enough for us. Now, we would like to use the same temperature cement and slag, but warmer natural sand with a temperature of 12°C like the crushed sand. We can do it by deicing the natural sand which contains water. You can see the calculations as below:

The numerator:

$$0.22 \,((510 \times 15) + (260 \times 15) + (674.5 \times 12) + (330 \times 12) + (350 \times 30)$$
$$+ (90 \times 27)) + (164 \times 1) + (35.5 \times 12) = 8627.5$$

Denominator:

$$0.22 \,(510 + 260 + 674.5 + 330 + 350 + 90) + 164 + 35.5 = 686.7$$

$$T = 8627.5/686.7 = 12.5°C$$

Again, the effect of using warmer sand is about 2°C which is not enough for us. Also, if we use warmer binders and warmer sand the maximum effect will be increasing the temperature for about 5°C which should not be enough for us.

Now, we are going to use the same temperature cement, slag and natural sand, and warm water with a temperature of 40°C:

The numerator:

$$0.22 \,((510 \times 15) + (260 \times 15) + (674.5 \times 5) + (330 \times 12) + (350 \times 30)$$
$$+ (90 \times 27)) + (164 \times 40) + (35.5 \times 5) = 13746.3$$

Denominator:

$$0.22 \,(510 + 260 + 674.5 + 330 + 350 + 90) + 164 + 35.5 = 686.7$$

$$T = 13746.3/686.7 = 20°C$$

You can see that this is a good temperature for a concrete in winter. For the last calculation, we will use warmer binders and natural sand with warm water.

The numerator:

$$0.22 ((510 \times 15) + (260 \times 15) + (674.5 \times 12) + (330 \times 12) + (350 \times 42)$$
$$+ (90 \times 42)) + (164 \times 40) + (35.5 \times 12) = 16244.5$$

Denominator:

$$0.22 (510 + 260 + 674.5 + 330 + 350 + 90) + 164 + 35.5 = 686.7$$

$$T = 16244.5/686.7 = 23.6°C$$

You can see that the most effective way to increase concrete temperature is by warming up water. This is the same as the hot weather conditions that we needed cooler concrete and the best way to produce cooler concrete was using powdered ice instead of water or a part of the water in the production process. This is because of high heating capacity of water which is the case that we use water as a cooler liquid in many industries.

REFERENCES

"A blacktop road in a forest during snowfall." Retrieved from: https://www.rawpixel.com/image/3284925/free-photo-image-forest-trees-cc0.

Aitcin P.C, High Performance Concrete, E&FN SPON, 2004.

American Society for Testing and Materials, Standard Practice for Making and Curing Concrete Test Specimens in the Field, ASTM C31–00.

American Society for Testing and Materials, Standard Practice for Sampling Freshly Mixed Concrete, ASTM C172-99.

American Society for Testing and Materials, Standard Practice for Making and Curing Concrete Test Specimens in the Laboratory, ASTM C192-00.

American Society for Testing and Materials, Standard Practice for Capping Cylindrical Concrete Specimens, ASTM C617-98.

American Society for Testing and Materials, Standard Specification for Ready-Mixed Concrete, ASTM C94-00.

American Society for Testing and Materials, Standard Specification for Air-Entraining Admixture for Concrete, ASTM C260-00.

American Society for Testing and Materials, Standard Specification for Chemical Admixtures for Concrete, ASTM C494-99.

American Society for Testing and Materials, Standard Test Method for Air Content of Freshly Mixed Concrete by the Volumetric Method, ASTM C173-01.

American Society for Testing and Materials, Standard Test Method for Heat of Hydration of Hydraulic Cement, ASTM C186-98.

American Society for Testing and Materials, Standard Test Method for Air Content of Freshly Mixed Concrete by the Pressure Method, ASTM C231-97.

American Society for Testing and Materials, Standard Test Method for Density, Absorption and Voids in Hardened Concrete, ASTM C642-97.

Bradley, Bill (Billbee), "Column formwork." Retrieved from: https://commons.wikimedia.org/wiki/File:Column-formwork.jpg.

European Standard Organization, Concrete-Part 1: Specification, Performance, Production and Conformity, EN206-1, 2000.

European Standard Organization, Methods of Testing Cement, EN196 Series.

European Standard Organization, Testing Fresh Concrete, EN12450 Series.

European Standard Organization, Testing Hardened Concrete, EN12390 Series.

Galuzzi, Luca, "Leaving traces on soft sand dunes in Tadrart Acacus a desert area in western Libya, part of the Sahara." Retrieved from: https://en.wikipedia.org/wiki/Sahara_desert_(ecoregion).

Gjorv E.Odd, *Durability Design of Concrete Structures in Severe Environments*, Taylor & Francis, 2009.

Harvey, John, "Concrete cracked." Retrieved from: https://pixnio.com/textures-and-patterns/concrete-texture/concrete-cracked#.

Hauschild Michael, Rosenbaum Ralph K, Olsen Sting Irving, *Life Cycle Assessment, Theory and Practice*, Springer, 2018.

Heinrichs Harald, Martens Pim, Michelsen Gerd, Wiek Arnim, *Sustainability Science, An Introduction*, Springer, 2016.

Iranian Institute for Research on Construction Industry, 9th topic of National Rules for Construction, "Concrete Structures", 2009.

Iranian Institute for Research on Construction Industry, National Concrete Mix Design Method, 2015.

Iranian National Management and Programming Organization, National Handbook of Concrete Structures, 2005.

Iranian Standard Organization, Concrete Admixtures, Specification, ISIRI2930, 2011.

Iranian Standard Organization, Concrete Specification of Constituent Materials, Production and Compliance of Concrete, ISIRI2284-2, 2009.

Iranian Standard Organization, Standard Specification for Ready Mixed Concrete, ISIRI6044, 2015.

Lamond F.Joseph, Pielert H.James, *Significance of Tests and Properties of Concrete and Concrete Making Materials*, ASTM International, 2006.

Mahmood Zadeh Amir, Iranpoor Jafar, Concrete Technology and Test (Farsi), Golhaye Mohammadi, 2007.

Nawy G.Edward, *Concrete Construction Engineering Handbook*, CRC Press, 2008.

Newman John, Choo Ban Seng, *Advanced Concrete Technology, Concrete Properties*, Elsevier, 2003.

Popovics Sandor, *Concrete Materials, Properties Specification and Testing*, NOYES Publications, 1992.

Ramachandran V.S, Beaudion James, *Handbook of Analytical Techniques in Concrete Science and Technology, Principles, Techniques and Applications*, William Andrew Publishing, 2001.

Ramachandran V.S, *Concrete Admixtures Handbook, Properties, Science and Technology*, NOYES Publications, 1995.

Ramachandran, Paroli, Beaudion, Delgado, *Handbook of Thermal Analysis of Construction Materials*, NOYES Publications, 2002.

Richardson M, *Fundamentals of Durable Reinforced Concrete*, SPON Press, 2004.

Safaye Nikoo Hamed, Introduction to Concrete Technology (Farsi), Heram Pub, 2008.

Sęk, Matylda, "Liquid nitrogen tanks and storage containers." Retrieved from: https://commons.wikimedia.org/wiki/File:Liquid_nitrogen_tanks-01.jpg.

Shekarchizade Mohammad, Liber Nicolas Ali, Dehghan Solmaz, Poorzarrabi Ali, Concrete Admixtures Technology and Usages (Farsi), Elm & Adab, 2012.

TSGT Walter Perkins Jr, "Airmen unfold a membrane tarp that will be used to cover a runway crater repaired during a quick reaction runway repair test." Retrieved from: https://nara.getarchive.net/media/airmen-unfold-a-membrane-tarp-that-will-be-used-to-cover-a-runway-crater-repaired-bc839a.

Zandi Yousof, Advanced Concrete Technology (Farsi), Forouzesh Pub, 2009.

13 Concrete for Special Purposes

For some types of projects, you may define special properties for concrete. In this chapter, we discuss this subject. Special concrete could be any type of concrete instead of normal. For each case, you may need a special mix design and considerations to get the best result. As the usage of concrete in different structures is going to increase day by day, the usage of special concretes with special properties will increase also. So, this chapter could be one of the most important chapters of this book.

To study about the special types of concrete, you need to know about concrete technology, and this is the reason that we are talking about this important subject in this chapter. We are not going to talk about all special types of concrete. But we are trying to talk about the most important and useful ones that you may need in different structures.

For each type of special concrete, after introduction, first, we define why it could be a special concrete. Then we will talk about the considerations for the production and implementation of the concrete. After that, we will talk about the tests to check the properties of the special concrete and finally, we will give you an example of the mix design and implementation of the special concrete.

13.1 SELF-COMPACTING CONCRETE

The compaction process is very important for any type of concrete because the quality of concrete elements refers to the compaction method and the quality of compaction. For normal concrete, the compaction is possible by using the vibrator machines that we talked about them in the previous chapters. But the need for better and easier compaction pushed us to the newly made type of concrete named self-compacting concrete (SCC). The need for this type of concrete refers to the earliest years of using concrete in structures. But making such a concrete is possible now only after the invention of high-performance super-plasticizers based on polycarboxylate ether.

13.1.1 Definitions of SCC

SCC is a type of concrete with the flowability and easy moving through the structural elements and between the rebars. So, it can flow easily inside the form even in congested sections. Using this type of concrete means that, there is no need for extra compaction with the vibrators.

The specification of an SCC concrete is not only the flowability. The aggregates gradation, amounts of filler and binders, and concrete weight are the other important specifications of SCC concrete which let it flow through the bars and forms. In fact, we can make a high-flowable concrete that is not SCC. We can call that easy compacting concrete. But to make a real SCC concrete, you need to consider some special

DOI: 10.1201/9781003384243-13

FIGURE 13.1   Self-compacting concrete. (Photograph by the author.)

features that we discuss them in this part of the book. Don't forget that the important specification for SCC concrete is flowing through the structural elements without the need for vibration. You can see a picture of an SCC concrete in Figure 13.1.

We can name the most important specifications of SCC as below:

- High flowability: It means a non-slump concrete with a flowing diameter of more than 600 mm.
- High moving capability: Instead of high flowing, the concrete should move easily through the rebars. It means that you can make this type of concrete only by using polycarboxylate super-plasticizers. You can make a concrete with high flowability with other types of super-plasticizers. But it doesn't have a high moving capacity.
- Controlled segregation and bleeding: To move good through the rebars and for good surface type, you should control the segregation and bleeding. Only a little bleeding in concrete will decrease the quality of SCC concrete very much. You can see a picture of segregated concrete in Figure 13.2.
- Softer aggregate texture: SCC should have a softer texture compared with the normal concrete because there is a need to move through the congested elements and a coarse concrete cannot move from the little spaces between the rebars in a congested element. On the other hand, a coarse concrete has more potential for segregation and bleeding.

13.1.2   Considerations for SCC Production and Implementation

According to the above specifications, we can name the most important considerations for SCC concrete production and implementation as below:

- For high flowability, you should use a strong polycarboxylate ether super-plasticizer with high dosage. It is recommended to use a super-plasticizer with lower solid content for better controlling of the dosage. But as you are going to use high dosage of the super-plasticizer, you should take care of

FIGURE 13.2 The result of segregation and bleeding in concrete. ("Concrete segregation".)

the retardation effect. It is better to use slump retention type PCE polymers without any retardation effect.

- For high moving capability, you should use strong PCE super-plasticizers. It is recommended to use the slump retention type polymers because this type of polymer will give you better moving ability compared with the water reducer types.
- To control the segregation and bleeding, it is recommended to use a slump retention type PCE because this type of polymer will be helpful for the control of segregation and bleeding. In fact, the segregation and bleeding probability is higher for the water reducer type of polycarboxylates.
- To control the segregation and bleeding, you should make a concrete with more than 1.5% of air entrapped. Concrete with lower amount of air bubbles has high capability for segregation and bleeding. On the other hand, it is better to make smaller air bubble in concrete and remove the coarser ones. You can do it only by modifying the formulation of the super-plasticizer. As the super-plasticizer is very important for SCC concrete, you should be in contact with the super-plasticizer manufacturer during the trials and production process.
- To prevent segregation and bleeding, you should use more fillers in the aggregates. As mentioned before, fillers are passed by sieve No.100. You can use the recommendations of Table 13.1 for the amount of fillers in SCC concrete.
- To prevent segregation and bleeding, if you couldn't use high amount of filler as mentioned, you should use viscosity modifier admixtures. This type of admixture will increase the viscosity of concrete. So, it can control the segregation and bleeding. Some of the PCE super-plasticizers contain VMA admixture, which can be suitable for the production of SCC concrete. Whereas if you use VMA in the formulation of a PCE admixture that you are going to use for the production of normal concrete, the user

TABLE 13.1

Recommendations for the Amounts of Fillers (Passing of Sieve No. 100) for the Production of SCC Concrete

Maximum Size of Coarse Aggregate (mm)	Minimum Amount of Filler in 1 m³ of SCC Concrete (kg)
25	80–100
19	100–120
12.5	120–150

found weaker admixture compared with the ones without VMA in their formulations.

- You can use any size of 25, 19, or 12.5 mm as the maximum size of coarse aggregate for SCC concrete production. But you should count the distances between the rebars in the structure. For more congested sections, it is recommended to use a smaller maximum size for the coarse aggregates.
- For the mix design of SCC concrete, it is recommended to use n of Table 9.4 less than 0.3 because the texture of an SCC concrete should be finer than normal. You can see Table 13.2 just as a recommendation for the mixture of the aggregates in SCC concrete.
- For w/b of an SCC concrete, you should use the same methods of Chapter 9. But it is recommended to produce a concrete with maximum w/b = 0.42 as an SCC concrete.
- For the free water, you should use more water than the values of Tables 9.10 and 9.11. For the super-plasticizer as mentioned before, you should use a higher dosage and you will see a higher water reduction rate. So, the high dosage of the super-plasticizer will recover the excess amount of water you assumed.
- For the amount of total binder, you should consider Table 13.3.
- It is recommended to use supplementary cementitious materials in the mix design of SCC concrete. They could be helpful for the prevention of the segregation and bleeding.

TABLE 13.2

Recommendations for the Mixture of the Aggregates in SCC Concrete

Size Range of Aggregates (mm)	SCC Concrete With the Max Aggregates Size of 25 mm	SCC Concrete With the Max Aggregates Size of 19 mm	SCC Concrete With the Max Aggregates Size of 12.5 mm
0–4.75	60%	60%	65%
4.75–12.5	15%	15%	35%
12.5–19	-----	25%	-----
12.5–25	25%	-----	-----

TABLE 13.3

Minimum Amount of Total Binder for the Production of SCC Concrete

Maximum Size of Coarse Aggregate (mm)	Minimum Amount of Binder in 1 m³ of Concrete (kg)
25	400–430
19	430–450
12.5	450–480

13.1.3 Tests for Checking the Properties of SCC

Now, it is time to talk about the special tests that can show us the quality of SCC concrete. As the self-compacting behavior refers to the specifications of the fresh concrete, all of the tests that are special for the SCC concrete are the tests for the fresh concrete. The hardened concrete tests for the SCC are the same as the normal concrete. Below, you can find brief descriptions about the special tests for SCC concrete. For more information, you can see the special texts about the SCC concrete.

- Slump flow test (Figure 13.3): This is the simplest test for the SCC concrete. There are no many differences between the slump test and slump flow test. You can use the slump cone on its normal set or you can use it in reverse set (The bigger circle on the top). In this way you can pour the cone with a high flow concrete simpler. The other difference is no need for tamping the concrete because of high flowability. Instead of measuring the height of collapsed concrete, you should measure the flowing circle. There are different suggestions for the minimum acceptable flow diameter for the SCC concrete. But the most acceptable in the world is 600 mm. So, if you produce a concrete with the flow diameter less than 600 mm, we can say that this is not an SCC concrete and there is no need for the other tests. This is the reason that we mentioned

FIGURE 13.3 Slump flow test for SCC concrete. ("Flow test after".)

this test as the first test for the SCC concrete. There is no maximum diameter for the slump flow. But we can say that most of the time, you cannot make a stable concrete mix with the flow diameter of more than 850 mm.

This test can only show us the flowing ability of concrete mix. But it is not showing us the ability of concrete for passing through the rebars or between the forms.

- T50 test: The second test which will give you data about the viscosity of SCC concrete is the T50 test. This is in fact the same test as the slump flow. But you should measure the time that the flow circle will touch the 500 mm diameter. For this case, as you can see in Figure 13.4, you should sign the 500 mm diameter on the plate of the slump flow test. Then you can easily measure the time in seconds.

  Best SCC concretes will reach the 500 mm diameter in 3–5 seconds. It is not easy to make a stable SCC concrete with T50 of less than 3 seconds and an SCC concrete with T50 of more than 5 seconds is a high viscous concrete. Most of the time, it is because of high amounts of binders or high amounts of fillers or using a strong VMA admixture. For better moving capacity of SCC concrete, you should modify the T50 to less than 5 seconds according to the above reason for the high viscosity of concrete.

- J-ring test: This is a test for measuring the passing ability of SCC concrete between the rebars. You can see a picture of the J-ring test in Figure 13.4.

You can see that this is the same test as the slump flow with only difference in the rebars set through the concrete passing. The most common instrument consists of the 16 mm rebars with the height of 100 mm and the mean distance of 40 mm. You should compare this test with the slump flow and T50 test. It means that you will have a slump flow diameter without any rebar and a flow with the rebars. Also, you will have a T50 with and without the rebars. Now, you should compare them together. You can find the ability to pass by this comparison. For this reason, you can use Table 13.4.

- V-funnel test: This test is another test for the evaluation of the concrete viscosity. You can see a picture of the V-funnel instrument in Figure 13.5.

FIGURE 13.4   J-ring test for SCC concrete. (Photograph by the author.)

TABLE 13.4
Passing Ability of SCC Concrete With J-Ring Test

Difference Between the Slump Flow With and Without the J-Ring (mm)	Passing Index	Description
0–25	0	High passing ability
25–50	1	Moderate passing ability
More than 50	2	Low passing ability

FIGURE 13.5   V-funnel instrument. (Photograph by the author.)

For this test, you need at least 12 L of SCC concrete. First, you should pour the funnel with the SCC concrete. After 10 seconds you should open the valve under the funnel and let the concrete discharge from the funnel and take the time for complete discharging. You can do this test again without any cleaning of the funnel with 5 minutes of concrete remaining inside the funnel before discharging. The best time for the V-funnel test is between 6 and 12 seconds. In the case of discharging time less than 6 seconds, you will have a concrete with very low viscosity and high risk of segregation and bleeding. For the discharging time of more than 12 seconds, you will have a concrete with high viscosity and low capability to pass through the rebars and forms.

FIGURE 13.6 The U-box test. (Photograph by the author.)

- U-box test: This is a special test for the evaluation of the pouring ability of SCC concrete. In this test, you should pour the instrument that you can see in Figure 13.6. After 1 minute, you should open the valve and let the concrete go through the other side of the U-box. Then you should measure the difference between the height of the concrete in two columns. You should know that there are rebars with a distance of 4 cm in the space between the two columns for better evaluation of the concrete pouring ability.

If the difference between the height of concrete will be near zero, this would be the best result for SCC concrete. More difference in the height means more viscosity and lower pouring ability for concrete. The maximum amount of height difference in this test should be 3 cm.

- L-box test (Figure 13.7): This is another test for the evaluation of concrete viscosity and passing ability through the rebars and forms. In this test, you should pour the vertical part of the instrument with the SCC concrete.

FIGURE 13.7   The L-box test. ("Workability test for SCC concrete" by Amit Kenny.)

After 1 minute, you should open the valve and let the concrete flow through the instrument. After stability of concrete, you should measure the height difference of two edges of the L-box and calculate the H2/H1. This is the blockage proportion of the SCC concrete which should be between 0.8 and 1.0. If the blockage proportion will be less than 0.8, it shows that the viscosity of concrete is so high and the passing ability through the rebars and forms is very low.

• Visual stability (VSI) index: The easiest and simplest test for SCC concrete is the VSI index evaluation. In fact, the evaluation of the concrete stability with the eye is the test procedure of the VSI index. You can search the internet to view the VSI index pictures.

As you can see, the best VSI is zero, which is the best stability for SCC concrete. On the other hand, the VSI of 3 is the worse one which is a concrete with segregation and bleeding. You cannot use a concrete with the VSI3 as an SCC concrete.

13.1.4   EXAMPLE FOR SCC MIX DESIGN AND IMPLEMENTATION

Now, we are going to design a C40 SCC concrete mix with below constituent materials:

• The SSD density for the natural sand is 2.73 kg/L and the SSD density for the 11–19 and 5–12 gravels is 2.79 kg/L.
• Portland cement type II with a density of 3.15 kg/L and minimum compressive strength of 430 kg/cm^2 at the age of 28 days.
• GGBS with a density of 2.85 kg/L
• PCE super-plasticizer with a density of 1.08 kg/L and a water reduction rate as you see in Figure 13.8.

TABLE 13.5
Sieve Analysis Test for the Natural Sand for SCC Production

Sieve Size (mm)	Weight of Aggregates Remained on Sieve (g)	Weight of Aggregates Passed by Sieve (g)	Percent of Aggregates Remained on Sieve (%)	Percent of Aggregates Passed by Sieve (%)
4.75	122	1595	7.1	92.9
2.36	329	1266	19.2	73.7
1.18	355	911	20.7	53.1
0.6	292	619	17.0	36.1
0.3	265	354	15.4	20.6
0.15	189	165	11.0	9.6
Total	1717	-----	100	-----

TABLE 13.6
Sieve Analysis Test for the 11–19 Gravel for SCC Production

Sieve Size (mm)	Weight of Aggregates Remained on Sieve (g)	Weight of Aggregates Passed by Sieve (g)	Percent of Aggregates Remained on Sieve (%)	Percent of Aggregates Passed by Sieve (%)
25	0	1770	0.0	100
19	110	1660	6.2	93.8
12.5	1056	604	59.7	34.1
9.5	469	135	26.5	7.6
4.75	135	0	7.6	0.0
Total	1770	-----	100	-----

TABLE 13.7
Sieve Analysis Test for the 5–12 Gravel for SCC Production

Sieve Size (mm)	Weight of Aggregates Remained on Sieve (g)	Weight of Aggregates Passed by Sieve (g)	Percent of Aggregates Remained on Sieve (%)	Percent of Aggregates Passed by Sieve (%)
19	0	1648	0.0	100
12.5	78	1570	4.7	95.3
9.5	734	836	44.5	50.7
4.75	647	189	39.3	11.5
2.36	189	0	11.5	0.0
Total	1648	-----	100	-----

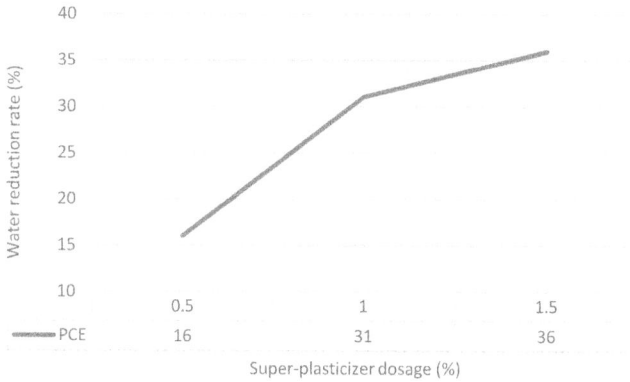

FIGURE 13.8 Water reduction curve for the PCE super-plasticizer for SCC production. (Graph created by the author.)

We should start with the step-by-step procedure mentioned in Chapter 9 by attention to the considerations of this chapter as below:

- Step 1: For this concrete, we use the standard deviation of 2.5 MPa.
- Step 2: Calculation of the mix design compressive strength as below:

$$F_{cm} = 40 + (1.34 \times 2.5) + 1.5 = 44.85 = 45\,MPa$$

- Step 3: For the percentage of each aggregate in concrete, you can see Tables 13.8 and 13.9. As mentioned before in this chapter, n should be less than 0.3 for SCC concrete. So, here we used n between 0.1 and 0.3.

TABLE 13.8

Max and Min Amounts for n = 0.1 to n = 0.3

Sieve size	19 mm	12.5 mm	9.5 mm	4.75 mm	2.36 mm	1.18 mm	0.6 mm	0.3 mm	0.15 mm
Min limit	100	85.4	76.8	58	42.6	30.2	20.3	12.1	5.4
Max limit	100	90.4	84.2	69.5	55.7	42.9	31.3	20.1	9.7

TABLE 13.9

Checking the Mix of Aggregates

Aggregate	19 mm	12.5 mm	9.5 mm	4.75 mm	2.36 mm	1.18 mm	0.6 mm	0.3 mm	0.15 mm
11–19 18%	16.9	6.1	1.4	0	0	0	0	0	0
5–12 10%	10	9.5	5.1	1.1	0	0	0	0	0
Sand 70%	72	72	72	66.9	53.1	38.2	26.0	14.8	6.9
Total	98.9	87.7	78.4	68.0	53.1	38.2	26.0	14.8	6.9

TABLE 13.10

Calculation of the Fineness Module of Total Aggregates

Sieve Size	37.5 mm	19 mm	9.5 mm	4.75 mm	2.36 mm	1.18 mm	0.6 mm	0.3 mm	0.15 mm	Total
Percent remained	0	1.1	20.4	10.4	14.9	14.9	12.2	11.1	7.9	-----
Cumulative percent remained	0	1.1	21.6	32	46.9	61.8	74	85.2	93.1	415.6

For the aggregates after trial and error, we will use 18% of gravel 11–19 and 10% of gravel 5–12 and 72% of the natural sand. If you compare them with Table 13.2, you will see that the concrete is finer than the recommendations of Table 13.2 which is better for the production of SCC concrete.

- Step 4: The fineness module of total aggregates as you can see in Table 13.10 is 4.16.
- Step 5: As the cement strength is 430 kg/cm², we assume that this is 425 kg/cm². So, we can use Tables 9.6 and 9.9. For $F_{cm} = 45$ MPa, we will have 0.46 and 0.42 from Table 9.6 and 0.38 from Table 9.9. The mean value for these three numbers is 0.42 which is the maximum amount of w/b recommended for the SCC concrete. So, we have:

w/b = 0.42

- Step 6: For the Portland cement of 425 kg/m², FM = 4.2, and slump of 90 mm we have the free water of 226 kg/m³ and for the slump of 150 mm we have the free water of 244 kg/m³. So, for the production of a flow concrete which is the slump of 270 mm the free water will be 280 kg/m³ with linear correlation.

As recommended for the SCC concrete, we should use a high dosage of a strong PCE super-plasticizer. So, we will use 1% of the PCE for this example. Its water reduction rate according to Figure 13.10 is 31%. So, the amount of reduced water will be 193 kg.

- Step 7: The calculations for the Portland cement and GGBS is as below:

b = 193/0.42 = 460 kg

For the GGBS we have a 15% of increasing rate according to Table 9.12. So, increased b = 460 × 1.15 = 530 kg.

We decided to use 20% of the GGBS. So, we will have:

GGBS = 105 kg

Portland cement = 425 kg

Total binder = 530 kg

Amount of the PCE super-plasticizer = 1% = 5.3 kg

As you can see, the amount of binder recommended is at least 430–450 kg/m³ from Table 13.3 which is 530 kg for this example. If you use a strong cement for

TABLE 13.11

Calculations for the Weight of SSD Aggregates for SCC Production

Type of Aggregate	Percent in Total Mixture (%)	Volume (L)	Density (kg/L)	Weight of SSD (kg)
Coarse 11–19	18	111	2.79	309
Coarse 5–12	10	62	2.79	173
Natural sand	71	443	2.73	1209
Total	100	616	-----	1691

FIGURE 13.9 A watertight structure. ("Water tank" by Abhinav Thakur.)

TABLE 13.12

Final Concrete Mix Design for SCC Concrete

Constituent Material	Mix Design for 1 m³ (kg)
Gravel 11–19	310
Gravel 5–12	175
Natural sand	1210
Cement Type II	425
GGBS	105
PCE super-plasticizer	5.3
Water	193
Total weight	2423

example with a compressive strength of more than 525 kg/cm², the amount of total binder will be within the recommendations.

- Step 8: Total volume of the aggregates will be:

$$V = 1000 - (425/3.15) - (105/2.85) - 193 - 15 - (5.2/1.08) = 615.5 = 616L$$

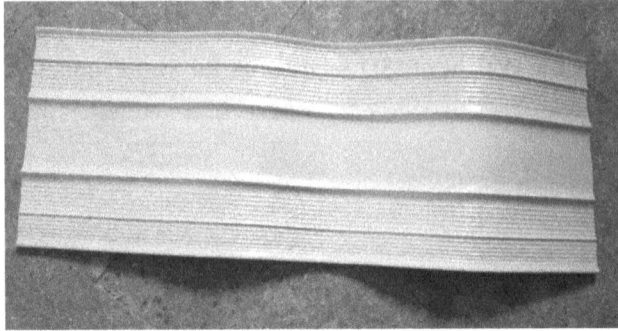

FIGURE 13.10 Using PVC water stops for watertightness of the joints. (Photograph by the author.)

As the recommendation for the amount of air in SCC concrete is a maximum of 1.5%, we used 1.5% of entrapped air in concrete which is 15 L.

- Step 9: The calculations for the weight of aggregates are as you can see in table 13.11

The SCC concrete mix design with SSD aggregates is as as you can see in table 13.12.

13.2 WATERTIGHT CONCRETE

There are too many structures that should be watertight (Figure 13.9). There are many considerations to build a watertight structure. Some of the considerations are very expensive. So, building water proof concrete structures is the best and cheaper choice to make a watertight structure.

Some of the most important structure that should be watertight and can be made with concrete is as below:

- Swimming and water storage pools
- Sewage settlement pools
- Water storage tanks
- Foundations and walls of structures under water pressure
- Some of the marine structures

If you couldn't build a watertight structure with concrete, the process of waterproofing by other methods will be very hard, time consumer, and expensive. In this chapter, we discuss the production of a water proof concrete and implement it in a watertight structure.

13.2.1 DEFINITIONS OF WATERPROOF CONCRETE AND WATERTIGHT STRUCTURE

A watertight structure is a structure that is in contact with water but without any infiltration or transition of water from concrete elements. For example, a swimming pool is made of concrete without any transition of water from floor, walls, or the joints between the walls and floor.

A water proof concrete is a concrete without any transition of water from the section. For example, the concrete used for the concreting of the above swimming pool is a waterproof concrete. Using a water proof concrete for a structure does not guarantee the watertightness of the total structure because the other problems and defections in the structure could cause the transition of water. But the first step to build a watertight structure is the production and use of a waterproof concrete. We can name below specifications for a watertight concrete structure:

• Concrete use for the structure should be a concrete with the ability to stop the transition of water.
• Implementation of concrete should be of the best quality without any defection or hole because the water could penetrate from the defections.
• The curing of concrete should be with the best performance to prevent any cracking because the water could penetrate the cracks.
• The joints of the structure should be watertight. You can do it by using the special PVC water stops at any place that you have any kind of joint (Figure 13.10) because water could penetrate from untightened joints.
• If you have any aggressive material in contact with concrete, like some types of sewage, you should cover the concrete surface for the best protection against the aggressive material by using special materials like epoxy or polyurea coatings.

13.2.2 Considerations for Watertight Concrete Production and Implementation

In this part, we discuss the most important considerations for the production of water proof concrete and for the implementation of a watertight structure.

• Produce concrete with a low water-to-binder ratio: To produce a water proof concrete, you should decrease the water-to-binder ratio as low as it possible because the capillary pores inside a concrete made with the minimum water-to-binder ratio are minimal and water cannot penetrate the concrete. We recommend to use a concrete with the maximum w/b = 0.4 as a watertight concrete.
• Using a strong super-plasticizer: To decrease the water-to-binder ratio, you should use strong super-plasticizers with optimum dosage. On the other hand, it is recommended to use supplementary cementitious materials, especially the silica fume instead of using only pure Portland cement.
• Using the silica fume: It is recommended to use silica fume for the production of water proof concrete. Many researches accomplished and show the good effect of the silica fume on the water proofing of concrete. Most of the recommendations asked us to use the silica fume more than 6% by weight of cement for this purpose. Using supplementary cementitious materials, especially the silica fume on the one hand, will cause decrease in the w/b, and on the other hand, decrease in the pores inside concrete.

- Complete aggregates gradation: Using a steady aggregates gradation is necessary for the production of a watertight concrete. It means that you should have all sizes of aggregates without any defection in the gradation system.
- Good compaction of concrete: To make a watertight concrete structure, it is very important to compact the concrete as good as it is possible to make a condensed element. To do that, you can make an easy compacting concrete by producing a high flow concrete and use good vibrators to get better results.
- Curing of concrete: As mentioned before, it is very important to prevent cracking in a watertight concrete structure because water can penetrate from the cracks. The best way to crack control is by curing concrete. For the watertight concrete, it is recommended to continue curing for longer time than normal.
- High-quality joints implementation: The joint implementation in watertight structures is very important because the joints are the weakest part of the structure from water penetration point of view. So, you should use PVC water stop tapes for the joints watertightness. On the other hand, you can use special chemical products (concrete adhesive) which is helpful for better bonding between the old and new concrete. For the width of the PVC water stops you should consider the manufacturers' brochure. Also, you should know that there are different types of water stops for different purposes as you can see in Figure 13.11.
- Repairing of cracks: Although you mentioned crack prevention, it is possible to see some cracks on the surface of the concrete elements. This could be very bad for a watertight structure. So, you should repair them by using special materials like polyurethane and epoxy fillers.
- Using water proofing admixtures: Water proofing admixtures that you may find in the market are in both liquid and powder forms. Some of them are special pore blocker chemicals that you can use with the dosage mentioned by the manufacturer for the production of water proof concrete. Some other types of them are special hydrophobia powders that can block some of the pores inside the concrete structure. Using these admixtures for the production of water proof concrete is beneficial, but they should be used with the other considerations that are mentioned in this part of the book. They cannot make a water proof concrete for you alone.
- Special surface protection: Sometimes, concrete surface will be exposed to a corrosive material. For example, some of the sewage filtration pools are in contact with some types of dangerous chemicals. In this case, you should cover the concrete surface with a protector chemical like epoxy coatings. they can help you to protect concrete surface and help to make a watertight structure. But the price of this kind of work is high. So, it is not economical to use these chemicals for the structures without any contact with the corrosive materials.

13.2.3 Tests for Checking the Properties of Watertight Concrete

In this part, we discuss the tests for checking the water proofing of concrete. We can test concrete specimens cured under the laboratory test. But sometimes according to the lack of good curing, the quality and permeability of concrete in the structure

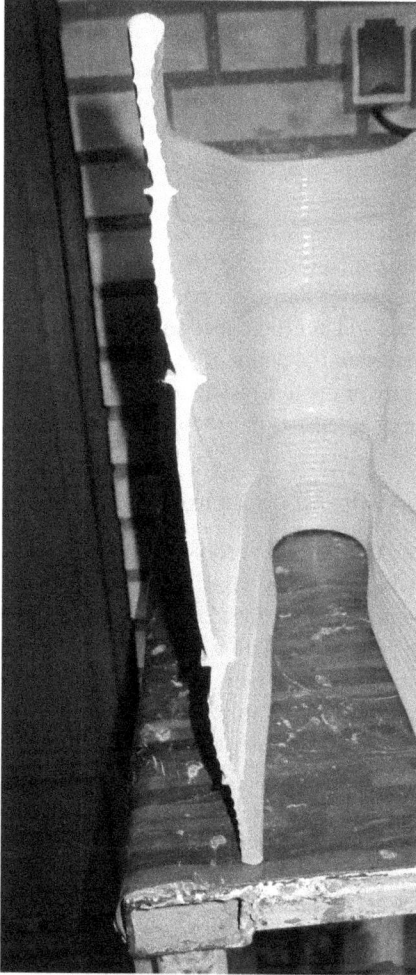

FIGURE 13.11 Section of types of PVC water stop tapes. (Photograph by the author.)

should be different from the laboratory specimens. So, for sensitive structures, you should try to cure concrete as well as it is possible in the structure to get good results nearer to the laboratory tests.

There are different test methods for the evaluation of concrete water proofing in the texts and standards. Here we discuss two methods. One of them is a very simple method with low accuracy about the watertightness and the other one is the most accurate and trustworthy test to check the water proofing of concrete.

The first test method is according to the ASTM C642. This test is for the evaluation of density, absorption, and voids of concrete specimens. Here, we would like to test the water absorption of concrete specimens. It does not exactly show us the water proofing of concrete. But it can be a point to obtain low-quality concretes with a high water absorption index.

You should test the concrete after the age of 28 days. First, you should oven dry the concrete specimen that you measured its dimensions and weight after cooling it to the room temperature. Then you should put the specimens under water for 24 hours and after drying with a napkin, you should weight the specimen again. Now, you can calculate the amount of water absorbed by the concrete specimen which could be a point for the water proofing of concrete. This test is simple but not accurate for the evaluation of concrete watertightness.

The second test which is the most accurate one is according to the EN12390-8 European standard that we talked about it before. In this test, you should put one face of the concrete specimen under the water pressure of 5 bars for 72 hours. Then you should dry the specimens and cut their section across the face which was under the pressure of water and measure the depth of water penetrated into the concrete specimen. You can see a picture of the test instrument in Figure 13.12. You should test the specimens after the age of at least 14 days to get meaningful results.

13.2.4   EXAMPLE FOR WATERTIGHT CONCRETE MIX DESIGN AND IMPLEMENTATION

Now, we would like to produce a water proof concrete for a special water storage with high rebar congestion which is a very sensitive structure by using the same constituent materials of this chapter example for the SCC concrete. The only material that we would like to add is the silica fume with a density of 2.2 kg/L.

As we should produce a concrete with the w/b less than 0.4 and we are not going to use pore blocker admixture as a help, it is better to produce a C45 concrete. Now, we start with the step-by-step procedure as below:

- Step 1: For this concrete, we use the standard deviation of 2.5 MPa.
- Step 2: calculation of the mix design compressive strength as below:

$$F_{cm} = 45 + (1.34 \times 2.5) + 1.5 = 49.85 = 50 \, MPa$$

FIGURE 13.12 Instrument for EN12390-8 test method for watertightness of concrete. (Photograph by the author.)

TABLE 13.13

Max and Min Amounts for n = 0.1 to n = 0.3

Sieve Size	12.5 mm	9.5 mm	4.75 mm	2.36 mm	1.18 mm	0.6 mm	0.3 mm	0.15 mm
Min limit	100	89.9	67.9	49.8	35.3	23.8	14.2	6.3
Max limit	100	91.7	72.5	55.7	41.3	28.9	17.9	8.3

- Step 3: For this concrete, because of the high congestion of the rebars, it is better to use only 5–12 and natural sand. So, we will have Table 13.13 for n between 0.2 and 0.3 for high slump concrete. Because of good compacting, you should use a high slump concrete for watertightness.

For the aggregates, after trial and error, we will use 27% of gravel 5–12 and 73% of the natural sand. If you see the gradation curve in Figure 13.16, you will see that this is a steady curve without any jump.

TABLE 13.14

Checking the Mix of Aggregates

Aggregate	12.5 mm	9.5 mm	4.75 mm	2.36 mm	1.18 mm	0.6 mm	0.3 mm	0.15 mm
5–12 27%	25.7	13.7	3.1	0	0	0	0	0
Sand 73%	73	73	67.9	53.8	38.7	26.3	15.1	7
Total	98.7	86.7	70.9	53.8	38.7	26.3	15.1	7

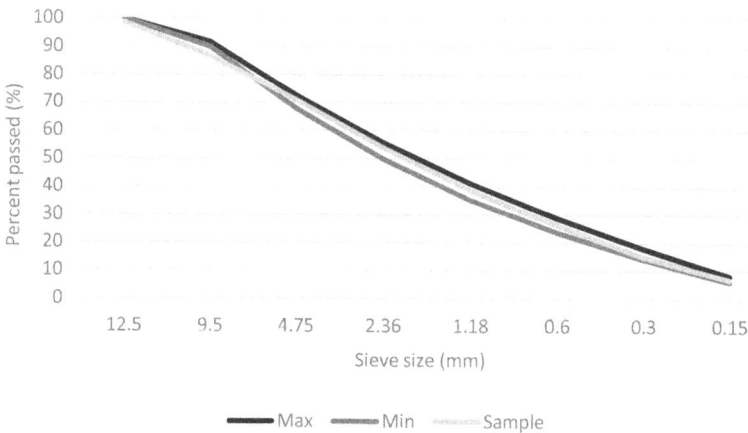

FIGURE 13.13 Total aggregates gradation curve for watertight concrete. (Graph created by the author.)

TABLE 13.15

Calculation of the Fineness Module of Total Aggregates for watertight concrete

Sieve Size	37.5 mm	19 mm	9.5 mm	4.75 mm	2.36 mm	1.18 mm	0.6 mm	0.3 mm	0.15 mm	Total
Percent remained	0	0	13.3	15.8	17.1	15.1	12.4	11.3	8	-----
Cumulative percent remained	0	0	13.3	29.1	46.2	61.3	73.7	84.9	93	401.5

- Step 4: The fineness module of total aggregates as you can see in Table 13.15 is 4.1.
- Step 5: As the cement strength is 430 kg/cm^2 we assume that this is 425 kg/cm^2. So, we can use Tables 9.6 and 9.9. For $F_{cm} = 50$ MPa, we will have 0.41 and 0.36 from Table 9.6 and 0.36 from Table 9.9. The mean value for these three numbers is 0.38 which is less than 0.4 recommendation for water proof concrete. So, we have:

w/b = 0.38

- Step 6: For the Portland cement of 450 kg/m^2, FM = 4.1 and slump of 90 mm, we have the free water of 234 kg/m^3 and for the slump of 150 mm, we have the free water of 252 kg/m^3. So, for the production of flow concrete with slump of 210 mm the free water will be 270 kg/m^3.

  For the production of high slump concrete, we should use a high dosage of super-plasticizer. We decided to use 0.8% by weight of the binder. Its water reduction rate according to Figure 13.10 is 25%. So, the amount of reduced water will be 202 kg.
- Step 7: The calculations for the Portland cement, GGBS, and silica fume are as below:

  b = 202/0.38 = 530 kg

For the GGBS we have a 15% of increasing rate and for the silica fume it is zero according to Table 9.12. So, increased b = 530 × 1.15 = 610 kg.

We decided to use 20% of the GGBS and 8% of silica fume which is more than the minimum recommendation for silica fume usage. This mixture will guarantee the watertightness of concrete. So, we will have:

GGBS = 120 kg
Silica fume: 50 kg
Portland cement = 440 kg
Total binder = 610 kg
Amount of the PCE super-plasticizer = 0.8% = 4.9 kg

- Step 8: Total volume of the aggregates will be:

TABLE 13.16

Calculations for the Weight of SSD Aggregates for Water Proof Concrete

Type of Aggregate	Percent in Total Mixture (%)	Volume (L)	Density (kg/L)	Weight of SSD (kg)
Coarse 5–12	27	155	2.79	432
Natural sand	73	419	2.73	1144
Total	100	574	-----	1576

$$V = 1000 - (440/3.15) - (120/2.85) - (50/2.2) - 202 - 15$$
$$- (4.8/1.08) = 574.1 = 574\,L$$

- Step 9: The calculations for the weight of aggregates are as you can see in table 13.16:

The watertight concrete mix design with SSD aggregates is as you can see in table 13.17:

13.3 HIGH-STRENGTH CONCRETE

We have many developments in concrete technology in the latest decades. So, the production of concrete with higher performance is simply possible. One of the most important mechanical properties of any type of concrete is the compressive strength. If you use a concrete with higher strength, you will have below advantages:

- More economical structure: It is because of decreasing the amount of rebars according to the higher strength of concrete and using less concrete in the structure because of lower dimensions of the elements.
- Higher durability: You will make a high-strength concrete with low w/b. So, the concrete permeability will be very low and this type of concrete will be more durable than normal concrete.

TABLE 13.17

Final Concrete Mix Design for Watertight Concrete

Constituent Material	Mix Design for 1 m³ (kg)
Gravel 5–12	440
Natural sand	1150
Cement Type II	440
GGBS	120
Silica fume	50
PCE super-plasticizer	4.8
Water	202
Total weight	2407

- Protection of the environment: It is because of using waste materials as the supplementary cementitious materials and using less cement for giving more compressive strength which means less air pollution. On the other hand, you will need less water for the production of high-strength concrete.
- Sustainable development: This is because of building a durable structure with using waste materials as the supplementary cementitious materials.
- Building special structures: You can build special structures from shape point of view by using high-strength concrete.

As you can see, building the structures with high-strength concrete is very common nowadays. So, in this part of the book, we discuss the high-strength concrete production and usage.

13.3.1 DEFINITIONS OF HIGH-STRENGTH CONCRETE

There are different definitions of the high-strength concrete in the text and papers. In fact, the important problem is the compressive strength border between the normal and high-strength concrete. For this purpose, you can see below considerations:

- Type of concrete failure: For normal concrete, the cracks of failure are in the paste or the boundary area of the aggregates and paste. But for high-strength concrete, the cracks will be in the aggregates (Figure 13.14). So, we can say that the paste in high-strength concrete should be very strong. The crushing of aggregates will be for the compressive strength of more than 50 MPa depending on the type of aggregates.
- The stress-strain curve: High-strength concrete is more brittle than the normal concrete. It means that the failure of high-strength concrete will be suddenly without high amount of strain as you can see it in Figure 13.15. The difference in the shape of stress-strain curve is for compressive strength of more than 60 MPa.
- Elastic modules and other calculations: As the stress-strain curve of high-strength concrete and normal concrete is different, the calculations for the elastic modules and other structural calculations will be different also. It means that, for the structural design, when you would like to use high-strength concrete, you should consider the equations and calculations of high-strength concrete. You cannot use the normal concrete techniques.
- Regional normal concrete definition: It is very important that which compressive strength is common in your region of the world. It can be a concept for the definition of high-strength concrete. For example, in some parts of the world, the most common concrete is C30, for these regions, if you produce a C40 concrete, it can be with special considerations and it can be a high-strength concrete. But for some other parts of the world with common compressive strength of 50 MPa, if you produce a C80 concrete, it can be a high-strength.

According to the above mentioned in this book, we define high-strength concrete as a concrete with compressive strength of more than 50 MPa.

FIGURE 13.14 Aggregates crushing in high-strength concrete. (Photograph by the author.)

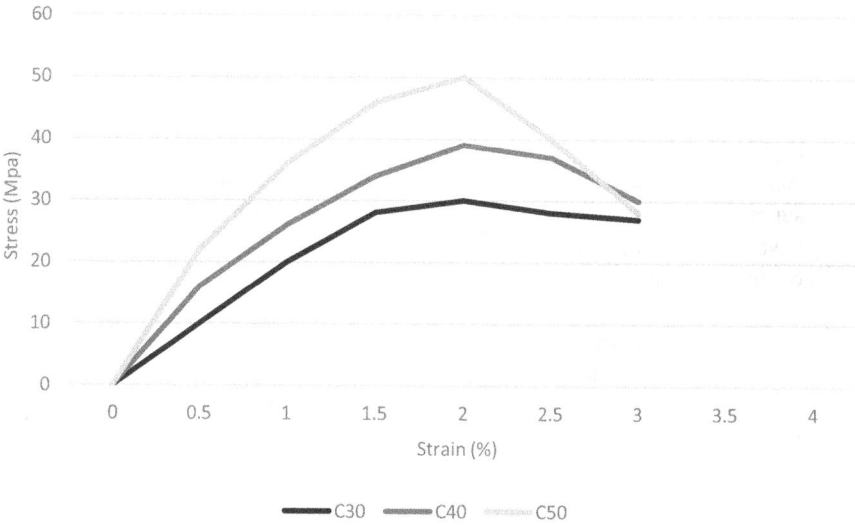

FIGURE 13.15 Stress-strain curve for different types of concrete. (Graph created by the author.)

13.3.2 Considerations for High-Strength Concrete Production and Implementation

Below considerations are the most important ones for the production and implementation of high-strength concrete:

- Using minimum water for the production of concrete: As you know, the most important parameter for the production of high-strength concrete is minimizing the w/b ratio. In this case, using minimum water is more important because you have limitations for the usage of binders, especially the Portland cement. For this purpose, you should use strong super-plasticizers with optimum dosage. The best choice is polycarboxylate type.
- Usage of high amount of binder: To minimize the w/b ratio, instead of minimizing water, you should use higher amounts of binders than normal. It is very important to use cement mixed with the supplementary cementitious materials. It will cause a higher powder amount which can control the high effect of super-plasticizers that can cause segregation and bleeding.
- Using supplementary cementitious materials: This is very important to use the supplementary cementitious materials for the production of high-strength concrete because you have limitations to use the Portland cement. So, if you would like to produce a concrete with lower w/b ratio, you should substitute some amount of Portland cement with the supplementary cementitious materials. On the other hand, because of higher amounts of binders in concrete, you need to control the hydration heat in concrete elements. You can do it, by substitution of the Portland cement with the supplementary cementitious materials. If you need more earlier strength, you should use the silica fume and for other purposes you can use GGBS, fly ash or the natural pozzolans. It is recommended to use mixture of Portland cement with the silica fume and GGBS or fly ash or natural pozzolan as a total binder because the silica fume can increase the compressive strength very much compared with the other supplementary cementitious materials. So, it is recommended to use it for all types of high-strength concrete.
- Adjustment of concrete flowability and viscosity: As you should use strong super-plasticizers for the production of high-strength concrete, you will have a high flowable concrete. On the other hand, as you use a high amount of binder in high-strength concrete, it is possible to have a high viscosity which will cause hard pumpability and concrete moving. So, you should adjust the flowability and viscosity by adjusting the amount of binder and check it with the amount of aggregates filler.
- Using stronger aggregates: As the failure of high-strength concrete will be in the aggregates, So, it is recommended to use stronger aggregates. If you use stronger aggregates with the same concrete mix design, you will give more compressive strength.
- Curing of concrete as well as it possible: Curing of concrete is more important for high-strength concrete. So, it is recommended to cure concrete as

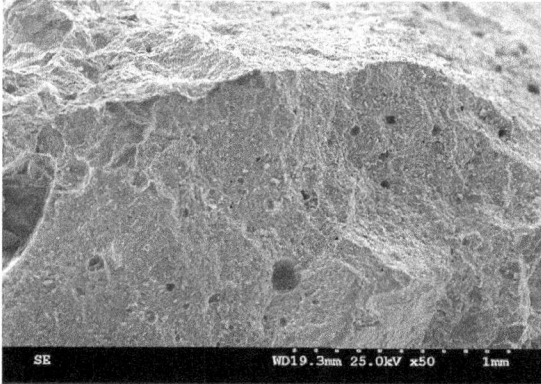

FIGURE 13.16   Air bubbles inside concrete. ("Pore structures in fresh concrete and air entrained concrete. Fresh concrete" by Fangzhi Zhu, Zhiming Ma, Tiejun Zhao.)

well as it possible and for longer time than normal. Plastic and chemical shrinkage are the most dangerous ones for high-strength concrete.

• Control the amount of air in concrete (Figure 13.16): It is recommended to produce a concrete with the amount of air between 1.8% and 2% as a high-strength concrete because more air will cause decreasing in the concrete strength and less air means higher viscosity and harder moving of concrete. You can do it by the adjustment of the super-plasticizer formulation.

13.3.3   Tests for Checking the Properties of High-Strength Concrete

As you can find from the name of high-strength concrete, the most important property for this type of concrete is the compressive strength. But you should know that making a concrete with low workability and high strength is not acceptable. So, testing the workability of fresh concrete should be very important. Most of the time, as you produce high-strength concrete with an optimum dosage of PCE super-plasticizers, the concrete will be a non-slump or high slump concrete. The best test is the slump test, but if the concrete was a high slump concrete you should report the slump of concrete and if the concrete was flowing concrete, you should report the flowing diameter of it.

The other important property that you should check in fresh concrete, is the segregation and bleeding. The concrete with segregation and bleeding is not an approved one as a high-strength concrete. So, you should check the stability of concrete visually as mentioned for the SCC concrete.

The most important test for high-strength concrete is the compressive strength test. This is the same test as the normal concrete. But the below considerations should be useful for the compressive strength test of high-strength concrete:

• It is recommended to use standard 15×30 cylinder or at least 10×20 cylinder specimens. Using cube specimens will not give you the exact result at higher strengths because the correlation between the compressive strength of cube and cylinder at higher strength could hardly interpret.

FIGURE 13.17   A capped cylinder specimens with melted sulfur and silica sand. (Photograph by the author.)

- The capping of the cylinders is very important for trustable results (Figure 13.17). You can use dental gypsum for a strength less than 40 MPa, melted sulfur with silica sand for a strength of less than 80 MPa, and for higher strength, it is recommended to cut the specimen accurately with special instruments. Sometimes, there is a need for capping the lower surface of the specimen, if the flatness of that wasn't acceptable.
- It is recommended to use testing machine with a nominal capacity of at least 300 tons because you should not use more than 80% of the nominal capacity. So lower capacity should not be useful for the high-strength concrete.

13.3.4   EXAMPLE FOR HIGH-STRENGTH CONCRETE MIX
DESIGN AND IMPLEMENTATION

In this part, we would like to design a C70 concrete mix with the constituent materials of this chapter. We should use the step-by-step procedures of Chapter 9 with the considerations of this chapter.

- Step 1: For this concrete, like the other ones in this chapter, we use the standard deviation of 2.5 MPa.
- Step 2: calculation of the mix design compressive strength as below:

$$F_{cm} = 70 + (1.34 \times 2.5) + 1.5 = 74.85 = 75\,MPa$$

- Step 3: According to the recommendations of Table 9.3, for C70 concrete, we should use the maximum size of coarse aggregate as 12.5 mm. So, we should use the mixture of the natural sand and 5–12 coarse aggregate and like the example of the water proof concrete, we will have Table 13.13 for the maximum and minimum values. Then we will have Table 13.18 for n between 0.2 and 0.3 for high slump concrete.

Also, for the aggregates mixture, we will use 27% of gravel 5–12 and 73% of the natural sand like Table 13.14 and Figure 13.16.

- Step 4: The fineness module of total aggregates as you can see in Table 13.15 is 4.1.
- Step 5: As the cement strength is 430 kg/cm² we assume that this is 425 kg/cm². So, we can use Table 9.9. For $F_{cm} = 75\,MPa$, we will have 0.29 from Table 9.9 which is the w/b for this example.
- Step 6: For the Portland cement of 500 kg/m², FM = 4.1, and slump of 90 mm, we have the free water of 242 kg/m³ and for the slump of 150 mm, we have the free water of 260 kg/m³. So, for the production of flow concrete with a slump of 210 mm the free water will be 278 kg/m³.

    For the production of high slump concrete with high strength, according to the recommendations of this chapter, we should use the highest dosage of super-plasticizer which is 1.5% by weight of the binder. Its water reduction rate according to Figure 13.10 is 36%. So, the amount of reduced water will be 178 kg.
- Step 7: As recommended in this chapter, we decided to use GGBS and silica fume for the production of high-strength concrete. So, the calculations for the Portland cement, GGBS, and silica fume are as below:

b = 178/0.29 = 610 kg
    For the GGBS we have 15% of increasing rate and for the silica fume, it is zero according to Table 9.12. So, increased b = 610 × 1.15 = 700 kg.
    We decided to use 20% of the GGBS and 10% of silica fume which is necessary for the production of high-strength concrete. So, we will have:
    GGBS = 140 kg

TABLE 13.18

Calculations for the Weight of SSD Aggregates for High-Strength Concrete

Type of Aggregate	Percent in Total Mixture (%)	Volume (L)	Density (kg/L)	Weight of SSD (kg)
Coarse 5–12	27	152	2.79	425
Natural sand	73	409	2.73	1116
Total	100	561	-----	1541

TABLE 13.19
Final Concrete Mix Design for C70 High Strength Concrete

Constituent Material	Mix Design for 1 m³ (kg)
Gravel 5–12	425
Natural sand	1120
Cement Type II	490
GGBS	140
Silica fume	70
PCE super-plasticizer	10.5
Water	178
Total weight	2433

Silica fume: 70 kg
Portland cement = 490 kg
Total binder = 700 kg
Amount of the PCE super-plasticizer = 1.5% = 10.5 kg

- Step 8: Total volume of the aggregates will be:

$$V = 1000 - (490/3.15) - (140/2.85) - (70/2.2) - 178 - 15$$
$$- (10.5/1.08) = 560.9 = 561 \, L$$

- Step 9: The calculations for the weight of aggregates are as you can see in table 13.18:

The C70 high-strength concrete mix design with SSD aggregates is as you can see in table 13.19.

13.4   ULTRA-HIGH-STRENGTH CONCRETE

We cannot produce a high-strength concrete with unlimited strength by the normal aggregates and techniques. The maximum compressive strength that you can achieve with the normal size aggregates and normal mix design method mentioned in Chapter 9 is between 80 and 100 MPa. It depends on the Portland cement's compressive strength. If you use cement with higher strength, you can achieve a higher strength for concrete. You can see Table 13.20 which is the compressive strength classes for normal density concrete derived from the European standard EN206. You can see in the table, the maximum strength defined for normal density concrete is 100 MPa.

It doesn't mean that you cannot produce a concrete with higher compressive strength. But you cannot achieve these strengths with normal aggregates and techniques. So, we should define concrete with compressive strength of more than 100 MPa as the ultra-high-strength concrete. You can produce concrete with the compressive strength of more than 400 MPa. But you should make this concrete with special techniques and methods that we discuss in this part of the book.

TABLE 13.20

Compressive Strength Classes of Normal Weight Concrete According to EN206

Compressive Strength Class	Minimum Characteristic Strength in Cylinder (MPa)	Minimum Characteristic Strength in Cylinder (MPa)
C8/10	8	10
C12/15	12	15
C16/20	16	20
C20/25	20	25
C25/30	25	30
C30/37	30	37
C35/45	35	45
C40/50	40	50
C45/55	45	55
C50/60	50	60
C55/67	55	67
C60/75	60	75
C70/85	70	85
C80/95	80	95
C90/105	90	105
C100/115	100	115

13.4.1 DEFINITIONS OF ULTRA-HIGH-STRENGTH CONCRETE

Ultra-high-strength concrete is a concrete with a compressive strength of more than 100 MPa. This is a special concrete that should produce with a special method. There are different methods for the production of ultra-high-strength concrete. In this book, we discuss the most common one with the name of reactive powder concrete. You can produce concrete with the compressive strength of more than 400 MPa by this method.

We can divide ultra-high-strength concrete into two types:

- Ultra-high-strength concrete without micro-steel fibers: You can produce a concrete with compressive strength of less than 250 MPa without using the micro-steel fibers. On the other hand, the tensile and flexural strength of this type of ultra-high strength concrete is lower than the other type because the steel fibers cause highly increase of the tensile and flexural strength in concrete.
- Ultra-high-strength concrete with the micro-steel fibers: You can produce concrete with a compressive strength of more than 250 MPa by using the micro-steel fibers. On the other hand, the tensile and flexural strength of this type of ultra-high-strength concrete is higher than the other type.

13.4.2 CONSIDERATIONS FOR ULTRA-HIGH-STRENGTH CONCRETE PRODUCTION AND IMPLEMENTATION

The production and mix design method of ultra-high-strength concrete is different from the other types of concrete. As we discuss the reactive powder concrete as the most common type of ultra-high-strength concrete, we can name below specifications and considerations for that:

- Very fine aggregates (Figure 13.18): This is a very fine structure concrete. The aggregates are quartz sand with the particle size of less than 0.5 mm and silica powder as the finer aggregates. So, the maximum size of coarse aggregate in this concrete is about 0.5 mm. The best choice for the aggregates of reactive powder concrete is the quartz sand and silica powder because the compressive strength of each particle in the reactive powder concrete is very important.
- Very low water-to-binder ratio: For the production of reactive powder concrete, we need very low water-to-binder ratio. It is recommended to produce a concrete with w/b less than 0.22 as the ultra-high-strength concrete. On the other hand, it is very hard to produce a concrete with w/b less than 0.1. So, minimum amount for the w/b could be 0.1.
- High amount of Portland cement: For the production of reactive powder concrete, you should use Portland cement about 30% to 40% by weight of the concrete. To decrease the w/b ratio less than 0.22 you need higher amount of Portland cement than normal.
- High amount of the silica fume: For the production of the reactive powder concrete, you should use silica fume of about 10% by weight of concrete. To decrease w/b ratio less than 0.22 you need high amount of binder. So, you should use high amount of silica fume. On the other hand, silica fume is a very active powder for increasing the compressive strength of concrete and the particles' compressive strength is very high.

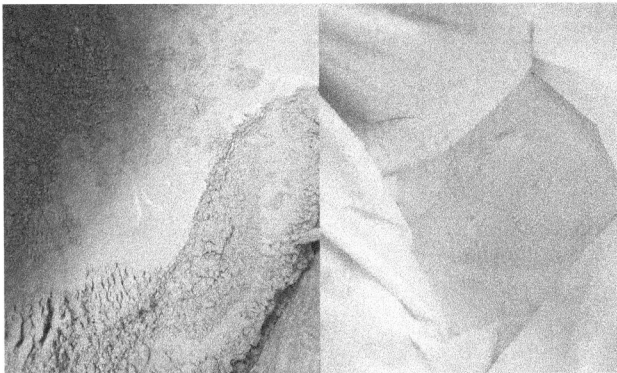

FIGURE 13.18   Silica powder left and stone powder right. (Photograph created by the author.)

- Using other supplementary cementitious materials: As mentioned before, the best type of supplementary cementitious material for RPC concrete is the silica fume. Using other types of supplementary cementitious materials, especially fly ash is under research. So, you can use them but with care.
- Very high dosage of strong PCE super-plasticizers: For the production of RPC concrete, you should use very high dosage of PCE super-plasticizer. You should use PCE between 3% and 10% by weight of binder. As you use this high amount of PCE, the product that you are using should be a pure PCE product without any retardation effect. Most of the times, super-plasticizer manufacturers mix retarders with the super-plasticizers specially polycarboxylates to improve the slump retention effect. In the case of using the PCE product for the production of RPC, you should mention this case to the manufacturer because high dosage of a retarded PCE will destroy the setting time and strength of concrete.
- Portland cement with minimum C3A content and high compressive strength: In RPC concrete, the most important problem is flowing the concrete with low amount of water and minimum dosage of the super-plasticize. If you use a cement with minimum C3A content, you will get better flowability with lower dosage of PCE. The Portland cement with minimum C3A is ASTM type V. On the other hand, to get better compressive strength, you should use a cement with higher compressive strength. Most of the time, compressive strength of type V cement is lower than the other types. So, you should use –I cement. So, the best that you can do is consulting your needs with the cement manufacturer.
- Using stone powder (Figure 13.18): Instead of the silica powder, you can use other types of the stone powders. It can decrease the compressive strength of your concrete because the water absorption of the stone powders is much higher than the silica powder. So, it will need more super-plasticizer or more water. For the lower compressive strength near 100 Mpa, you can use stone powder instead of the silica powder according to the economic considerations.
- Using the micro-steel fibers (Figure 13.20): As mentioned before, you can use the micro-steel fibers for increasing the compressive strength of RPC concrete. You can increase the compressive strength of RPC about two times, if you use micro-steel fibers. It is very important to disperse the micro-steel fibers into the concrete volume uniformly. You can do it, by designing a soft texture concrete with optimum dosage of the super-plasticizer and by using a suitable mixer. The most suitable mixers for the production of RPC are pan mixers type as you can see in Figure 13.19.

13.4.3 Tests for Checking the Properties of Ultra-High-Strength Concrete

Like the high-strength concrete, the most important tests for RPC are testing the flowability of fresh concrete and testing the compressive strength of hardened concrete. As the texture of RPC is very soft, the slump test is not a suitable test for the flowability of RPC. It is better to use a special test method designed for testing the

FIGURE 13.19 Pan mixer, suitable for the production of RPC. (Photograph by the author.)

FIGURE 13.20 Micro-steel fibers. (Photograph by the author.)

flowability of cement paste or mortar like the test defined in ASTM C1437. The flow table that you would like to use, should be in accordance with ASTM C230 as you can see in Figure 13.21.

In the flow table test, you should pore the mold with the RPC concrete and take the mold up and let the concrete flow through the table. Then you should measure the flowing circle in two perpendicular lines. Good RPC concrete should flow at

FIGURE 13.21 Flow table test instruments. (Photograph by the author.)

least 260 mm in flow table test. Lower flowability means very high viscosity and very hard working with concrete. Higher flowability should be good if the segregation and bleeding controlled.

For the compressive strength test of RPC concrete, according to the higher strength of this concrete compare with the other types, you should use smaller specimens. It is recommended to use 10×20 cylinder specimens. As mentioned before, about the compressive strength test, it is very important to have a very soft and steady surface on both sides of the specimens. It will be more important for higher strength concrete. So, for RPC it is recommended to cut the surface of the specimens with a very accurate instrument (Figure 13.22) because trustable results are only possible by this method. Using any type of capping is not acceptable. It is more important when you would like to test RPC concrete produced with micro-steel fibers because the compressive strength of this concrete could be more than 250 MPa and any type of capping never be acceptable for this high amount of compressive strength.

FIGURE 13.22 Cutting instrument for the surface of cylinder specimens ("Concrete cutter")

The crush of the cylinder specimen is acceptable only when the specimen divides into two equal cones. From the crushing pattern of the specimens, you can find the accuracy of compressive strength test.

13.4.4 EXAMPLE FOR ULTRA-HIGH-STRENGTH CONCRETE MIX DESIGN AND IMPLEMENTATION

The mix design technique mentioned in Chapter 9 cannot use for the RPC concrete because this is accepted and tested only for concrete with normal sized aggregates. For RPC concrete, there are too many methods for mix design that you can find in text and lectures. We are not going to talk about these methods here and we cannot say which method is better and more accurate for the design of RPC concrete. So, here we only give you examples of RPC mix design with the compressive strength achieved that you can see in Tables 13.21 and 13.22.

TABLE 13.21
RPC Mix Design Without Using the Micro-Steel Fibers

Material	Amount
Portland cement	850 kg
Silica fume	230 kg
Quartz sand	1020 kg
Silica powder	120 kg
PCE super-plasticizer	3.5% = 38 kg
Water	160 kg
Total	2418

w/b for this concrete is 0.15 and the 28 days compressive strength achieved is 226 MPa

TABLE 13.22
RPC Mix Design by Using the Micro-Steel Fibers

Material	Amount
Portland cement	750 kg
Silica fume	210 kg
Quartz sand	990 kg
Silica powder	130 kg
PCE super-plasticizer	4% = 38 kg
Micro-steel fibers	150 kg
Water	160 kg
Total	2428

w/b for this concrete is 0.17 and the 28 days compressive strength achieved is 340 MPa

13.5 MASS CONCRETE

There are different types of structural elements that you should pour with concrete. Some of them are thin elements with low amount of concrete like columns or thin beams. Some other types are mid-size concrete elements like roofs, walls, or large beams. The last type is huge elements that need high amount of concrete. As the concrete hydration reaction is an exothermic reaction, you should take care of huge concrete elements about the temperature increasing. In this part of the book, we discuss the huge elements concreting.

13.5.1 DEFINITIONS FOR MASS CONCRETE

Mass concrete refers to the concrete that you should use in huge structural elements (Figure 13.23). The most important huge elements are:

- Big foundations
- Huge walls
- Concrete dams

The most important parameter that you should control in mass concrete is the concrete temperature. As the hydration reaction grows, the temperature will increase. If you do not control the heating release, the temperature inside the element will raise and can cause many problems as below:

- Problems in the setting time of concrete
- Problems in strength grow of concrete
- Decreasing the final compressive strength and loading capacity of the structure
- Temperature gradient between inside and outside of the element which can cause undefined stress on the structure

FIGURE 13.23 A huge structure which is the case of mass concreting. (Photograph by the author.)

- High shrinkage and cracking
- Damaging the other structural parts near the huge element according to the high temperature

13.5.2 Considerations for Mass Concrete Production and Implementation

Now, we discuss the most important considerations that you should mention for the production of mass concrete as below:

- Minimum Portland cement usage: For the production of mass concrete, you should use a minimum amount of Portland cement to control the hydration heat and temperature. Most of the time, the compressive strength of huge structure is not too high. So, you can design a concrete with low amount of Portland cement without any problem.
- Using low heating release cement: For huge structures, it is recommended to use Portland cement with lower hydration heat like type IV, V, or at least II. By using these cements, you can control temperature increases inside the huge element.
- Using supplementary cementitious materials: Another way for controlling concrete temperature is the usage of supplementary cementitious materials. As you should substitute these materials with pure Portland cement, you can decrease the hydration heat by using these materials. On the other hand, you can increase the durability and compressive strength of concrete.
- Design coarser concrete: Mass concrete should be coarser than the other types of concrete. It means that you should use more coarse aggregates that normal and it is recommended to use bigger size as the maximum size of coarse aggregate. It can be helpful for controlling the shrinkage and cracks in huge structures. On the other hand, it will be more economical for huge structures to use coarser concrete because coarser concrete will need less cement for the same compressive strength.
- Controlling the temperature of fresh concrete: As you would like to control the temperature of huge elements, you should control the temperature of fresh concrete that you are going to use. It is recommended to use a concrete with temperature of less than 20° in huge structures.
- Good curing of concrete: Curing is very important for the huge structures. The best type of curing in this case is water curing because you can control the temperature of elements by using cold water. On the other hand, curing is the best way to control shrinkage and cracking.
- Using retarder admixtures or retarded super-plasticizers: It is recommended to use a retarded super-plasticizer or a retarder admixture separately to control the hydration reaction and increase the temperature in the structure. In this case, you should give the support of admixture manufacturers because optimizing the dosage of retarders is very difficult and depends to several factors like ambient and concrete temperature, cement type and amount and others.

13.5.3 Tests for Checking the Properties of Mass Concrete

The most important tests for checking the properties of mass concrete are checking the temperature of fresh concrete and checking the compressive strength for the compatibility with designed strength. Sometimes, for very sensitive structures, you can put a thermometer inside the concrete element to control the temperature growth during the hydration reaction. For the compressive strength, you can take a core from the element to ensure the compressive strength of the element in the structure because in huge elements, uncontrolled temperature can cause decreasing the compressive strength. So, you should design a special testing procedure for the huge elements according to the project needs.

13.5.4 Example for Mass Concrete Mix Design and Implementation

In this part, we would like to design a C30 concrete mix for a mass foundation with all of the constituent materials in this chapter. The only difference is using a 12–25 coarse gravel instead of the 11–19 one. You can see the sieve analysis test in Table 13.23. The SSD density of this gravel is 2.75 kg/L.

Now, we should start the step-by-step procedure mentioned in Chapter 9 with the considerations of mass concrete design. The first thing that we should mention here is the low design compressive strength of this foundation which is normal for mass foundations. It can help us to control the amount of cement and heat of hydration.

- Step 1: For this concrete, we use the standard deviation of 2.5 MPa.
- Step 2: calculation of the mix design compressive strength as below:

$$F_{cm} = 30 + (1.34 \times 2.5) + 1.5 = 34.85 = 35 \, MPa$$

- Step 3: For the percentage of each aggregate in concrete, you can see Tables 13.24 and 13.25. For a pumpable low slump mass concrete which is suitable for a foundation according to Table 9.4, n should be between 0.3 and 0.4.

For the aggregates, after trial and error, we will use 32% of gravel 12–25 and 18% of gravel 5–12 and 50% of the natural sand.

TABLE 13.23
Sieve Analysis Test for the 12–25 Gravel for Mass Concrete Production

Sieve Size (mm)	Weight of Aggregates Remained on Sieve (g)	Weight of Aggregates Passed by Sieve (g)	Percent of Aggregates Remained on Sieve (%)	Percent of Aggregates Passed by Sieve (%)
25	12	1620	0.7	99.3
19	465	1155	28.5	70.8
12.5	832	323	51.0	19.8
9.5	221	102	13.5	6.3
4.75	102	0	6.3	0.0
Total	1632	-----	100	-----

TABLE 13.24
Max and Min Amounts for n = 0.3 to n = 0.4

Sieve size	25 mm	19 mm	12.5 mm	9.5 mm	4.75 mm	2.36 mm	1.18 mm	0.6 mm	0.3 mm	0.15 mm
Min limit	100	88.5	73.2	64.4	46.2	32.3	21.8	14.1	8.0	3.5
Max limit	100	90.4	77.2	69.5	52.4	38.5	27.3	18.4	10.9	4.9

TABLE 13.25
Checking the Mix of Aggregates

Aggregate	25 mm	19 mm	12.5 mm	9.5 mm	4.75 mm	2.36 mm	1.18 mm	0.6 mm	0.3 mm	0.15 mm
12–25 32%	31.8	22.6	6.3	2.0	0	0	0	0	0	0
5–12 18%	18	18	17.1	9.1	2.1	0	0	0	0	0
Sand 50%	50	50	50	50	46.4	36.9	26.5	18	10.3	4.8
Total	99.8	90.6	73.5	61.1	48.5	36.9	26.5	18	10.3	4.8

- Step 4: The fineness module of total aggregates as you can see in Table 13.26 is 5.
- Step 5: As the cement is a type II cement with strength of 430 kg/cm² it could be suitable for the production of mass concrete. We can use Tables 9.6 and 9.7. For $F_{cm} = 35$ MPa, we will have 0.54 and 0.51 from Table 9.6 and 0.53 from Table 9.7. The mean value for these three numbers is 0.52 that we should use it as the w/b for mass concrete.
- Step 6: For the Portland cement of 300 kg/m² which is the minimum in Table 9.11, FM = 5 and slump of 150 mm which is suitable for a foundation concrete, we have the free water of 196 kg/m³.

To decrease the amount of Portland cement and as we assume that the PCE is a retarded one, we decided to use high dosage of this super-plasticizer like 0.8%. So, we will use 0.8% of the PCE for this example. Its water reduction rate according to Figure 13.10 is 25%. So, the amount of reduced water will be 147 kg.

TABLE 13.26
Calculation of the Fineness Module of Total Aggregates

Sieve Size	37.5 mm	19 mm	9.5 mm	4.75 mm	2.36 mm	1.18 mm	0.6 mm	0.3 mm	0.15 mm	Total
Percent remained	0	9.4	29.5	12.6	11.6	10.3	8.5	7.7	5.5	-----
Cumulative percent remained	0	9.4	38.9	51.5	63.1	73.5	82	89.7	95.2	503.2

- Step 7: According to the economic considerations and low strength of concrete, we decide to use only GGBS. So, the calculations for the Portland cement and GGBS is as below:

$b = 147/0.52 = 285\,kg$

For the GGBS we have a 15% of increasing rate according to Table 9.12. So, increased $b = 285 \times 1.15 = 330\,kg$.

We decided to use 20% of the GGBS. So, we will have:

GGBS $= 65\,kg$
Portland cement $= 265\,kg$
Total binder $= 330\,kg$
Amount of the PCE super-plasticizer $= 0.8\% = 2.6\,kg$

- Step 8: Total volume of the aggregates will be:

$$V = 1000 - (265/3.15) - (65/2.85) - 147 - 18 - (2.6/1.08) = 725.6 = 726\,L$$

- Step 9: The calculations for the weight of aggregates are as you can see in table 13.27:

The mass concrete mix design with SSD aggregates is as you can see in table 13.28:

TABLE 13.27

Calculations for the Weight of SSD Aggregates for Mass Concrete Production

Type of Aggregate	Percent in Total Mixture (%)	Volume (L)	Density (kg/L)	Weight of SSD (kg)
Coarse 12–25	32	232	2.75	638
Coarse 5–12	18	131	2.79	365
Natural sand	50	363	2.73	990
Total	100	726	-----	1993

TABLE 13.28

Final Concrete Mix Design for Mass Concrete

Constituent Material	Mix Design for 1 m³ (kg)
Gravel 12–25	640
Gravel 5–12	365
Natural sand	990
Cement Type II	265
GGBS	65
PCE super-plasticizer	2.6
Water	147
Total weight	2474

13.6 PRECAST CONCRETE

Precast concrete segments are the concrete elements produced and cured in a factory with high-quality control system. They are made to use in prefabricated structures like:

- Some parts of bridges
- Tunnel segments (Figure 13.24)
- Precast walls for prefabricated structures
- Precast roofs for prefabricated structures
- Cement base artificial stones

13.6.1 DEFINITIONS OF PRECAST CONCRETE

Precast concrete is a type of concrete that we should use for the production of precast segments. This type of concrete should be designed for this purpose to have below considerations:

- Rapid setting time
- High initial strength
- High-quality finished surface
- High durability

In this part of the book, we discuss the consideration for making a high-quality concrete to use in precast products.

13.6.2 CONSIDERATIONS FOR PRECAST CONCRETE PRODUCTION AND IMPLEMENTATION

You can see below the considerations for the production and implementation of precast concrete:

- Use more Portland cement: As we need higher initial strength, it is recommended to use Portland cement more than normal.

FIGURE 13.24 Tunnel precast segments. (Photograph by the author.)

- Using cement with higher strength: For precast concrete, it is recommended to use ASTM type I or III Portland cement. If you don't reach these types or according to the durability considerations, you can use type II cement. The importance is high initial and final strength for the cement for higher initial strength of concrete. For example, it is better to use a I-525 cement instead of I-425.
- Using new and high-quality forms: As we need high-quality final surface for the precast segments, it is recommended to use new and high-quality forms.
- Using high-quality mold releasing agent: For the best finishing surface quality, you should use a high-quality mold releasing agent. Using used car oil or some other types of low-quality and cheap releasing agents will cause very low quality of final surface.
- Consider high durability for the segments: Like any other type of concrete, durability considerations are very important for the precast segments. So, you should use supplementary cementitious materials, high-quality super-plasticizers, and good aggregate to produce a durable concrete.
- Concrete curing: Like any other type of concrete, curing is very important. The difference is rapid mold releasing which is helpful for the speed of segment production. So, you can steam cure the segments for the best result. In this process, you can produce high-quality segments and you can release the molds as soon as it is possible.
- Using air-entraining admixtures: For some of the segments like New Jersey, you may need air-entraining admixture in the production process of the concrete to increase freeze-thaw resistance of the New Jersey. But for all types of the precast segments, there is no need to use air-entraining admixtures. Controlling the amount of air in precast factories is simpler than any other project because quality control of the production process in a precast factory is simpler.

13.6.3 Tests for Checking the Properties of Precast Concrete

There is no special test for precast concrete. For this type of concrete, tests are similar to normal concrete. We can name the most important tests as below:

- Fresh concrete temperature and slump: This test should be done the same as normal concrete.
- Air content of fresh concrete: It is very important when you would like to produce a segment like New Jersey which needs entrained air. So, you should check the amount of air before pouring the forms.
- Compressive strength test: The test is the same as normal concrete. If you need a special compressive strength at any age for mold release, you should test the concrete specimens at that age. You can put the specimens at the same place that you cure the segments to get more trustable results for this purpose.

13.6.4 EXAMPLE FOR PRECAST CONCRETE MIX DESIGN AND IMPLEMENTATION

In this part, we would like to design a concrete mix with the constituent materials of this chapter for a C40 concrete that we would like to use it for the production of pre-cast segments. You need to release the molds at the age of 24 hours when the concrete achieved the compressive strength of 20 MPa. On the other hand, as the segments are exposed to freezing and thawing, you need to use a concrete with 6% of total air.

Like before, we should start with the step-by-step procedure mentioned in chapter 9 by attention to the considerations of this chapter as below:

- Step 1: For this concrete, we use the standard deviation of 2.5 MPa.
- Step 2: We need 40 MPa concrete with 6% of air. It means that if we assume that the entrapped air in a concrete without using air-entraining admixture will be 1.5%, the excess air is 4.5%. Each 1% of excess air will decrease the compressive strength by about 5%. So, you should increase the compressive strength $(4.5 \times 5 = 22.5)$ percent. It means that the compressive strength that we should use for concrete mix design should be 49 MPa that we use 50 MPa. Now, the calculation of the mix design compressive strength is as below:

$$F_{cm} = 50 + (1.34 \times 2.5) + 1.5 = 54.85 = 55 \, MPa$$

- Step 3: For the percentage of each aggregate in concrete, as we don't need pumping for a precast segment and we have high-quality vibrating tables, we should use a concrete with the slump of 150 mm. So, you can see Table 13.29 and 13.30 for n between 0.35 and 0.45.

For the aggregates after trial and error, we will use 22% of gravel 11–19 and 28% of gravel 5–12 and 50% of natural sand.

- Step 4: The fineness module of total aggregates as you can see in Table 13.31 is 4.9.
- Step 5: As the cement strength is 430 kg/cm², we assume that this is 425 kg/cm². So, we can use Tables 9.6 and 9.9. For $F_{cm} = 55$ MPa, we will have 0.36 and 0.34 from Table 9.6 and 0.35 from Table 9.9. The mean value for these three numbers is 0.35.
- Step 6: For the Portland cement of 475 kg/m³, FM = 4.9, and slump of 150 mm, we have the free water of 227 kg/m³.

TABLE 13.29
Max and Min Amounts for n = 0.35 to n = 0.45

Sieve size	19 mm	12.5 mm	9.5 mm	4.75 mm	2.36 mm	1.18 mm	0.6 mm	0.3 mm	0.15 mm
Min limit	100	81.3	70.8	49.4	33.6	22.2	14	7.8	3.3
Max limit	100	84.1	74.8	55.1	39.5	27.3	18	10.5	4.6

TABLE 13.30

Checking the Mix of Aggregates

Aggregate	19 mm	12.5 mm	9.5 mm	4.75 mm	2.36 mm	1.18 mm	0.6 mm	0.3 mm	0.15 mm
11–19 22%	20.6	7.5	1.7	0	0	0	0	0	0
5–12 28%	28	26.7	14.2	3.2	0	0	0	0	0
Sand 50%	50	50	50	46.4	36.9	26.5	18	10.3	4.8
Total	98.6	84.2	65.9	49.7	36.9	26.5	18	10.3	4.8

For this concrete, as we need a 150 mm slump we decided to use 0.6% of the PCE super-plasticizer. Its water reduction rate according to Figure 13.10 is 19%. So, the amount of reduced water will be 184 kg.

• Step 7: For this concrete, as we need high initial strength, we decided to use only Portland cement and 8% of silica fume for increasing the durability. So, the calculations for the Portland cement and silica fume are as below:

b = 184/0.35 = 525 kg

For the silica fume, we have 0% of increasing rate according to Table 9.12. So, we will have:

Silica fume = 8% = 45 kg

Portland cement = 480 kg

Total binder = 525 kg

Amount of the PCE super-plasticizer = 0.6% = 3.2 kg

According to the manufacturer's recommendations, we will use 0.2% of an air-entraining admixture for 6% of the air. You should make trials to ensure the amount of air. The density of this admixture is 1 kg/L.

Amount of air-entraining admixture = 0.2% = 1 kg

• Step 8: Total volume of the aggregates will be:

TABLE 13.31

Calculation of the Fineness Module of Total Aggregates

Sieve size	37.5 mm	19 mm	9.5 mm	4.75 mm	2.36 mm	1.18 mm	0.6 mm	0.3 mm	0.15 mm	Total
Percent remained	0	1.4	32.8	16.2	12.8	10.3	8.5	7.7	5.5	–
Cumulative percent remained	0	1.4	34.1	50.3	63.1	73.5	82	89.7	95.2	489.3

TABLE 13.32

Calculations for the Weight of SSD Aggregates for Precast Concrete

Type of Aggregate	Percent in Total Mixture (%)	Volume (L)	Density (kg/L)	Weight of SSD (kg)
Coarse 11–19	22	127	2.79	354
Coarse 5–12	28	162	2.79	452
Natural sand	50	290	2.73	792
Total	100	579	-----	1598

TABLE 13.33

Final Concrete Mix Design for Precast Concrete

Constituent Material	Mix Design for 1 m³ (kg)
Gravel 11–19	360
Gravel 5–12	450
Natural sand	790
Cement Type II	480
Silica fume	45
PCE super-plasticizer	3.2
Air-entraining admixture	1
Water	184
Total weight	2313

$$V = 1000 - (480/3.15) - (45/2.2) - 184 - 60 - (3.2/1.08) - (1) = 579.2 = 579\,L$$

- Step 9: The calculations for the weight of aggregates are as you can see in figure 13.32:

The precast concrete mix design with SSD aggregates is as you can see in figure 13.33:

REFERENCES

Aitcin P.C, High Performance Concrete, E&FN SPON, 2004.

American Society for Testing and Materials, Standard Practice for Making and Curing Concrete Test Specimens in the Field, ASTM C31-00.

American Society for Testing and Materials, Standard Practice for Capping Cylindrical Concrete Specimens, ASTM C617-98.

American Society for Testing and Materials, Standard Specification for Concrete Aggregates, ASTM C33-01.

American Society for Testing and Materials, Standard Specification for Ready-Mixed Concrete, ASTM C94-00.

American Society for Testing and Materials, Standard Specification for Portland Cement, ASTM C150-00.

American Society for Testing and Materials, Standard Specification for Chemical Admixtures for Concrete, ASTM C494-99.

American Society for Testing and Materials, Standard Specification for Use of Silica Fume as a Mineral Admixture in Hydraulic Cement Concrete, Mortar and Grout, ASTM C1240-00.

American Society for Testing and Materials, Standard Test Method for Compressive strength of Cylindrical Concrete Specimens, ASTM C39-01.

American Society for Testing and Materials, Standard Test Method for Organic Impurities in Fine Aggregates for Concrete, ASTM C40-99.

American Society for Testing and Materials, Standard Test Method for Surface Moisture in Fine Aggregates, ASTM C70-94.

American Society for Testing and Materials, Standard Test Method for Materials Finer than 75 μm in Aggregates by Washing, ASTM C117-95.

American Society for Testing and Materials, Standard Test Method for Specific Gravity and Absorption of Coarse Aggregates, ASTM C127-88.

American Society for Testing and Materials, Standard Test Method for Specific Gravity and Absorption of Fine Aggregates, ASTM C128-97.

American Society for Testing and Materials, Standard Test Method for Sieve Analysis of Fine and Coarse Aggregates, ASTM C136-01.

American Society for Testing and Materials, Standard Test Method for Slump of Hydraulic Cement Concrete, ASTM C143-00.

American Society for Testing and Materials, Standard Test Method for Density of Hydraulic Cement, ASTM C188-95.

American Society for Testing and Materials, Standard Test Method for Compressive Strength of Hydraulic Cement Mortars, ASTM C109-99.

Bertolini L, Elsener B, Pedeferri P, Polder R, *Corrosion of Steel in Concrete, Prevention, Diagnosis, Repair*, WILEY-VCH, 2004.

Bicanski, "Concrete cutter." Retrieved from: https://pixnio.com/media/concrete-cutting-saw-cutter-industrial#.

Connor Jerome J, Faraji Susan, *Fundamentals of Structural Engineering*, Springer, 2016.

Ervanne Heini, Hakanen Martti, Analysis of Cement Super-plasticizer and Grinding Aids: A Literature Survey, Posiva Oy, 2007.

European Standard Organization, Admixtures for Concrete Mortar and Grout, EN934 Series.

European Standard Organization, Admixtures for Concrete, Mortar and Grout Test Methods, EN480 Series.

European Standard Organization, Cement Composition, Specifications and Conformity Criteria for Common Cements, EN197-1: 2000.

European Standard Organization, Concrete-Part 1: Specification, Performance, Production and Conformity, EN206-1, 2000.

European Standard Organization, Methods of Testing Cement, EN196 Series.

European Standard Organization, Testing Fresh Concrete, EN12450 Series.

European Standard Organization, Testing Hardened Concrete, EN12390 Series.

European Standard Organization, Tests for General Properties of Aggregates, EN932 Series.

European Standard Organization, Tests for Geometrical Properties of Aggregates, EN933 Series.

Gjorv E.Odd, *Durability Design of Concrete Structures in Severe Environments*, Taylor & Francis, 2009.

Gjorv Odd E, *Durability Design of Concrete Structures*, Taylor & Francis, 2009.

Hauschild Michael, Rosenbaum Ralph K, Olsen Sting Irving, *Life Cycle Assessment, Theory and Practice*, Springer, 2018.

Heinrichs Harald, Martens Pim, Michelsen Gerd, Wiek Arnim, *Sustainability Science, An Introduction*, Springer, 2016.

Iranian Institute for Research on Construction Industry, 9th topic of National Rules for Construction, "Concrete Structures", 2009.

Iranian Institute for Research on Construction Industry, National Concrete Mix Design Method, 2015.

Iranian National Management and Programming Organization, National Handbook of Concrete Structures, 2005.

Iranian Standard Organization, Concrete Admixtures, Specification, ISIRI2930, 2011.

Iranian Standard Organization, Concrete Specification of Constituent Materials, Production and Compliance of Concrete, ISIRI2284-2, 2009.

Iranian Standard Organization, Standard Specification for Ready Mixed Concrete, ISIRI6044, 2015.

Janamian Kambiz, Aguiar Jose, *A Comprehensive Method for Concrete Mix Design*, Materials Research Forum LLC, 2020.

Kenny, Amit, "workability test for SCC concrete. the end of a L test." Retrieved from: https://commons.wikimedia.org/wiki/File:SCC_test_L_end.jpg.

Knipptang, "Flow test after" Retrieved from: https://commons.wikimedia.org/wiki/File:Flow_Test_after.jpg.

Lamond F.Joseph, Pielert H.James, *Significance of Tests and Properties of Concrete and Concrete Making Materials*, ASTM International, 2006.

Mahmood Zadeh Amir, Iranpoor Jafar, Concrete Technology and Test (Farsi), Golhaye Mohammadi, 2007.

Mostofinejad Davood, Concrete Technology and Mix Design (Farsi), Arkane Danesh, 2011.

Nawy G.Edward, *Concrete Construction Engineering Handbook*, CRC Press, 2008.

Newman John, Choo Ban Seng, *Advanced Concrete Technology, Concrete Properties*, Elsevier, 2003.

Popovics Sandor, *Concrete Materials, Properties Specification and Testing*, NOYES Publications, 1992.

Ramachandran V.S, Beaudion James, *Handbook of Analytical Techniques in Concrete Science and Technology, Principles, Techniques and Applications*, William Andrew Publishing, 2001.

Ramachandran V.S, *Concrete Admixtures Handbook, Properties, Science and Technology*, NOYES Publications, 1995.

Ramachandran, Paroli, Beaudion, Delgado, *Handbook of Thermal Analysis of Construction Materials*, NOYES Publications, 2002.

Ramezanianpoor Aliakbar, Arabi Negin, Cement and Concrete Test Methods (Farsi), Negarande Danesh, 2011.

Richardson M, *Fundamentals of Durable Reinforced Concrete*, SPON Press, 2004.

Richardson M, *Fundamentals of Durable Reinforced Concrete*, Spon Press, 2002.

Safaye Nikoo Hamed, Introduction to Concrete Technology (Farsi), Heram Pub, 2008.

Shekarchizade Mohammad, Liber Nicolas Ali, Dehghan Solmaz, Poorzarrabi Ali, Concrete Admixtures Technology and Usages (Farsi), Elm & Adab, 2012.

Thakur, Abhinav, "Water tank." Retrieved from: https://pixahive.com/photo/water-tank-2/.

Tux-Man, "Concrete segregation." Retrieved from: https://commons.wikimedia.org/wiki/File:S%C3%A9gr%C3%A9gation_b%C3%A9ton.jpg.

Zandi Yousof, Advanced Concrete Technology (Farsi), Forouzesh Pub, 2009.

Zandi Yousof, Concrete Tests and Mix Design (Farsi), Forouzesh Pub, 2007.

Zhu, Fangzhi, Zhiming Ma, Tiejun Zhao, "Pore structures in fresh concrete and air entrained concrete. Fresh concrete." Retrieved from: https://commons.wikimedia.org/wiki/File:Pore-structures-in-fresh-concrete-and-air-entrained-concrete.jpg.

Index

Note: **Bold** page numbers refer to tables and *italic* page numbers refer to figures.

abrasion resistance 116, **117**
absorption 97, **117**, 119, 121, 124, **126**, 128, 195, 196, 239, 245, 246, 247, 252–254, 259, 343, 357
accelerating admixtures 318, 319, 321
accelerator 9, 12, 46, 133–138, 142, 143, 145, 321
admixtures 2, 3, 4, 8, 9, 10, 12, 17, 20, 21, 27, 30, 46, 48, 97, 116, 133–149, 156–166, 187, 188, 217, 225, 238, 271, 283, 284, 305, 307, 312, 318–321, 329, 342, 362, 367
aggregates 2, 5, 21, 22, 23, 97, 101–130, 133, 149, 157, **158**, 159, 161, 163, 164, 175, *176*, 184, 185, 187, 195, 198, 202, 225, 228, 229, **231**, 232, 233, **234**, **235**, 236–239, **240**, 242–245, **247**, **248**, 249–252, **253**, **254**, 255, 256, *257*, 258–260, 264, 266–269, 271, 272, 282, 286, 292, 294, 295, 301, 307, 308, 314, 315, 320, 327, 329, 330, **336**, **337**, 338–340, 342, **345**, 346–348, *349*, 350–354, 356, 360, 362–365, 368–370
air entraining admixtures 30, 48, 134, 156–158, 187, 188, 217, 238, 367
alkali aggregate reaction 22, *23*, 44, 68, 81, 85, 88, 91, **93**, 97, 98, 173, 174
alkali carbonate reaction 22
alkali silica reaction 22, *23*
analysis of cement 64, 67
ASTM 6, 12–14, 23, 29, 35, 44, 47, 51, 54, 56, 57, 59, 61, 63, 67, 108, **109**, 116, **117**, 118, 124, 128, 134, 139, 143, 152, 180, 185–189, 193, 195–197, 215, 230, 343, 357, 358, 367
autoclave 44, **45**, **46**, **47**, 57, 58, *59*, 65

batching plant 71, 94, 121, 123, 124, 164, 225, 246, 261, 264, 267, 271, 272, 276, 286, 292, 306, 313, 319, 322
binder 1–4, 8, 16, 17, 20, 28, 29, 35–38, 77, 79, 81, 84, 85, 87, 88, 90, 91, 127, 209–211, 225, 230, 231–233, 236, 237, 243–245, 250–252, 256–258, 261, 264, 266–272, 314, 319, 323, 324, 330, 332, 338, 339, 341, 346, 350, 353, 354, 356, 357, 365, 369
bleeding 80, 84, 87, 91, **93**, 114, **123**, 127, 148, 163, 164, 229, 328, 329, 330, 333, 335, 350, 351, 359
blended cement 23, 51, 52, 63, 318

calcium chloride 136, 172
calcium silicate hydrate 42, **50**, 75
capillary pores 158, 160, 161, 197, 310, 341
carbonation 24–26, 211, 212, 218, 220, **221**
C3A 43, 45, 46, **47**, 49, **50**, **65**, **66**, **67**, 69, 357
C4AF 44, 46, **47**, 49, **50**, **65**, **66**, **67**, 69
cement 1–9, 12, 16, 17, 20–23, 28, 29, 35–71, 76–81, 83–90, 94, **117**, 125, 127, 133–136, 139, 141, 144, 145, 148–150, 152–155, 158, 160, 161, 165, 171–173, 179, 185, 186, 210, **212**, 214, 215, **216**, **217**, 219–221, 225, **228**, 229–233, **234**, **235**, 236, 237, 240, 243–245, 250–252, 257, 258, 260, 261, 264, 266–272, 286, 287, 301, 304, 307–312, **313**, 314–316, 318–320, 323, 330, 335, 338, **339**, 341, 346, **347**, 348, 350, 353, 354, 356, 357, **360**, 362–369, **370**
cement paste 59, 60, 80, 149, 150, 152, 153
cement type 5, 44, 45–48, **51**, 52, 61, 62, 139, 215, **216**, 221, 231, 232, **233**, 240, **246**, 247, **253**, 255, **260**, 335, **339**, **347**, **354**, 362, **365**, **370**
chemical attack 49, 215, **216**
chemical shrinkage 309, 310, 351
C-H-S 7, 8, 20, 42, 43, 49, 50, 56, 75, 77, 80, 85, 88, 91, 160, 161, 162, 210
chloride attack 25–27, 81, 92, 213
clay 37, 38, 88, 104, *105*, 114, 129, 159
CLC 165, 166, **297**
clinker 39, *40*, 41, 43, 51, 52, 54, 63, 68, 85, 86
coarse aggregates 2, 109, 111, 116, **117**, 118, 119, **121**, 124, 127, 159, 198, 202, **229**, 232, 239, 282, 292, 294, 295, 314, 330, 362
cold weather concreting 185, 301, 318, 320, 321
compaction 4, 13, 20, 186, 198, 210, 280, 327, 342
compositions 7, 41, 44, 45, 46, 47, 48, 49, 69
compressive strength 1, 4, 6, 8, 11–19, 44, 54, 59–62, 66, 77, 78, 80, 85, 88, 91, **93**, 97, 102, 104, 110, 111, 114, 115, 121, 128, 129, 133, 137, 138, 156, 158, 161, 174, 175, 180, 184, 185, 187, 188–191, 193, 195, 198, 199, 201, 202, 204, 211, 212, **214**, **215**, **216**, **217**, 219–221, 226–232, 237, 240, 241, 243, 244, 248, 250, 251, 255, 256, 257, 260, 261, 264, 266, 298, 304, 305, 309, 317–319, 321, 335, 337, 339, 344, 347, 348, 350, 351, 352, 354–357, 359–363, 367, 368

concrete admixtures 133, 143, 165
concrete mix design 37, 60, 61, 63, 81, 97, 109,
116, 119, 123, 124, 154, 158, 163, 188,
225–228, 230–232, 236, 239–241,
246, 247, **253**, 254, 255, 260, 261, 264,
290, 292, 294, 295, 298, 301, 302, 313,
317, 319, 322, **339**, 340, 344, 347, 350,
352, 354, 360, 363, 365, 368, 370
concrete production 4, 8, 16, 17, 22, 23, 28, 30,
52, 55, 61, 67, 75, 83, 86, 93, 199, 116,
119, **123**, 130, 133, 141, 146, 148, 158,
171, 172, 174–176, 197, 210, 226, 227,
231, 236, 240, 263, 264, 267, 286, 287,
294, 301, 304, 306–308, 317, 319, 320,
328, 330, 341, 348, 350, 356, 362, **363**,
365, 366
concrete pump 114, 164, 165, 277, 278, *279*,
280, 286
concrete temperature 6, 8, 159, 161, 184, 185,
301, 302, 304, 306–308, 313–317, 320,
322–324, 361, 362, 367
conveyor belt 271, 277
core test 198, 202, 204
corrosion 24–29, 129, 136, 172, 173, 175, 196,
209–211, **212**, 213, 214, **215**, **216**,
217–219, 304
cracking 4, 10, 16, 20, 22, 24, 58, 80, 85, 88,
93, 138, 162, 171, 282, 284, 291, 292,
294, 296, 298, 305, 309–312, 341,
342, 362
crushed aggregates 5, 101, 113–115, 159, **231**,
232, 233, 250, 251
C3S 42, **47**, 49, **50**, **65**, **66**, **67**, 69
C2S 42, **47**, 49, **50**, **65**, **66**, **67**, 69
cube molds 189
curing 9–11, 20, 26, 30, 61, 80, 141, 160–163, 171,
174–176, 190, 195, 282, 284, *285*,
308–312, 317, 319, 320, 322, 323, 340,
342, 350, 362, 367
curve 17, 18, *19*, 49, 56, *57*, 193, 194, *202*, **230**,
236, *241*, *248*, 255, *337*, *345*, 348, *349*
cylinder 12, *13*, 14, 15, 18, 41, 116, 186, 189–191,
192, 193, 195–197, 202, 204, 266, 278,
351, 352, **355**, 359, 360

density of fresh concrete 186, 187, 189
density of hardened concrete 189, 195, 196
destructive tests 198, 200, 202
dicalcium silicate 42, **50**
dosage 6, 17, 26, 48, 54, 81, 94, 128, 139–141,
149, 153–155, *156*, *157*, 159, 163, 236,
241, 244, *248*, 252, 258, 283, 292, 305,
328, 330, *337*, 338, 341, 342, 346, 350,
351, 353, 357, 362, 364
drying shrinkage 310, 311
ductility 18

durability 2, 19, 21, 22, 25, 29, 48, 49, 158, 161,
189, 196, 209, 211, 213, 214, 216–220,
221, 226, 237, 282, 284, 347, 362, 366,
367, 369

elastic modules 189, 192, 193, *194*, 348
EN 206, 209, 354, **355**
entrained air *157*, **158**, 185, 367
entrapped air 80, 84, 88, 91, **93**, 128, 157, 180,
185, 216, 238, 340, 368
ettringite 28, 43, **50**
environment 21, 29, 35–37, 43, 49, 52, 75, 77, 86,
114, 115, **158**, 209, 210–213, 215–219,
287, 320, 348
exposure classes **212**, **213**, **214**, **215**, **216**

fibers 1, 18, 19, 289–292, **293**, 294–298, 355, 357,
358, 359, **360**
fiber reinforced concrete *289*, 290, 292, *293*
fillers 5, 37, 105, 114–116, 125, 127, **128**, 160,
161, 163, 164, 233, 269, 329, **330**,
332, 342
fine aggregates *2*, 5, 97, 109, 111, 115, **117**, 118,
122, 124, 125, 127, 159, 232, 356
fineness modules 118
fire resistance 296
flexural strength 1, 11, 17, 291, 294, 296, 298, 355
flowability 5–7, 154, 155, 161, 179–182, 229, 233,
236, 327, 328, 331, 350, 357–359
flow table spread 182
flow table test 179, 182, 226, 358, 359
fly ash 8, 16, 49–51, 76, 82, 83–86, 88, 89, 91, **92**,
93, **237**, 350, 357
foaming agent admixtures 165, 166
formwork 321
freeze thaw 29, 30, 80, 85, 88, 91, **93**, 134,
156–158, 187, 216, 217, 219, 221, 367
fresh concrete 1, 4, 58, 80, 84, 87, 90, 92, **93**,
97, 114, 125, 129, *157*, 172, 173, 179,
180, 182, 184–189, 195, 260, 282,
304, 305, 307, 315, 331, 351, 357, 362,
363, 367

GFRC 292, 294, *295*
GGBS 16, 29, 50, 51, 76–78, 85–89, 92, **93**, 159,
237, 247, 252, **253**, 255, 258, **260**, **313**,
322, 335, 338, **339**, 346, 347, 350, 353,
354, 365
glass fibers 292, **293**, 294
grading 127, **128**, 161
gravel 101, 109, *115*, **120**, *121*, **156**, 239, **240**, 242,
243, 245–247, **248**, 249, 252–254, **255**,
256, 259, **260**, 335, **336**, 338, **339**, 345,
347, 353, 354, 363, **365**, 368, **370**
ground granulated blast furnace slag 16, 50, 76,
85, 247

hardened concrete 1, 4, 13, 59, 75, 80, 85, 88, 91, 92, **93**, 97, 129, 171–173, 175, 179, 185, 188, 189, 192, 195–197, 225, 260, 289, 290, 294, 302, 304, 309, 331, 357

hardening accelerator 133, 134, 137, 138, 321

harmful materials 114, 129, 130

heat of hydration 8, 46, 54, 63, 69, 80, 84, 88, 91, 138, 321, 363

heavy weight aggregates 105, *107*, 108

heavy weight concrete 105, *107*, 186

high strength concrete 15, 16, 18, 61, 97, 115, 141, *193*, 194, 201, 258, 290, 292, 309, 347, 348, *349*, 350–357, 360

hot weather concrete 6, 11, 133, 134, 138, 139, 163, 176, 184, 185, 284, 301, 302, 304–309, 312, 313, 315–318, 320, 329, 324

hydration heat 7, 8, 16, 42, 56, **93**, 138, 319, 350, 362

hydration reaction 2, 4, 6, 7–11, 36, 42, 43, 49, 50, 54, 58, 63, 75, 138, 139, 171, 304, 305, 307, 309, 310, 312, 314–319, 323, 361–363

igneous rocks 98

impermeability 42, 162

impurities 102, **117**, 129, 171, 172, 174, 175

initial setting **45**, **46**, **47**, 58–60, **143**, 179, 305

joints 20, 36, 139, 140, 160, 283, 340–342

laboratory 52, 54, 61, 134, *140*, 175, 183, 199, 225, 246, 260, 261, 264, *265*, 266, 284, 342, 343

L-box 183, 334, 335

Le Chatelier flask 63, 64

light weight aggregates 104

light weight concrete 103–105, *106*

lignosulfonate 141, 144–148, **155**

Los Angeles test 116, *117*

low alkali cement 23, 44, 68

manufacture 44, 48, 49, 133, 134, 139, 140, 142, 145, 156, 159, 161, 165, 187, 200, 202, 236, 260, 290, 308, 319, 320, 329, 342, 357, 362, 369

mass concrete 8, 361–365

maximum size of coarse aggregates 109, 111, 198, 202, **229**, 282

metakaolin 49, 76, 88, 89, 90

metamorphic rocks 99, 100

mineral additives 12, 16, 20, 21, 23, 26, 29, 35, 49–51, 75, 77, 93, 94

mini slump test 149, 150, *151*, 152, 155

mix design 37, 60, 61, 63, 81, 97, 109, 112, 116, 119, 123, 124, **152**, **153**, 154, 158, 163, 185, 188, 195, 199, 225–228, 230–232, 236, 239–241, 243, 245–248, 250–261, 264, 266, 290, 292, 294–296, 298, 301, 302, 304, 313, 317, 319, 322, 327, 330, 335, 337, **339**, 340, 344, 347, 350, 352, 354, 356, 360, 363, 365, 368, 370

mixing time 260, 292

moisture 8, 10, 11, 22, 24, 28–30, 61, 69–71, 80, 81, 85, 88, 91, 121, 123–125, **126**, 156, 158, 195, 211, **228**, 238, 239, 245, 246, 252–254, 259, 260, 266, 272, 284, 308, **313**, 314, 320, **322**

molds 12–14, 58, 59, 61, 133, 137, 174, 182, 189–191, *192*, 229, 266, 320, 358, 367, 368

mortar 36–38, 44, 45, 48, 57, 58, 61, 62, 69, 70, 111, 125, 149, 155, *156*, 165, 203, 230, 358

natural aggregates 5, 113–115, **231**, 232, 233

natural pozzolans 16, 51, 76, 88, 89–91, **92**, **93**, 350

non-destructive tests 202

organic impurities 117, 174, 175

particle size distribution 41, 49, 55, 56, *57*, 63

passing by sieve 125, 127, 128, 159, 161, 233

PCE 147, 148, 150, **152**, **153**, *154*, *156*, *157*, 163, 164, 248, 249, 252, **253**, 255, 258, 259, **260**, 298, 329, 335, *337*, 338, **339**, 347, 351, 354, 357, **360**, 364, 365, 369, **370**

perlite 104, *106*

permeability 4, 19, 20, 24–26, 29, 30, 42, 80, 81, 85, 88, 91, **93**, 97, 162, 175, 189, 196, 197, 204, 211, 226, 284, 342, 347

plasticizers 4–6, 8, 17, 20, 21, 77, 133, 134, 141, 142, 144, 154, 236

plastic shrinkage 282, 309

poly carboxylate ether **155**, **238**

poly melamine sulfonate *147*, **155**

polymers 145, 148, 307, 329

poly naphthalene sulfonate 145, *146*, 147, **155**

polypropylene fibers 295, 296

portland cement 1–4, 7, 8, 16, 17, 21, 23, 28, 29, 35–55, 57, 61, 63, 64, 75, 78, 79, 82–91, 94, 141, 173, 185, 210, 221, 231, 236, 237, 240, 245, 247, 252, 255, 258, 264, 311, **313**, 319, **322**, 335, 338, 341, 346, 350, 353, 354, 356, 357, 360, 362, 364–369

pozzolan 16, 50, 51, **52**, 75, 76, 78, 80, 82, 83, 85, 88–91, **92**, **93**, 159, 237, 350

precast 30, 59, 133, 137, 156, 366–368, 370

pumping concrete **230**

quality control 35, 52, 54, 60, 67, 97, 115, 116,
121, 179, 183, 188, 198, 226, 227, **228**,
240, 261, 264, 366, 367

raw materials 23, 35, 38–41, 43, 44, 48, 75
reactive powder concrete 355, 356
ready mixed concrete 61, 138, 225, 276, 283
rebound 136, 200–202
rebound hammer 200
recycling *176*, 286
retarder 8, 12, 133, 134, 138–143, 145, 146, 148
retarding 3, 46, 139–141, 145, 307, 312, 318, 320
retarding admixtures 307, 312, 320
rheometer 179, 183, 184
rocks 98–100, 102, 103, 113, 114, 116, **117**, 123
roller compacted concrete 277

sample 12, 17, 44, 45, 46, **47**, 54, 56, 65, 66, *120*,
121, *122*, 175, 181, 197, 202, *243*, *250*,
257, 264, *345*
sand 25, 37, 58, 61, 101, *102*, 111–116, 118, 119,
122, **123**, 124, 127, **128**, *129*, 155, 157,
159, 161, 163–165, 191, 232, 239, **242**,
243, **245**, 246, **247**, 249, **250**, 253, 254,
256, 259, **260**, 269, 287, 292, 294, 295,
303, 313, 314, 320, 322–324, 335, **336**,
337, 338, **339**, 345, 346, 352, 353, **354**,
356, **360**, 363, **364**, **365**, 368, **369** **370**
SCC concrete 164, 182, 183, **230**, 280, *281*,
327–335, 337, 338, **339**, 340, 344, 351
Schmidt hammer 198–202
sea water 26, 28, 174, 175, 213–215, 218–220,
304, 316
sedimentary rocks 99
segregation 80, 84, 87, 91, **93**, 114, **123**, 127, 148,
163, 164, 229, 328–330, 333, 335, 350,
351, 359
self-compacting concrete 7, **128**, 163, 164, 182,
183, 229, 327, *328*
set accelerator 134–136
setting time 43, **45**, 46, **47**, 48, 49, 56, 58–60, 80,
85, 88, 91, **93**, 129, 133–135, 137, 138,
141, **143**, 173, 174, 179, 282, 304, 305,
307, 316, 357, 361, 366
shotcrete 36, 136
shrinkage 4, 48, 80, 85, 88, 91, 162, 171, 282, 297,
309–312, 351, 362
sieve size 108, 109, 116, **120**, *121*, *122*, **239**, **240**,
242, *243*, **244**, **247**, **248**, **249**, **250**,
251, **254**, **255**, **256**, **257**, **337**, **338**, **345**,
346, **364**, **368**, **369**
silica fume *3*, 16, 20, 26, 50, 51, 76–82, 86, 88,
92, **93**, *94*, **237**, 255, 258, **260**, 341,
344, 346, **347**, 350, 353, 354, 356, 357,
360, 369, 370
slag 8, 16, 50, **51**, **52**, 76, 85, 86, *87*, 247, 314, 323
slump flow test 182, *183*, 331, 332

slump test 6, *7*, 124, 179–182, 184, 186, 226, 266,
331, 351, 357
specific gravity 62, 63, *64*, 79, 83, 87, 89, **92**, 97,
116, **117**, 165
specimens 13, 52, 61, 189–191, *192*, 195–197,
200, 202–204, 227, 266, 342–344, 351,
352, 359, 360, 367
SSD conditions 245, 246, 252, 253, 259
standard 12, *13*, *14*, 15, 18, 29, 35, 44, 47, 51, 52,
53, 56–61, 63, 65, 66, 97, 108, 116,
117, 118, **121**, **122**, 134, 143, 152, 158,
172, 179, 180, 182, 183, 185–187, 189,
190, 193, 195–197, 204, 209, 211, 215,
217, 227–230, 240, 241, 248, 255, 337,
343, 344, 351, 352, 354, 363, 368
standard deviation 60, 61, 226–228, 240, 241,
248, 255, 337, 344, 352, 363, 368
steam curing 284, 309
steel fibers 1, 18, 19, 289–292, 355, 357, *358*,
359, **360**
storage 19, 35, 63, 70, 71, 103, 267, *268*, *308*, 319,
340, 344
strength development 305, 318
sulfate attack 27–29, 43, 45, 46, 88, 91, 215
super-plasticizer 4–6, 8, 17, 20, 21, 54, 77, 81,
94, 97, 110, 116, 128, 133, 134, 138,
140–150, 152–155, **156**, *157*, 160,
161, 163–165, 175, 185, 187, 210, 229,
236–238, 240, *241*, 244, 245, **246**,
247, *248*, 249, 252, **253**, 255, 258,
259–261, 264, 266, 272, 290, 292, 294,
295, 307, 312, 313, 320, 322, 327–330,
335, *337*, 338, **339**, 341, 346, 350, 351,
353, 354, 357, **360**, 362, 364, 365, 367,
369, **370**
supplementary cementitious materials 50, 75–78,
82, 92, 93, 141, 160, 161, 163, 210, 212,
231, 237, 301, 319, 330, 341, 348, 350,
357, 362, 367
sustainable development 21, 37, 348

temperature 6, 8–11, *12*, 29, 30, 41, 49, 58, 59, 61,
97, 134, 138, 139, 140, 156, 158, 159,
161, 162, 176, 179, 181, 184, 185, 196,
213, 216, 260, 284, 301, 302, 304–309,
311, 313–324, 344, 361–363, 367
tensile strength 17, 291, **293**, **296**
test machine 14, *15*, 61, *62*, 116, *117*, 202, 203
tetracalcium alumino ferrite 44
thermal shrinkage 311, *312*
transportation 6, 35, 54, 69, 116, 138, 147, 175,
225, 260, 263, 270, 276, 277, 286, 287,
292, 301, 304–306, 323
tricalcium aluminate 43, 50
tricalcium silicate 42, 50
truck mixer 12, 58, *142*, 159, 179, 181, 273, 276,
277, 286, 292, 306

ultra-high strength concrete 15, 290, 292, 354–357, 360
ultrasonic 198–201
under water **212**, **213**, 340, 344

V-funnel test 183, 332, 333
vibration 20, 182, 269, 271, 280, *281*, 282, 328
VMA 130, 163, 164, 329, 330, 332
voids 62, 160, 196, 343
volume 1–5, 57, 97, 124, **126**, 158, 181, 185–187, 195, 196, **228**, 237, 238, 245, 252, **253**, 258, 271, 272, *273*, 278, 286, 296, 308–310, 339, 347, **353**, 354, 357, 365, 369, **370**

washed sand 124, 320
water penetration 134, 197, 342
water proofing 159, 341–344
water reducer admixtures 143

water reduction rate 143, 145–148, 150, 153, 154, 155, **156**, *157*, 164, 236, 240, *241*, 244, 247, *248*, 252, 258, 260, 264, 312, 330, 335, *337*, 338, 346, 353, 364, 369
water-tightness *340*, 341–346
water tight structure 19, 159, *160*, *339*, 340–342
water to binder ratio 77, 209–211, 230, 231, **232**, 233, 236, 243, 244, 250, 251, 256, 257, 341, 356
water to cement ratio 16, 17, 20, 133, 134, 141, 160
workability 3–6, 48, 54, 80, 84, 87, 90, 93, 97, 114–116, 128, 133, 134, 141, 144, 145, 157, 171, 172, 179–184, 211, 226, 228, 229, 233, 264, 290, 294, 295, 298, *335*, 351

x-ray 67, *68*, 105, *107*

For Product Safety Concerns and Information please contact our EU
representative GPSR@taylorandfrancis.com
Taylor & Francis Verlag GmbH, Kaufingerstraße 24, 80331 München, Germany

www.ingramcontent.com/pod-product-compliance
Lightning Source LLC
Chambersburg PA
CBHW060754220326
41598CB00022B/2427